Analysis, Modeling and Control of Rotating Electrical Machines

Analysis, Modeling and Control of Rotating Electrical Machines

Editor: Elsa Hughes

MURPHY & MOORE
www.murphy-moorepublishing.com

www.murphy-moorepublishing.com

MURPHY & MOORE

Cataloging-in-Publication Data

Analysis, modeling and control of rotating electrical machines / edited by Elsa Hughes.
p. cm.
Includes bibliographical references and index.
ISBN 978-1-63987-747-8
1. Electric machinery. 2. Electric machines. 3. Electric motors. 4. Turbomachines.
4. Rotors. 5. Electric controllers. I. Hughes, Elsa.
TK2000 .A53 2023
621.310 42--dc23

Murphy & Moore Publishing
1 Rockefeller Plaza,
New York City,
NY 10020, USA

ISBN 978-1-63987-747-8

Contents

Preface

This book was inspired by the evolution of our times; to answer the curiosity of inquisitive minds. Many developments have occurred across the globe in the recent past which has transformed the progress in the field.

Rotating electric machines are electric machines that are used to convert mechanical energy into electrical energy. There are three basic types of rotating electric machines, namely, the DC machine, the polyphase synchronous machine, and the polyphase induction machine. Steel, copper and aluminum are the three materials commonly used in manufacturing these machines. Rotating electric machines consists of two parts, the rotor and the stator. The rotor is the cylindrical rotating component whereas the stator is the annular stationary component. These components are made of a magnetic material, which conducts magnetic flux necessary for energy conversion process. In the case of synchronous and DC machines, the magnetic field is generated by field poles induced by the use of direct current. Rotating motors can be controlled by simple start/stop functions or may involve complex controls that regulate motor output parameters such as shaft speed and acceleration. This book includes some of the vital pieces of work being conducted across the world, on various topics related to the analysis, modeling and control of rotating electric machines. It aims to serve as a resource guide for students and experts alike and contribute to the growth of study on this subject.

This book was developed from a mere concept to drafts to chapters and finally compiled together as a complete text to benefit the readers across all nations. To ensure the quality of the content we instilled two significant steps in our procedure. The first was to appoint an editorial team that would verify the data and statistics provided in the book and also select the most appropriate and valuable contributions from the plentiful contributions we received from authors worldwide. The next step was to appoint an expert of the topic as the Editor-in-Chief, who would head the project and finally make the necessary amendments and modifications to make the text reader-friendly. I was then commissioned to examine all the material to present the topics in the most comprehensible and productive format.

I would like to take this opportunity to thank all the contributing authors who were supportive enough to contribute their time and knowledge to this project. I also wish to convey my regards to my family who have been extremely supportive during the entire project.

Editor

1

Fault Diagnosis of Rotating Electrical Machines using Multi-Label Classification

Adrienn Dineva [1], Amir Mosavi [1,2,*], Mate Gyimesi [1], Istvan Vajda [1], Narjes Nabipour [3,*] and Timon Rabczuk [4]

1 Institute of Automation, Kalman Kando Faculty of Electrical Engineering, Obuda University, 1034 Budapest, Hungary; dineva.adrienn@kvk.uni-obuda.hu (A.D.); mate.gyimesi@generalmechatronics.com (M.G.); vajda@uni-obuda.hu (I.V.)
2 Institute of Structural Mechanics, Bauhaus Universität-Weimar, D-99423 Weimar, Germany
3 Institute of Research and Development, Duy Tan University, Da Nang 550000, Vietnam
4 Department of Computer Engineering, College of Computer and Information Sciences, King Saud University, Riyadh 12372, Saudi Arabia; timon.rabczuk@uni-weimar.de
* Correspondence: amir.mosavi@kvk.uni-obuda.hu (A.M.); narjesnabipour@duytan.edu.vn (N.N.)

Abstract: Fault Detection and Diagnosis of electrical machine and drive systems are of utmost importance in modern industrial automation. The widespread use of Machine Learning techniques has made it possible to replace traditional motor fault detection techniques with more efficient solutions that are capable of early fault recognition by using large amounts of sensory data. However, the detection of concurrent failures is still a challenge in the presence of disturbing noises or when the multiple faults cause overlapping features. Multi-label classification has recently gained popularity in various application domains as an efficient method for fault detection and monitoring of systems with promising results. The contribution of this work is to propose a novel methodology for multi-label classification for simultaneously diagnosing multiple faults and evaluating the fault severity under noisy conditions. In this research, the Electrical Signature Analysis as well as traditional vibration data have been considered for modeling. Furthermore, the performance of various multi-label classification models is compared. Current and vibration signals are acquired under normal and fault conditions. The applicability of the proposed method is experimentally validated under diverse fault conditions such as unbalance and misalignment.

Keywords: multiple fault detection; rotating electrical machines; drive systems and power electronics; multi-label classification; machine learning; data science; big data; energy conversion; soft computing; fault severity; fault classifiers; electric machine

---

1. Introduction

Rotating electrical machines are responsible for converting a great amount of worldwide energy into mechanical energy [1]. Mobility, transportation, logistics, construction, production, agriculture, food, automation, and basically any economic activity and industry directly or indirectly depend on rotating electrical machines, as discussed in literatures, e.g., [2–6]. The rapidly evolving industries have suggested that we will be witnessing a further increase in this rate [7]. Furthermore, the increasing demand for hybrid and electric vehicles, the rapid transition toward automated systems and micro and nano mechatronics devises, increasing interests for more efficient energy conversion systems, and emerging new robotics machines have been motivating further advancement in the rotating electrical machines reported in the works of scholars, e.g., [7–18].

One of the key factors of overall efficiency maximization covers the well-sized and high-efficient components [19–22]. Therefore, the reduction and prediction of faults occurring in electrical machines and drive systems such as electrical, thermal, and mechanical faults of electrical machines are strongly suggested to be essential [23–29]. Classical solutions of Fault Detection and Diagnosis (FDD) [30] are based on the complex mathematical models, e.g., supervised diagnosis [25–29,31], or dynamic models [32–37] of the processing system. Intelligent modernization has contributed to the widespread use of Machine Learning (ML) techniques in industrial applications [38–43]. As a result, the latest FDD systems demand more artificial intelligent solutions to incorporate multiple fault events or dynamically changing load profiles in case of incomplete or noisy measurements [44–49]. Commonly, the diagnosis and predictions are calculated through motor current signature analysis (MCSA) [50,51], i.e., examining the output signals of the motor stator's current while running on a steady-state operating mood [52–56]. MCSA analyses the time-frequency decomposition of the current signals or by faults' frequencies in the frequency domain. MCSA works based on a single input source and representing a simple, low-cost and non-invasive monitoring method [50,57,58]. An enhanced method of MCSA in the case of multiphase electrical machines is called electrical signature analysis (ESA) [59].

A timely diagnosis of the complete rotating machinery system contributes to avoiding overpriced reparations and unexpected breakdowns. According to [60], the great majority of recent electric motor condition monitoring methods can be classified into three main categories. The first-class includes the detection of single faults by analyzing one or multiple parameters; the second class covers the detection of different faults with multiple parameters and processing techniques, and the last one contains the mixed techniques of various computing-intensive approaches to analyze different electrical and mechanical parameters in order to detect multiple faults [61–64]. In contrast to conventional signal processing based fault detection techniques [65], recently a few attempts were made for the application of intelligent algorithms [66,67] including new approaches to Fault Detection and Isolation (FDI) [68] based on fuzzy logic, decision trees, neural networks, and further machine learning techniques [69–73]. However, most of them rely on the measurement and processing of vibration signals, which require at least one vibration sensor, which demands extra costs for its proper installation and maintenance [74–77]. In addition, a technician needs knowledge and a good amount of experience to correctly use such sensors [78–84]. However, the ESA reported being able to reveal a large number of relations between the machine parameters [85–88]. Therefore, the ML techniques are highly suitable to support the processing of such extracted information.

There are some intelligent methods that have evolved in recent years with the purpose of improving fault diagnosis methods of electrical motor and drive systems associated with various fault events on the basis of current or vibration signature analysis. For instance, in [89] it has been shown that motor current signature analysis based Support Vector Machine classifier achieves better accuracy in case of gearbox multiple fault detection than decision trees. In [90], the authors state that hybridization of machine learning methods, e.g., adaptive neuro-fuzzy inference system (ANFIS) in combination with Classification and Regression Tree (CART) has the potential for fault diagnosis of induction motors. The proposed method also employs vibration and current signals and achieves nice performance; however, the method lacks the flexibility to easily fit with specific types of machines. Random Forest (RF) classifier was developed for multi-class bearing faults in [91]. This work also employs input features extracted from vibration signals. Worth mentioning that the bearing fault is an important source of faults in the rotating electrical machines that has been modeled by machine learning methods. Wardle [92] and Lynagh et al. [93] provide extended description on the causes of the bearing fault.

A number of studies report the advantages of the wavelet transform in signal detection and fault feature identification (see, e.g., [94–96]). The decision tree's capability, in combination with the well-studied wavelet technique, also provides a possible tool for fault detection [97]. In [97], a large number of possible wavelets are analyzed in order to find the best match. The selection of a suitable wavelet function is still a challenge for the users for specific applications. The investigation of mechanical fault signatures of misalignment and unbalance is carried out in [97], where the authors found that a multilayer perceptron model only with three layers, can classify the faults. High accuracy is reported that is achieved by training the network on a large dataset. In [98] a Convolutional Neural Network (CNN) is successfully applied for fault classification. This approach relies on the S-transform of vibration signals into images displaying the time-frequency patterns in which the pattern recognition is performed. We can clearly observe that the common pattern recognition methods are relying on the assumption of a single fault scenario, and only a limited number of work proposes flexible and comprehensive solutions for multiple fault detection. In the presence of multiple faults, the single fault recognition techniques' performance may be degraded. In addition, the other machines operating in the motor's environment or the coupled subsystems, etc., may introduce further noise components in the measured signal. As a result, it may occur that one of the faults or noise components obscures a particular fault feature and makes it impossible to recognize it. Its further consequence is that the fault isolation operations may become rather difficult.

However, the above brief literature review shows that only a few studies have been published on the investigation of the multilabel prediction performance of recent classifier algorithms in the presence of disturbing noises in the case of electrical signature analysis. One of the major drawbacks from adopting these algorithms to ESA based motor diagnoses is that they require a large training set, which is difficult and time-consuming to collect. In addition, each individual electrical machine and drive system would require their own type- and operation-specific pattern fault library. The feature extraction step raises further questions. The performance of a classifier strongly depends on the effectiveness of the features representing the patterns of different fault events. In the case of fault diagnosis of electrical machines, the knowledge about the physical system may limit the number of necessary features. Furthermore, in engineering practice, there is still considerable ambiguity concerning the selection of the most suitable classifier.

Based on the above briefly summarized antecedents and state-of-the-art applications, this research on the development of new condition monitoring and diagnostic methodology for electrical machines and drives is focused into the applicability of improved versions of multi-label classification machine learning approaches for multiple fault detection under noisy conditions and also the simultaneous determination of fault severity. The contribution of this work is to reveal the most suitable multi-label classification methodology for simultaneously diagnosing multiple faults and evaluating the fault severity under noisy conditions. For this purpose, the performance of various multi-label classification models is compared. Some observations indicate that numerous attributes result in wide parameter space. In our view of the physical model approach, we highlight the important features to be extracted. Current and vibration signals are acquired under normal and fault conditions on an experimental set-up. To enable fast training set preparation, we propose a semi-synthetic database based on a small number of measurements.

The rest of this paper is organized as follows: in Section 2, we briefly introduce the latest multi-label classification methods, and we derive a new methodology for multi-fault diagnosis and severity assessment for rotating electrical machines and drive systems. Section 3 is devoted to the description of the experimental setup as well as the description of feature extraction and dataset preparation. Section 4 presents the results of our investigations on the multi-label classification methods. After, in Section 5 we draw some pertinent conclusions.

2. Multiple Fault Classification and Fault Severity Determination

2.1. Brief Introduction of Multi-Label Classification Methods

Modern industry requires data mining algorithms that are able to efficiently cope with the growing amount of information and large datasets. Specialized processing tasks of various practical application fields deal with common characteristics of the stored data that can be assigned to multiple categories [99,100]. Therefore, multi-label classification algorithms have gained increasing interest in recent years. Specialized techniques for learning such type of data, which have the capability of predicting a set of relevant labels for new species, are still the focus of researches. Currently, three main groups of newly developed multi-label classification methods are proposed in the literature, namely the data transformation methods, adaptation methods, and ensemble of classifiers [101,102]. Early solutions for multi-label classification methods cover the data transformation techniques. The concept is to turn the original multi-label set into binary sets or multi-class sets that adequately can be processed with the classical algorithms. Besides binarization with the widely applied binary relevance technique, also the voting methods and divide-and-conquer approaches are applied to accomplish multi-label transformation [100,103,104]. Such separated sets are learned by single-label classifiers, such as decision trees. This group also includes the Classifier Chains that are similar to the binary relevance technique but perform the binarization in consecutive classifiers, since the models' order has importance [105,106]. The output of one response variable for a sub-classifier is used as an additional feature in the next sub-classifier. The optimal order of the classification models can be enhanced by the Naïve Bayes method.

Adaptation methods are based on the adaptation of conventional classification methods to multi-label versions without problem transformation. The adaptation methods are extensions of well-founded automatic classification algorithms. For instance, the support vector machine (SVM) classifiers [107], the k-nearest neighbors (KNN) [108], or temporal classifiers [109] are able to predict binary or multiclass outputs simultaneously [110,111]. KNN is a non-parametric method used for classification and regression [112,113]. In KNN, an object is classified through a plurality vote of its neighbors while the object assigned to the class of the most common k nearest neighbors. K is a positive and small number, to be able to reduce the number of calculations, and thus the process duration. Similarly, the neural networks and decision trees have such abilities, but the adaptation of the algorithm may become difficult. Ensemble classifiers are also held notable interest. The KNN employs an euclidean distance function (D_{rt}) as follows:

$$D_{rt} = \sqrt{\sum_{i=1}^{m} W_i (x_{ir} - x_{it})^2}, \; t = 1, 2, \ldots, n \tag{1}$$

where $W_i(i = 1,2, \ldots ,m)$ is the weight of predictors. D_{rt} estimates the neighborhood of the prediction value $X_r = \{x_{1n}, x_{2n}, x_{3n}, \ldots ,_r\}$ in real-time using former observation value of $X_t = \{x_{1b}, x_{2b}, x_{3b}, \ldots ,x_{mr}\}$. The classification ensemble approach is based on the aggregation of the outputs of a number of the individual classifier by weighted or unweighted averaging [114]. According to the 'no-free-lunch' theory [115,116], it is assumed that the weaker classifiers with different biases can achieve better performance than the better ones [114,117,118].

2.2. The Scheme of the Proposed Method

We mainly focus on diagnostics based on Electrical Signature Analysis, but we also utilize traditional vibration data. The analysis starts with the elimination of the components that do not contribute to the system from the acquired stator phase current signals. After, the filtered signal is used for extracting the fault features affecting the machine health. Then, the trained multi-label classifier receives the new feature vectors and performs the multi-label prediction. The flowchart of the method is depicted in Figure 1 below.

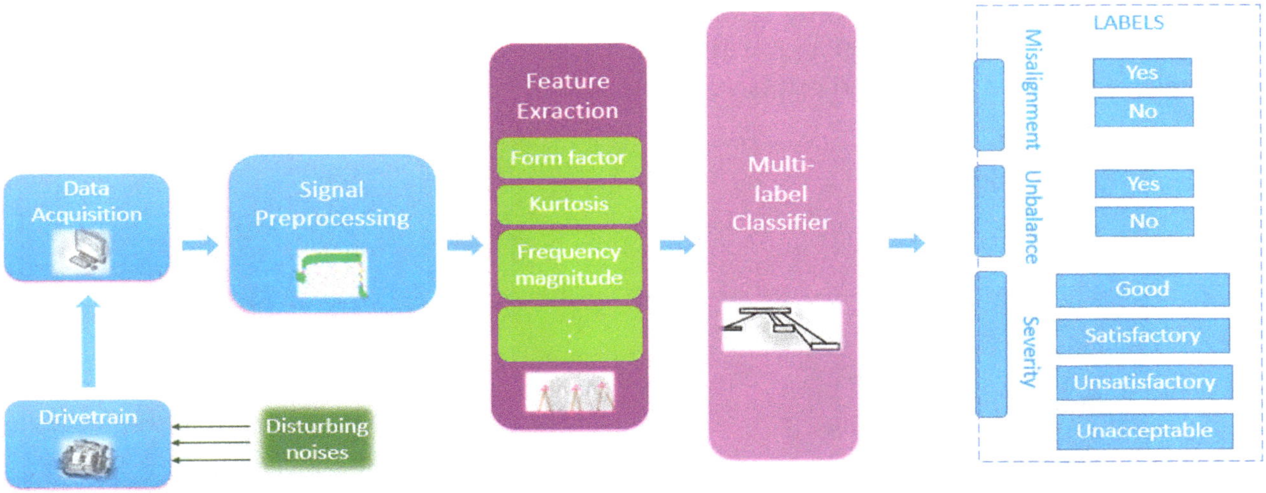

Figure 1. Flowchart of the proposed multi-label fault diagnosis system.

For both of the misalignment and unbalance faults, the two targets can be treated as logical labels, i.e., each output is a binary value, indicating whether a label is associated with the fault or not.

For evaluating the fault severity, we can inject additional classes or alternatively we can generate a second classifier that is executed parallel. Acquired vibration data is sufficient for proper evaluation. The fault severity can be determined easily according to the ISO 2372 Standard [119,120] for vibration severity of machines operating between 600 to 1200 rpm range. The severity classes are shown in Table 1.

Table 1. Vibration severity per ISO (International Organization for Standardization) 10816.

Vibration Velocity [mm/s]	Class I. Small Machines	Class II. Medium Machines	Class III. Large Rigid Foundation	Class IV. Large Soft Foundation
0.28	GOOD	GOOD	GOOD	GOOD
0.45	GOOD	GOOD	GOOD	GOOD
0.71	GOOD	GOOD	GOOD	GOOD
1.12	SATISFACTORY	GOOD	GOOD	GOOD
1.80	SATISFACTORY	SATISFACTORY	GOOD	GOOD
2.80	UNSATISFACTORY	SATISFACTORY	SATISFACTORY	GOOD
4.50	UNSATISFACTORY	UNSATISFACTORY	SATISFACTORY	SATISFACTORY
7.71	UNACCEPTABLE	UNSATISFACTORY	UNSATISFACTORY	SATISFACTORY
11.20	UNACCEPTABLE	UNACCEPTABLE	UNSATISFACTORY	UNSATISFACTORY
18.00	UNACCEPTABLE	UNACCEPTABLE	UNACCEPTABLE	UNSATISFACTORY
28.00	UNACCEPTABLE	UNACCEPTABLE	UNACCEPTABLE	UNACCEPTABLE
45.90	UNACCEPTABLE	UNACCEPTABLE	UNACCEPTABLE	UNACCEPTABLE

3. System Description and Experiment

3.1. System Description

The outlined method requires precise measurements and an appropriate dataset. The theoretical considerations and their usability are validated by simulation investigations by using the collected data on the test bench. The Laboratory of Electrical Machines of the Institute of Automation provides equipment for motor diagnostics and expert system development that allows the integration of measurement, computing, and communication. The laboratory has a wind power simulation system, where the wind power is simulated by an inverter-driven cage induction machine (Leroy Somer 3fFLSE225M-TC; No28221L12001/2012; IP55IK08; P = 30 kW; n = 985 1/min; U = 230/400 V; f = 50 Hz; cos φ = 0.82). Three different types of motors are available on the test bench: one brushless synchronous machine of 40 kW, one double-fed asynchronous machine, and one permanent magnet synchronous machine (PMSM) (Leroy Somer 3fLSRPM200L-T; No728333K12001/2012; IP55IK08; P = 40 kW; n = 1500 1/min; U = 400 V; f = 100 Hz; I = 83 A; 95.2%; Imax/In = 145%) on which the measurements and tests are carried out (see, Figure 2).

Figure 2. The test bench; (1) brushless motor, (2) permanent magnet synchronous motor, (3) double fed asynchronous motor, (4) coupling; Bellows shaft coupling connects the generator and the motor, (5) inverter-driven cage induction machine.

The system is capable of simulating fault events. Misalignment and unbalance are introduced in the system, and a simple NI data acquisition system for data collection purposes is installed. The laboratory has the latest NI LabVIEW Software. The data acquisition system consists of the National Instruments PCI 6013 B-Series 16-Analog-Input multifunction DAQ board and the SC-2345 signal conditioning connector block with various modules and sensors connected to the test bench. The representation of the data acquisition system through the visualization of block schematic illustration is available in Figure 3.

The measuring and controlling LabVIEW software collaborates with the above-presented devices and additional hardware components and conducts the measurements. The measurement data are processed simultaneously on each channel, which takes a relatively short time. The LabVIEW software is developed in order to inspect the proper configuration of the test bench. The software displays the amplitude-time signals of the three-phase stator current and vibration signals of the motor and the generator in real-time. The second tab of the software views the Park Vector's spectrums calculated from the current signals. Frequency-amplitude diagrams can be used to predict machine errors during measurement. The third part of the program saves the measurement results in the Excel.csv format, so the measured values can be evaluated and processed later.

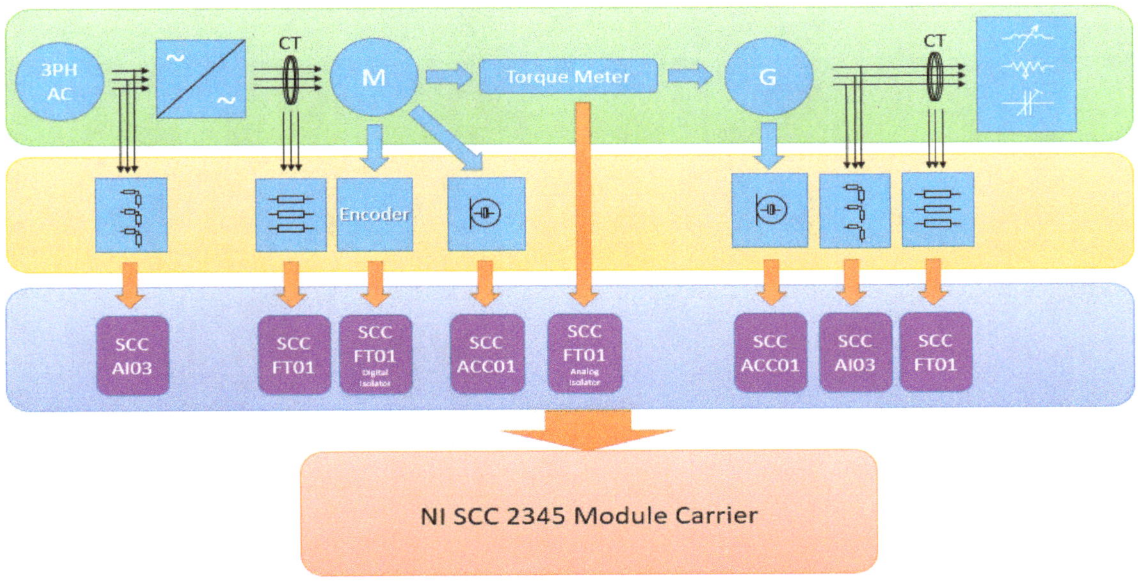

Figure 3. Block scheme of the data acquisition system.

3.2. *Feature Extraction and Dataset Preparation for Training*

Stator current signals and vibration data picked for fault-free no load and 40% load of full load at the speed of 1500 rpm conditions with fs = 10000 sampling rate and t = 20 s duration in the presence of both mechanical faults. Unbalance results in high-frequency amplitudes at frequencies at once the rotational speed. The typical fault features of misalignment are dominant frequencies at one or two times the rotational speed depending on the degree of angular misalignment and the type of the couplings. Spectral images also display sub-harmonic multiples of 1/2xRPM. However, the different speeds, loads, motor parameters, and operational setups, etc., may affect the fault frequencies. Differences between faulty and healthy conditions can immediately be observed from vibration spectra during the measurement (see Figure 4).

It is clearly visible in the spectrum that the machine vibrates strongly around 25 Hz, which indicates a shaft misalignment. From the generator current spectrum, a 50 Hz component appeared in the spectra, while in the motor current spectrum a 75 Hz component (see Figure 5). The reason is that a generator is a 4-pole machine, while the motor has six poles. This indicates exactly that the machine vibrates at 25 Hz.

The unbalancing load is placed at the generator's side, so the flexible bellows coupling may attenuate the fault signature in its spectra. For fault extraction, we can also apply the Thomson multitaper spectral estimates that are proven to be efficient in the case of weaker signals [26] and combine the beneficial properties of high resolution and low variance. Figure 6 displays the Thomson multitaper spectrum of the generator's current signal in which we can observe how the distinguished magnitudes are emphasized. This allows a fast, automatic extraction of the peaks.

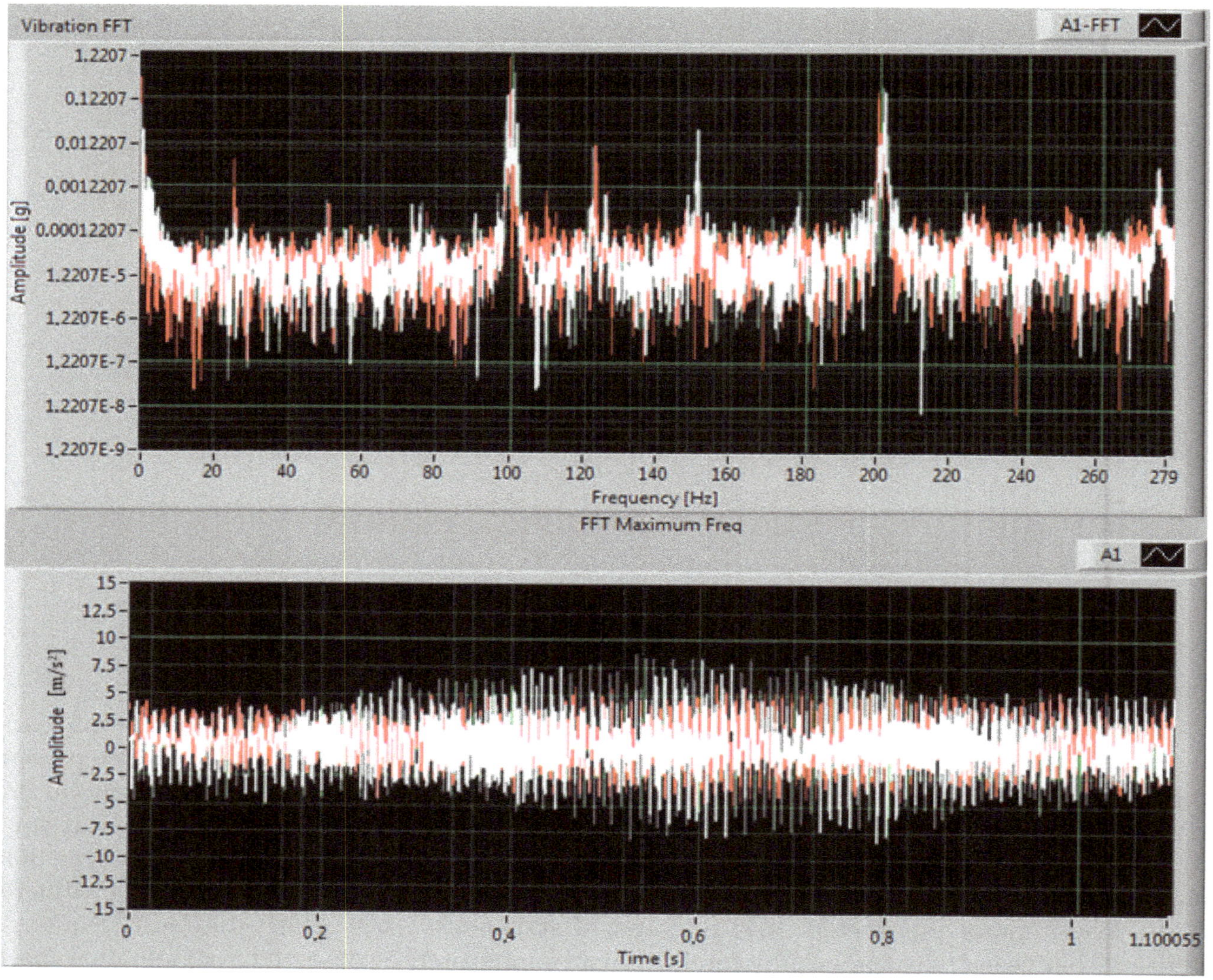

Figure 4. Fast Fourier Transform spectrum (upper chart) and time-domain signal (lower chart) of vibration data displayed online by the LabVIEW software during the measurement. A1 (white line) denotes the generator's vibration signal, and A2 (red line) stands for the motor's vibration signal.

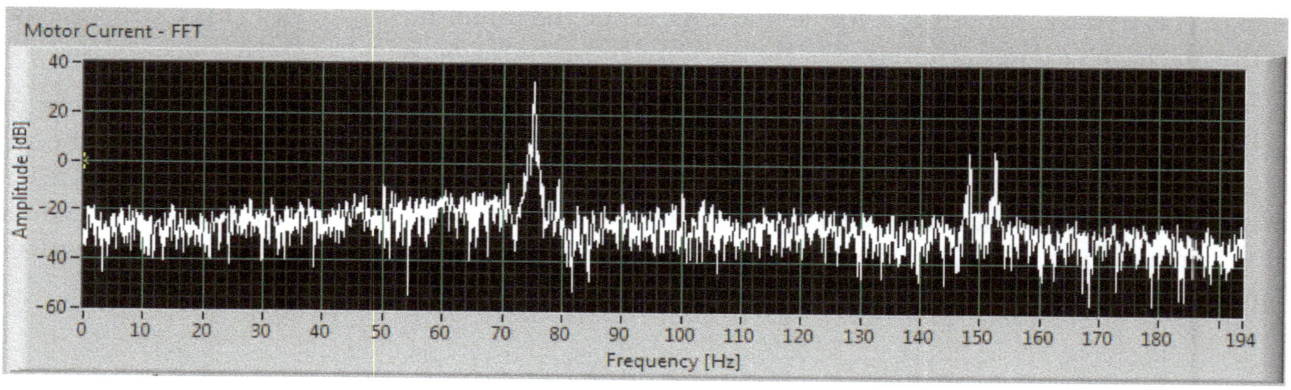

Figure 5. Fast Fourier Transform (FFT) spectra of the motor current (*Im*) stands for the motor's current component *a*.

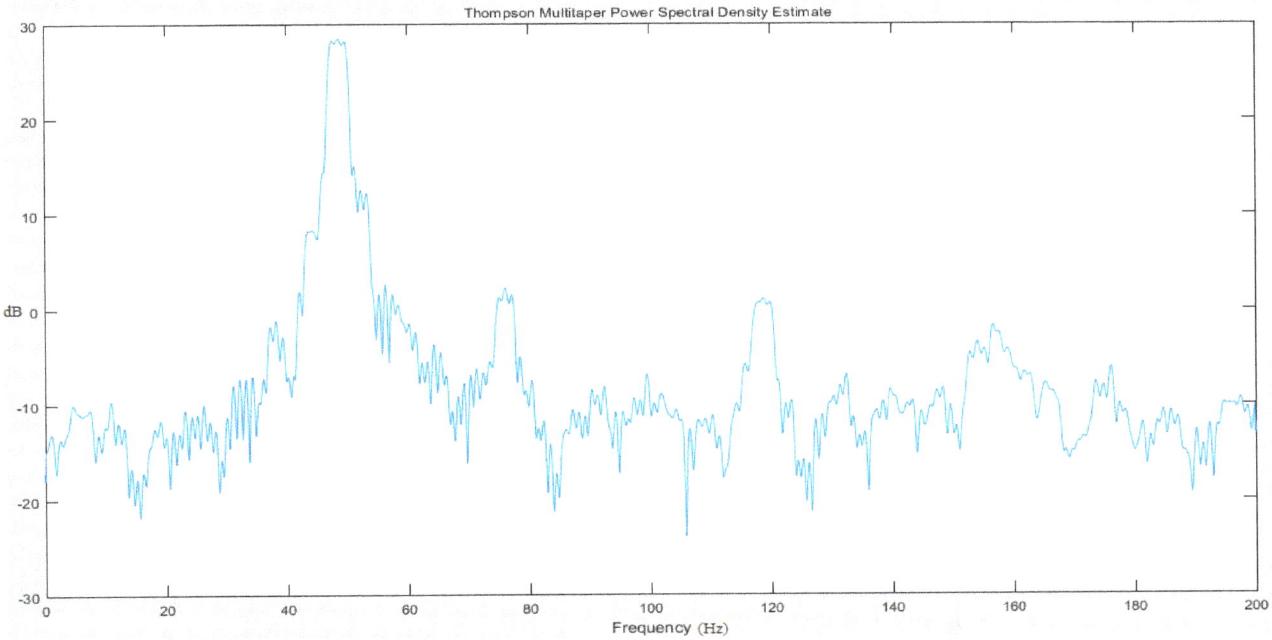

Figure 6. Multitaper power spectral density estimation of generator current signal for fault frequency magnitude detection.

The features carrying most of the information of interest are mainly in the frequency domain. However, we have inspected a few time-domain features also. The feature vector is composed of the magnitudes of the distinguished sideband frequencies and RMS variance frequencies from the spectra for all three phases of both machines. The time-domain features cover the form factor, kurtosis, and entropy deviation from the fault-free sample. The fault feature vector also includes the distance calculated between the observed signature and each pure signature associated with the identified fault [121–123].

It is worth noting that many attempts have been made with the purpose of improving performance through sophisticated feature extraction methods. In his seminal paper on recognition of diagnostic signals based on acoustic signals, Glowacz and Glowcz [124] have introduced a shortened method of frequencies selection. In [125], a novel method of selection of amplitudes of frequencies ratio 50-s frequency coefficient was introduced in order to build the feature vector. Recently, the new Method of Selection of Amplitudes of Frequency Ratio of 24% Multiexpanded Filter 8 Hz and k-means clustering approach has been presented in [126]. This technique draws our attention to investigate the applicability of acoustic-based fault feature extraction techniques to ESA in our future work.

The dataset is built by synthetically adding a random number of ten to twenty of external disturbing frequency components for the faulty and fault free signals. Fifteen different samples are generated with random contaminating frequencies for all of the fault-free, multiple-fault, and single fault (unbalance or misalignment) cases and additionally the data collected by the accelerometers at both sides for the severity evaluation. The training set consists of 64 feature vectors, including the original ones and each of them containing 27 features.

In our qualitative models *'isUnbalance'* and *'isMisalignment'* (Table 2) are the fault label columns which consist of values 0 (No) and 1 (Yes) corresponding to the presence of the symptom of a fault in a given sample. For the fault severity classifier, we have specified the labels according to the previously presented ISO standard [25] as *'Good'*, *'Satisfactory'*, *'Unsatisfactory'*, *'Unacceptable'*. The dataset is divided into a training set 70% of the vectors to train the classifier and a testing set of 15 % and 15% for validating the accuracy of the classifier on new data. When most of the samples have the same labels in the training set, the classifier will achieve a high accuracy value. In order to avoid such a bias, half of the dataset composed of samples including the fault features. Further metrics are applied for the proper evaluation of the performance.

Table 2. Structure of the dataset.

id	Ia_motor	Ib_motor	...	isUnbalance	isMisalignm	Severity
0	0.687150	0.028560	...	0	1	'Good'
1	0.019699	0.388238	...	1	0	'Good'

4. Results

The experiments with three different algorithms are performed by applying the Python Scikit-multi-learn library, which offers various classification approaches that are suitable for predicting simultaneously multiple outputs. At first, a binarized Decision Tree has been implemented for predicting the labels. As a criterion, which defines the function to measure the quality of a split, both criteria for the Gini Index and entropy for Information Gain were used. We have also modified the maximum of tree depth in order to compare the performance using various maximum depths of the trees. Various tree models were also built, including and excluding the error attribute values. According to initial tests, we have found that applying the error feature attributes in the training does not result in significant improvement in the accuracy. The best accuracy score we have achieved is 0.7333. The accuracy score is not the most representative metric in the case of multi-label classification. Therefore, further metrics are applied. The results are collected in Table 3.

Table 3. Performance evaluation of the tested models.

Prediction Method and Label (0/1)		Precision	Recall	f1-Score
Binarized Decision Tree	0	0.79	0.88	0.83
	1	0.85	0.94	0.89
Classifier Chain	0	0.79	0.88	0.83
	1	0.78	1.00	0.88
KNN	0	0.76	0.76	0.76
	1	0.82	1.00	0.90

The precision also shows the accuracy of the model by the ration of total predicted positive and the number of true positive ones. Recall or sensitivity is calculated by the number of true positives divided by the number of true positives plus the number of false negatives. The F1 score covers the weighted average of the sensitivity and precision values that are suitable to characterize the test performance. After, a Classifier Chain multi-label method using the Gaussian Naïve Bayes approach was tested. We found it to be more efficient. Its performance can be seen in Table 3. The Classifier Chain's score accuracy resulted in 0.8333. Subsequently, we have tested the multi-label KNN method whose best percentage accuracy resulted in 0.7 after testing the algorithm with various k values.

It can be seen from Table 3 that the classification performance has been the most enhanced by using Classifier Chains and KNN. The accuracy may be further improved by training a model with a larger training set and further tuning the algorithms. Its excellent pattern recognition capabilities can be effectively utilized for the fault classification of electrical machines in the presence of disturbing noises. The prediction performance of the parallel severity classification tree resulted in 99% accuracy because most of the vibration data is labeled *'good'*.

5. Discussion

In modern industrial systems, complexity is increasing as multi-sensor network systems are expanding towards largescale systems. The intelligent solution of Electrical Signature Analysis allow simplifying the fault identification processes because it does not require a large number of sensors, which results in remarkable cost reduction, and supports sensorless and largescale technologies. Furthermore, a well-developed theoretical and practical methodology could serve as a basis for reliable remote diagnostic tasks of electrical machines, especially important diagnostics carried out in hazardous environments (for instance, in nuclear plants). The development of appropriate diagnostic-prediction methods, which are capable of reliably and timely evaluating the health status of the system on the basis of representative parameters acquired directly or indirectly, is an important part of the modern electric drivetrain monitoring system. The state-of-health of a certain dynamic system and its possible initial failures can be addressed by a number of approaches published in the literature. The strengths and weaknesses of conventional fault diagnosis methods have already been proven. In recent years, diagnostics research has focused on the prediction algorithms that can identify the fault features of progressive malfunctions. As such algorithms are highly technology-dependent, it is important to define the method as a function of a large number of parameters of the dynamic system for fitting to the system under consideration, taking into account boundary conditions. Data-driven technologies aim such difficulties.

Fault classification methods have already been established for the investigation of the relations between the symptom and the fault feature. It is clear that binary relationships can be easily represented by such systems. Early versions of these algorithms are used rather for visualizing diagnostic reasoning. However, practical engineering problems demanded the development of new automatic methods that are able to deal with multiple fault scenarios and noisy or uncertain fault features. A possible solution is the extension of the most commonly applied classification methods. Recently, binary trees, Chain Classifiers, Neural Networks, and other multi-label and multi-class techniques were introduced in the literature. However, these algorithms need more development and improvement, especially in terms of robustness. In addition, the application of Machine Learning techniques on sensory data is still not well-established and requires the synergy of classical signal processing and intelligent data analysis. As a conclusion from all cases studied, the analysis of motor current signatures in combination with machine learning-based processing and evaluation algorithms is a viable methodology for fault detection and prediction in electrical machine and drive systems that has the following additional advantages; the measurement or monitoring does not interfere with the production process; it is well suited for any machine with speed, performance, and power; it is universal because it is able to detect the quality of electricity supply, disturbances, faults in a motor control, starter- and control-devices, the efficiency of the electric machine and its characteristics, errors in the electrical machine, plant-wide machine monitoring and its operating costs (energy consumption, repair, maintenance) meet the economy's expectations. The transfer of recent Machine Learning approaches to practical fault diagnosis problems of rotating electrical machines and drive systems may help to facilitate smart industrialization and intelligent modernization.

6. Conclusions

Automated electrical signature analysis-based fault detection and diagnosis of electrical machines is still an especially important issue in modern industrial applications. The recent data acquisition devices with high sample rate signal inputs have high levels of noise. The coupled machines may cause even heavy disturbances in the signal. A greater source of concern is the selection of the appropriate model and the unavailability of the dataset rich in fault signatures. With this in mind, this paper investigates the applicability of recent multi-label classification methods and suggests a new methodology for simultaneously diagnosing multiple faults and evaluating the fault severity under noisy conditions. Our study provides a new framework and strategy that has been tested on an experimental setup. According to the comparative investigations, we have found that the Classifier Chain strategy is the most suitable for issues of simultaneous noisy misalignment and unbalance fault. The fault signatures are composed of magnitudes of the fault and disturbing noise frequencies and some additional parameters. We have confirmed that relying on the knowledge of the physics of the fault event has twofold advantages. Obtaining the required training dataset is much shorter if we build a semi-synthetic database. A reliable model can be built by recording only a few measurements beforehand, which can be further refined during operation. On the other hand, the high dimension of attributes can be even simply reduced. Our investigations into this area are still ongoing and seem likely to confirm our hypothesis. To further our research, we intend to carry out more extensive investigations regarding the feature extraction techniques. For instance, extending recent findings of motor fault feature extraction, e.g., [126], based on acoustic signals to electrical signature analysis, also taking into account sensor fusion, can be undertaken.

Author Contributions: Principal investigator, lead author, writing, machine learning expertise, electric machines expertise, A.D.; data collection, data curation, data analysis, technical expertise, laboratory expertise, data acquisition system implementation, LabVIEW programming, M.G., N.N.; machine learning modeling expertise, writing, revision support, A.M., N.N.; electric machines expertise, conceptualization, supervision, resources, software, control and verification of the results, I.V., T.R.

Nomenclature

FDD	Fault detection and diagnosis
ML	Machine learning
MCSA	Motor current signature analysis
ESA	Electrical signature analysis
FDI	Detection and isolation
ANFIS	Adaptive neuro-fuzzy inference system
CART	Classification and regression Tree
CNN	Random forest classifier convolutional neural network
SVM	Support vector machine
KNN	k-nearest neighbors
PMSM	Permanent magnet synchronous machine
RF	Random Forest
ISO	International Organization for Standardization

References

1. Dineva, A.; Mosavi, A.; Ardabili, S.; Vajda, I.; Shamshirband, S.; Rabczuk, T.; Chau, K.W. Review of soft computing models in design and control of rotating electrical machines. *Energies* **2019**, *12*, 1049. [CrossRef]
2. Jokanović, B.; Bebić, M.; Kartalović, N. The influence of combined strain and constructive solutions for stator insulation of rotating electrical machines on duration of their reliable exploitation. *Int. J. Electr. Power Energy Syst.* **2019**, *110*, 36–47. [CrossRef]
3. Roubache, L.; Boughrara, K.; Dubas, F.; Ibtiouen, R. Elementary subdomain technique for magnetic field calculation in rotating electrical machines with local saturation effect. *COMPEL Int. J. Comput. Math. Electr. Electron. Eng.* **2019**, *38*, 24–45. [CrossRef]
4. Boughrara, K.; Dubas, F.; Ibtiouen, R. 2-D exact analytical method for steady-state heat transfer prediction in rotating electrical machines. *IEEE Trans. Magn.* **2018**, *54*. [CrossRef]
5. Caruso, M.; Di Tommaso, A.O.; Miceli, R.; Rizzo, R. The use of slightly asymmetrical windings for rotating electrical machines. *Int. Trans. Elecr. Energy Syst.* **2018**, *28*. [CrossRef]
6. Kande, M.; Isaksson, A.J.; Thottappillil, R.; Taylor, N. Rotating electrical machine condition monitoring automation-A review. *Machines* **2017**, *5*, 24. [CrossRef]
7. Byerly, K.; Ohodnicki, P.R.; Moon, S.R.; Leary, A.M.; Keylin, V.; McHenry, M.E.; Simizu, S.; Beddingfield, R.; Yu, Y.; Feichter, G.; et al. Metal Amorphous Nanocomposite (MANC) Alloy Cores with Spatially Tuned Permeability for Advanced Power Magnetics Applications. *JOM* **2018**, *70*, 879–891. [CrossRef]
8. Fu, Y.; Huang, X.; Xu, Y.; Bai, C. Controllable inertial control strategy of rotating motor in DC distribution network. *Dianli Zidonghua Shebei Electr. Power Autom. Equip.* **2018**, *38*, 32–38. [CrossRef]

9. Guerroudj, C.; Saou, R.; Boulayoune, A.; El-hadi Zaïm, M.; Moreau, L. Performance analysis of Vernier slotted doubly salient permanent magnet generator for wind power. *Int. J. Hydrogen Energy* **2017**, *42*, 8744–8755. [CrossRef]
10. Song, Z.; Li, Y.; Zhang, C.; Liu, Y. Rotating core loss model for motor considering skin effect and dynamic hysteresis effect. *Nongye Gongcheng Xuebao* **2019**, *35*, 74–80. [CrossRef]
11. Zhang, J.; Yu, H.; Chen, Q.; Hu, M. Dynamic characteristics and experiment analysis of a single phase permanent magnet linear generator for wave energy conversion. *Diangong Jishu Xuebao* **2013**, *28*, 110–116.
12. Mosavi, A.; Faizollahzadeh Ardabili, S.; Shamshirband, S. Demand prediction with machine learning models: State of the art and a systematic review of advances. **2019**. Available online: https://eprints.qut.edu.au/129249/ (accessed on 20 November 2019). [CrossRef]
13. Akhtar, M.J.; Behera, R.K. Optimal design of stator and rotor slot of induction motor for electric vehicle applications. *IET Electr. Syst. Transp.* **2019**, *9*, 35–43. [CrossRef]
14. Deng, W.; Zuo, S. Electromagnetic vibration and noise of the permanent-magnet synchronous motors for electric vehicles: An overview. *IEEE Trans. Transp. Electrif.* **2019**, *5*, 59–70. [CrossRef]
15. Fan, Y.; Qu, G.; Tan, C.; Zhang, Q.; Wang, W. Design and analysis of a new five-phase dual-stator consequent-pole brushless hybrid excitation machine. *IEEE Trans. Magn.* **2019**, *55*. [CrossRef]
16. Ulu, C.; Korman, O.; Kömürgöz, G. Electromagnetic and thermal design/analysis of an induction motor for electric vehicles. *Int. J. Mech. Eng. Robot. Res.* **2019**, *8*, 239–245. [CrossRef]
17. Mosavi, A.; Salimi, M.; Faizollahzadeh Ardabili, S.; Rabczuk, T.; Shamshirband, S.; Varkonyi-Koczy, A.R. State of the art of machine learning models in energy systems, a systematic review. *Energies* **2019**, *12*, 1301. [CrossRef]
18. Baranyai, M.; Mosavi, A.; Vajda, I.; Varkonyi-Koczy, A.R. Optimal Design of Electrical Machines: State of the Art Survey. In Proceedings of the International Conference on Global Research and Education, Iasi, Romania, 25–28 September 2017; pp. 209–216.
19. Aljehaimi, A.M.; Pillay, P. Operating Envelopes of the Variable-Flux Machine with Positive Reluctance Torque. *IEEE Trans. Transp. Electrif.* **2018**, *4*, 707–719. [CrossRef]
20. Catuogno, G.R.; Garcia, G.O. Conversion of Three-phase Commercial Machines into Six- phase Machines for Didactic and Research Purposes. *IEEE. Lat. Am. Trans.* **2018**, *16*, 467–475. [CrossRef]
21. Morozov, A.; Humphries, K.; Zou, T.; Rahman, T.; Angeles, J. Design, Analysis, and Optimization of a Multi-Speed Powertrain for Class-7 Electric Trucks. *SAE Int. J. Altern. Powertrains* **2018**, *7*. [CrossRef]
22. Palanivel, A.; Padmanabhan, S. Software-based performance estimation and real-time thermal analysis of brushless direct current motor with corroded permanent magnets. *Comput. Electr. Eng.* **2018**, *71*, 938–952. [CrossRef]
23. Praveenkumar, T.; Saimurugan, M.; Ramachandran, K.I. Comparison of vibration, sound and motor current signature analysis for detection of gear box faults. *Int. J. Progn. Health Manag.* **2017**, *8*, 132–151.
24. Bessous, N.; Sbaa, S.; Megherbi, A.C. Mechanical fault detection in rotating electrical machines using MCSA-FFT and MCSA-DWT techniques. *Bull. Pol. Acad. Sci. Tech. Sci.* **2019**, *67*, 571–582. [CrossRef]
25. Corne, B.; Vervisch, B.; Derammelaere, S.; Knockaert, J.; Desmet, J. Emulating single point bearing faults with the use of an active magnetic bearing. *IET Sci. Meas. Technol.* **2018**, *12*, 39–48. [CrossRef]
26. Glowacz, A.; Glowacz, W.; Glowacz, Z.; Kozik, J.; Gutten, M.; Korenciak, D.; Khan, Z.F.; Irfan, M.; Carletti, E. Fault diagnosis of three phase induction motor using current signal, MSAF-Ratio15 and selected classifiers. *Arch. Metall. Mater.* **2017**, *62*, 2413–2419. [CrossRef]
27. Martín-Díaz, I.; Duque-Perez, O.; Romero-Troncoso, R.; Morinigo-Sotelo, D. Supervised diagnosis of induction motor faults: A proposed methodology for an improved performance evaluation. In Proceedings of the IEEE 10th International Symposium on Diagnostics for Electrical Machines, Power Electronics and Drives, Guarda, Portugal, 1–4 September 2015; pp. 359–365.
28. Irhoumah, M.; Pusca, R.; Lefevre, E.; Mercier, D.; Romary, R.; Demian, C. Information Fusion with Belief Functions for Detection of Interturn Short-Circuit Faults in Electrical Machines Using External Flux Sensors. *IEEE Trans. Ind. Electron.* **2018**, *65*, 2642–2652. [CrossRef]
29. Lopez-Perez, D.; Antonino-Daviu, J. Application of Infrared Thermography to Failure Detection in Industrial Induction Motors: Case Stories. *IEEE Trans. Ind. Appl.* **2017**, *53*, 1901–1908. [CrossRef]
30. Ma, Y.; Wu, X. Discriminant sparse and collaborative preserving embedding for bearing fault diagnosis. *Neurocomputing* **2018**, *313*, 259–270. [CrossRef]

31. Sapena-Bano, A.; Martinez-Roman, J.; Puche-Panadero, R.; Pineda-Sanchez, M.; Perez-Cruz, J.; Riera-Guasp, M. Induction machine model with space harmonics for fault diagnosis based on the convolution theorem. *Int. J. Electr. Power Energy Syst.* **2018**, *100*, 463–481. [CrossRef]
32. Antonino-Daviu, J.; Popaleny, P. Detection of Induction Motor Coupling Unbalanced and Misalignment Via Advanced Transient Current Signature Analysis. In Proceedings of the 2018 XIII International Conference on Electrical Machines (ICEM), Alexandroupoli, Greece, 3–6 September 2018; pp. 2359–2364.
33. Bessous, N.; Chemsa, A.; Sbaa, S. New Vision about the Mixed Eccentricity Fault Causes in Induction Motors and its relationship with the Rolling Element Bearing Faults: Analytical model dedicated to the REB faults. In Proceedings of the 2018 International Conference on Communications and Electrical Engineering (ICCEE), El Oued, Algeria, 17–18 December 2018.
34. Brandt, M.; Gutten, M.; Koltunowicz, T.; Zukowski, P. Analysis of winding fault in electric machines by frequency method. In Proceedings of the 2018 ELEKTRO, Mikulov, Czech Republic, 21–23 May 2018; pp. 1–4.
35. Nikita, T.; Manickavasagam, K.; Sachin, S. Magnetostriction analysis on doubly fed induction generator under normal and low voltage ride through (LVRT) condition. In Proceedings of the 2017 International Conference on Technological Advancements in Power and Energy (TAP Energy), Kollam, India, 21–23 December 2017; pp. 1–5.
36. Ugale, R.T.; Gore, S.N.; Chaudhari, B.N. Web based remote and locally operated all in one electrical machine laboratory with data acquisition, fault diagnosis and protection. In Proceedings of the 2018 XIII International Conference on Electrical Machines (ICEM), Alexandroupoli, Greece, 3–6 September 2018; pp. 1979–1985.
37. Ullah, S.; McDonald, S.P.; Martin, R.; Benarous, M.; Atkinson, G.J. A permanent magnet assist, segmented rotor, switched reluctance drive for fault tolerant aerospace applications. *IEEE Trans. Ind. Appl.* **2019**, *55*, 298–305. [CrossRef]
38. Wang, X.B.; Zhang, X.; Li, Z.; Wu, J. Ensemble extreme learning machines for compound-fault diagnosis of rotating machinery. *Knowl. Syst.* **2019**. [CrossRef]
39. Kankar, P.K.; Sharma, S.C.; Harsha, S.P. Fault diagnosis of ball bearings using machine learning methods. *Expert Syst. Appl.* **2011**, *38*, 1876–1886. [CrossRef]
40. Pöyhönen, S.; Negrea, M.; Jover, P.; Arkkio, A.; Hyötyniemi, H. Numerical magnetic field analysis and signal processing for fault diagnostics of electrical machines. *COMPEL Int. J. Comput. Math. Electr. Electron. Eng.* **2003**, *22*, 969–981. [CrossRef]
41. Ijadi Maghsoodi, A.; Ijadi Maghsoodi, A.; Mosavi, A.; Rabczuk, T.; Zavadskas, E. Renewable energy technology selection problem using integrated h-swara-multimoora approach. *Sustainability* **2018**, *10*, 4481. [CrossRef]
42. Papoutsidakis, M.; Kalovrektis, K.; Drosos, C.; Stamoulis, G. Design of an Autonomous Robotic Vehicle for Area Mapping and Remote Monitoring. *Int. J. Comput. Appl.* **2017**, *167*, 36–41. [CrossRef]
43. Zekveld, M.; Hancke, G.P. Vibration Condition Monitoring Using Machine Learning. In Proceedings of the IECON 2018-44th Annual Conference of the IEEE Industrial Electronics Society, Washington, DC, USA, 21–23 October 2018; pp. 4742–4747.
44. Aydemir, G. Deep learning based spectrum compression algorithm for rotating machinery condition monitoring. In Proceedings of the ASME 2018 Conference on Smart Materials, Adaptive Structures and Intelligent Systems, San Antonio, TX, USA, 10–12 September 2018.
45. Dos Santos, T.; Ferreira, F.J.T.E.; Pires, J.M.; Damasio, C. Stator winding short-circuit fault diagnosis in induction motors using random forest. In Proceedings of the 2017 IEEE International Electric Machines and Drives Conference (IEMDC), Miami, FL, USA, 21–24 May 2017.
46. Ferreira, J.G.; Warzecha, A. An application of machine learning approach to fault detection of a synchronous machine. In Proceedings of the 2017 International Symposium on Electrical Machines (SME), Naleczow, Poland, 18–21 June 2017.
47. Senanayaka, J.S.L.; Kandukuri, S.T.; Khang, H.V.; Robbersmyr, K.G. Early detection and classification of bearing faults using support vector machine algorithm. In Proceedings of the 2017 IEEE Workshop on Electrical Machines Design, Control and Diagnosis (WEMDCD), Nottingham, UK, 20–21 April 2017; pp. 250–255.
48. Senanayaka, J.S.L.; Van Khang, H.; Robbersmyr, K.G. Towards online bearing fault detection using envelope analysis of vibration signal and decision tree classification algorithm. In Proceedings of the

2017 20th International Conference on Electrical Machines and Systems (ICEMS), Sydney, NSW, Australia, 11–14 August 2017.

49. Zhang, J.; Zhan, W.; Ehsani, M. On-line fault diagnosis of electric machine based on the Hidden Markov Model. In Proceedings of the 2016 IEEE Transportation Electrification Conference and Expo (ITEC), Dearborn, MI, USA, 27–29 June 2016.
50. Abid, F.B.; Zgarni, S.; Braham, A. Distinct bearing faults detection in induction motor by a hybrid optimized SWPT and aiNet-DAG SVM. *IEEE Trans. Energy Convers.* **2018**, *33*, 1692–1699. [CrossRef]
51. Giantomassi, A.; Ferracuti, F.; Iarlori, S.; Ippoliti, G.; Longhi, S. Signal Based Fault Detection and Diagnosis for Rotating Electrical Machines: Issues and Solutions. In *Studies in Fuzziness and Soft Computing*; Springer: Berlin/Heidelberg, Germany, 2015; Volume 319, pp. 275–309.
52. Afrasiabi, S.; Afrasiabi, M.; Parang, B.; Mohammadi, M. Real-Time Bearing Fault Diagnosis of Induction Motors with Accelerated Deep Learning Approach. In Proceedings of the 2019 10th International Power Electronics, Drive Systems and Technologies Conference (PEDSTC), Shiraz, Iran, 12–14 February 2019; pp. 155–159.
53. He, D.; Fahimi, B. Power management of a self-powered multi-parameter wireless sensor for IoT application. In Proceedings of the 2018 IEEE Applied Power Electronics Conference and Exposition (APEC), San Antonio, TX, USA, 4–8 March 2018; pp. 1380–1385.
54. Martin-Diaz, I.; Morinigo-Sotelo, D.; Duque-Perez, O.; Romero-Troncoso, R.J. An Experimental Comparative Evaluation of Machine Learning Techniques for Motor Fault Diagnosis under Various Operating Conditions. *IEEE Trans. Ind. Appl.* **2018**, *54*, 2215–2224. [CrossRef]
55. Shamshirband, S.; Rabczuk, T.; Chau, K.W. A Survey of Deep Learning Techniques: Application in Wind and Solar Energy Resources. *IEEE Access* **2019**, *7*, 164650–164666. [CrossRef]
56. Senanayaka, J.S.L.; Van Khang, H.; Robbersmyr, K.G. Online Fault Diagnosis System for Electric Powertrains Using Advanced Signal Processing and Machine Learning. In Proceedings of the 2018 XIII International Conference on Electrical Machines (ICEM), Alexandroupoli, Greece, 3–6 September 2018; pp. 1932–1938.
57. Chahine, K. Rotor fault diagnosis in induction motors by the matrix pencil method and support vector machine. *Int. Trans. Elecr. Energy Syst.* **2018**, *28*. [CrossRef]
58. Kao, I.H.; Wang, W.J.; Lai, Y.H.; Perng, J.W. Analysis of Permanent Magnet Synchronous Motor Fault Diagnosis Based on Learning. *IEEE Trans. Instrum. Meas.* **2019**, *68*, 310–324. [CrossRef]
59. Mendonça, P.L.; Bonaldi, E.L.; de Oliveira, L.E.L.; Lambert-Torres, G.; Borges da Silva, J.G.; Borges da Silva, L.E.; Salomon, C.P.; Santana, W.C.; Shinohara, A.H. Detection and modelling of incipient failures in internal combustion engine driven generators using Electrical Signature Analysis. *Electr. Power Syst. Res.* **2017**, *149*, 30–45. [CrossRef]
60. Cabal-Yepez, E.; Romero-Troncoso, R.; Garcia-Perez, A.; Osornio-Rios, R. Single-parameter fault identification through information entropy analysis at the startup-transient current in induction motors. *Electr. Power Syst. Res.* **2012**, *89*, 64–69. [CrossRef]
61. Dehghani, M.; Riahi-Madvar, H.; Hooshyaripor, F.; Mosavi, A.; Shamshirband, S.; Zavadskas, E.K.; Chau, K.W. Prediction of hydropower generation using grey wolf optimization adaptive neuro-fuzzy inference system. *Energies* **2019**, *12*, 289. [CrossRef]
62. Delpha, C.; Diallo, D.; Al Samrout, H.; Moubayed, N. Multiple incipient fault diagnosis in three-phase electrical systems using multivariate statistical signal processing. *Eng. Appl. Artif. Intell.* **2018**, *73*, 68–79. [CrossRef]
63. Vas, P. *Parameter Estimation, Condition Monitoring, and Diagnosis of Electrical Machines*; Oxford University Press: Oxford, UK, 1993; Volume 27.
64. Tidriri, K.; Tiplica, T.; Chatti, N.; Verron, S. A generic framework for decision fusion in fault detection and diagnosis. *Eng. Appl. Artif. Intell.* **2018**, *71*, 73–86. [CrossRef]
65. Delgado-Arredondo, P.A.; Morinigo-Sotelo, D.; Osornio-Rios, R.A.; Avina-Cervantes, J.G.; Rostro-Gonzalez, H.; de Jesus Romero-Troncoso, R. Methodology for fault detection in induction motors via sound and vibration signals. *Mech. Syst. Signal Process.* **2017**, *83*, 568–589. [CrossRef]
66. Serdio, F.; Lughofer, E.; Pichler, K.; Pichler, M.; Buchegger, T.; Efendic, H. Fuzzy fault isolation using gradient information and quality criteria from system identification models. *Inf. Sci.* **2015**, *316*, 18–39. [CrossRef]

67. Serdio, F.; Lughofer, E.; Zavoianu, A.-C.; Pichler, K.; Pichler, M.; Buchegger, T.; Efendic, H. Improved fault detection employing hybrid memetic fuzzy modeling and adaptive filters. *Appl. Soft Comput.* **2017**, *51*, 60–82. [CrossRef]
68. Jafari, H.; Poshtan, J. Fault detection and isolation based on fuzzy-integral fusion approach. *IET Sci. Meas. Technol.* **2019**, *13*, 296–302. [CrossRef]
69. Chen, Z.; Li, X.; Yang, C.; Peng, T.; Yang, C.; Karimi, H.R.; Gui, W. A data-driven ground fault detection and isolation method for main circuit in railway electrical traction system. *ISA Trans.* **2019**, *87*, 264–271. [CrossRef]
70. Jung, D.; Frisk, E. Residual selection for fault detection and isolation using convex optimization. *Automatica* **2018**, *97*, 143–149. [CrossRef]
71. Jung, D.; Sundstrom, C. A Combined Data-Driven and Model-Based Residual Selection Algorithm for Fault Detection and Isolation. *IEEE Trans. Control. Syst. Technol* **2019**, *27*, 616–630. [CrossRef]
72. Kannan, R.; Solai Manohar, S.; Senthil Kumaran, M. Nominal features-based class specific learning model for fault diagnosis in industrial applications. *Comput. Ind. Eng.* **2018**, *116*, 163–177. [CrossRef]
73. Na, W.; Park, C.; Lee, S.; Yu, S.; Lee, H. Sensitivity-based fault detection and isolation algorithm for road vehicle chassis sensors. *Sensors* **2018**, *18*, 2720. [CrossRef] [PubMed]
74. El Bakri, A.; Koumir, M.; Boumhidi, I. Extreme learning machine-based non-linear observer for fault detection and isolation of wind turbine. *Aust. J. Electr. Electron. Eng.* **2019**, *16*, 12–20. [CrossRef]
75. Sarwar, M.; Mehmood, F.; Abid, M.; Khan, A.Q.; Gul, S.T.; Khan, A.S. High impedance fault detection and isolation in power distribution networks using support vector machines. *J. King Saud Univ. Eng. Sci.* **2019**. [CrossRef]
76. Shahnazari, H.; Mhaskar, P.; House, J.M.; Salsbury, T.I. Modeling and fault diagnosis design for HVAC systems using recurrent neural networks. *Comput. Chem. Eng.* **2019**, 189–203. [CrossRef]
77. Yang, J.; Guo, Y.; Zhao, W. Long short-term memory neural network based fault detection and isolation for electro-mechanical actuators. *Neurocomputing* **2019**. [CrossRef]
78. Abderrahmane, M.; Mohammed, B. Fault diagnosis of a wind turbine benchmark via statistical and support vector machine. *Int. J. Eng. Res. Afr.* **2018**, *37*, 29–42. [CrossRef]
79. Ait-Izem, T.; Harkat, M.F.; Djeghaba, M.; Kratz, F. On the application of interval PCA to process monitoring: A robust strategy for sensor FDI with new efficient control statistics. *J. Process. Control.* **2018**, *63*, 29–46. [CrossRef]
80. Fazai, R.; Ben Abdellafou, K.; Said, M.; Taouali, O. Online fault detection and isolation of an AIR quality monitoring network based on machine learning and metaheuristic methods. *Int. J. Adv. Manuf. Technol.* **2018**, *99*, 2789–2802. [CrossRef]
81. Khorasgani, H.; Biswas, G. A methodology for monitoring smart buildings with incomplete models. *Appl. Soft Comput. J.* **2018**, *71*, 396–406. [CrossRef]
82. Lindahl, P.A.; Green, D.H.; Bredariol, G.; Aboulian, A.; Donnal, J.S.; Leeb, S.B. Shipboard Fault Detection Through Nonintrusive Load Monitoring: A Case Study. *IEEE Sens. J.* **2018**, *18*, 8986–8995. [CrossRef]
83. Mohamed Syed Ali, A. Helmet deduction using image processing. *Indones. J. Electr. Eng. Comput. Sci.* **2018**, *9*, 342–344. [CrossRef]
84. Zhang, D.; Qian, L.; Mao, B.; Huang, C.; Huang, B.; Si, Y. A Data-Driven Design for Fault Detection of Wind Turbines Using Random Forests and XGboost. *IEEE Access* **2018**, *6*, 21020–21031. [CrossRef]
85. Cheng, F.; Wang, J.; Qu, L.; Qiao, W. Rotor current-based fault diagnosis for DFIG wind turbine drivetrain gearboxes using frequency analysis and a deep classifier. *IEEE Trans. Ind. Appl.* **2018**, *54*, 1062–1071. [CrossRef]
86. Mosavi, A.; Faizollahzadeh ardabili, S.; Várkonyi-Kóczy, A.R. List of Deep Learning Models. *Preprints* **2019**, 2019080152. [CrossRef]
87. Gao, Z.; Sheng, S. Real-time monitoring, prognosis, and resilient control for wind turbine systems. *Renew. Energy* **2018**, *116*, 1–4. [CrossRef]
88. Zhu, Y.; Zhu, C.; Song, C.; Li, Y.; Chen, X.; Yong, B. Improvement of reliability and wind power generation based on wind turbine real-time condition assessment. *Int. J. Electr. Power Energy Syst.* **2019**, *113*, 344–354. [CrossRef]

89. Vigneshkumar, S.; Shankar, V.K.; Krishna, P.N.; Supriya, P. Fault Detection in Gearbox Using Motor Electrical Signature Analysis. In Proceedings of the 2018 9th International Conference on Computing, Communication and Networking Technologies (ICCCNT), Bangalore, India, 10–12 July 2018; pp. 1–6.
90. Yang, B.-S.; Oh, M.-S.; Tan, A.C.C. Fault diagnosis of induction motor based on decision trees and adaptive neuro-fuzzy inference. *Expert Syst. Appl.* **2009**, *36*, 1840–1849.
91. Patel, R.K.; Giri, V. Feature selection and classification of mechanical fault of an induction motor using random forest classifier. *Perspect. Sci.* **2016**, *8*, 334–337. [CrossRef]
92. Wardle, F.P. Vibration forces produced by waviness of the rolling surfaces of thrust loaded ball bearings Part 1: Theory. *Proc. Inst. Mech. Eng. Part C J. Mech. Eng. Sci.* **1988**, *202*, 305–312. [CrossRef]
93. Lynagh, N.; Rahnejat, H.; Ebrahimi, M.; Aini, R. Bearing induced vibration in precision high speed routing spindles. *Int. J. Mach. Tools Manuf.* **2000**, *40*, 561–577. [CrossRef]
94. Zimroz, R.; Bartelmus, W.; Barszcz, T.; Urbanek, J. Diagnostics of bearings in presence of strong operating conditions non-stationarity—A procedure of load-dependent features processing with application to wind turbine bearings. *Mech. Syst. Signal Process.* **2014**, *46*, 16–27. [CrossRef]
95. Chen, J.; Pan, J.; Li, Z.; Zi, Y.; Chen, X. Generator bearing fault diagnosis for wind turbine via empirical wavelet transform using measured vibration signals. *Renew. Energy* **2016**, *89*, 80–92. [CrossRef]
96. Teng, W.; Ding, X.; Zhang, X.; Liu, Y.; Ma, Z. Multi-fault detection and failure analysis of wind turbine gearbox using complex wavelet transform. *Renew. Energy* **2016**, *93*, 591–598. [CrossRef]
97. Muralidharan, V.; Sugumaran, V. Feature extraction using wavelets and classification through decision tree algorithm for fault diagnosis of mono-block centrifugal pump. *Measurement* **2013**, *46*, 353–359. [CrossRef]
98. Wen, L.; Gao, L.; Li, X.; Wang, L.; Zhu, J. A Jointed Signal Analysis and Convolutional Neural Network Method for Fault Diagnosis. *Procedia CIRP* **2018**, *72*, 1084–1087. [CrossRef]
99. Tsoumakas, G.; Katakis, I. Multi-label classification: An overview. *Int. J. Data Warehous. Min. (IJDWM)* **2007**, *3*, 1–13. [CrossRef]
100. Herrera, F.; Charte, F.; Rivera, A.J.; Del Jesus, M.J. Multilabel classification. In *Multilabel Classification*; Springer: Berlin/Heidelberg, Germany, 2016; pp. 17–31.
101. Read, J.; Pfahringer, B.; Holmes, G. Multi-label classification using ensembles of pruned sets. In Proceedings of the 8th IEEE International Conference on Data Mining, Pisa, Italy, 15–19 December 2008; pp. 995–1000.
102. Read, J.; Pfahringer, B.; Holmes, G.; Frank, E. Classifier chains for multi-label classification. *Mach. Learn.* **2011**, *85*, 333. [CrossRef]
103. Fu, H.; Cheng, J.; Xu, Y.; Wong, D.W.K.; Liu, J.; Cao, X. Joint optic disc and cup segmentation based on multi-label deep network and polar transformation. *IEEE Trans. Med. Imaging* **2018**, *37*, 1597–1605. [CrossRef] [PubMed]
104. Rauber, T.W.; Mello, L.H.; Rocha, V.F.; Varejão, F.M. Multi-label fault classification experiments in a chemical process. In Proceedings of theIEEE Brazilian Conference on Intelligent Systems, Sao Paulo, Brazil, 18–22 October 2014; pp. 265–270.
105. Cakir, E.; Heittola, T.; Huttunen, H.; Virtanen, T. Polyphonic sound event detection using multi label deep neural networks. In Proceedings of the 2015 International Joint Conference on Neural Networks (IJCNN), Killarney, Ireland, 12–17 July 2015; pp. 1–7.
106. Yang, X.; Song, J.; Wu, X.; Xie, L.; Liu, X.; Li, G. Identification of unhealthy Panax notoginseng from different geographical origins by means of multi-label classification. *Spectrochim. Acta Part A Mol. Biomol. Spectrosc.* **2019**, *222*, 117243. [CrossRef] [PubMed]
107. Hsu, C.-W.; Chang, C.-C.; Lin, C.-J. *A Practical Guide to Support Vector Classification*; National Taiwan University of Science and Technology (Taiwan Tech): Taipei City, Taiwan, 2003.
108. Dudani, S.A. The distance-weighted k-nearest-neighbor rule. *IEEE Trans. Syst. Man Cybern.* **1976**, *SMC-6*, 325–327. [CrossRef]
109. Basu, K.; Debusschere, V.; Bacha, S.; Maulik, U.; Bondyopadhyay, S. Nonintrusive load monitoring: A temporal multilabel classification approach. *IEEE Trans. Ind. Inf.* **2014**, *11*, 262–270. [CrossRef]
110. Domeniconi, C.; Gunopulos, D. Adaptive nearest neighbor classification using support vector machines. In Proceedings of the Advances in Neural Information Processing Systems, Vancouver, BC, Canada, 3–8 December 2001; pp. 665–672.
111. Keskes, H.; Braham, A.; Lachiri, Z. Broken rotor bar diagnosis in induction machines through stationary wavelet packet transform and multiclass wavelet SVM. *Electr. Power Syst. Res.* **2013**, *97*, 151–157. [CrossRef]

112. Imandoust, S.B.; Bolandraftar, M. Application of k-nearest neighbor (knn) approach for predicting economic events: Theoretical background. *Int. J. Eng. Res. Appl.* **2013**, *3*, 605–610.
113. Hu, L.-Y.; Huang, M.-W.; Ke, S.-W.; Tsai, C.-F. The distance function effect on k-nearest neighbor classification for medical datasets. *SpringerPlus* **2016**, *5*, 1304. [CrossRef]
114. Torabi, M.; Hashemi, S.; Saybani, M.R.; Shamshirband, S.; Mosavi, A. A Hybrid clustering and classification technique for forecasting short-term energy consumption. *Environ. Prog. Sustain. Energy* **2019**, *38*, 66–76. [CrossRef]
115. Gómez, D.; Rojas, A. An empirical overview of the no free lunch theorem and its effect on real-world machine learning classification. *Neural Comput.* **2016**, *28*, 216–228. [CrossRef]
116. Wolpert, D.H.; Macready, W.G. No free lunch theorems for optimization. *IEEE Trans. Evolut. Comput.* **1997**, *1*, 67–82. [CrossRef]
117. Galar, M.; Fernandez, A.; Barrenechea, E.; Bustince, H.; Herrera, F. A review on ensembles for the class imbalance problem: Bagging-, boosting-, and hybrid-based approaches. *IEEE Trans. Syst. Man Cybern. Part C (Appl. Rev.)* **2011**, *42*, 463–484. [CrossRef]
118. Polikar, R. Ensemble based systems in decision making. *IEEE Circuit. Syst. Mag.* **2006**, *6*, 21–45. [CrossRef]
119. Stack, J.R.; Habetler, T.G.; Harley, R.G. Effects of machine speed on the development and detection of rolling element bearing faults. *IEEE Power Electr. Lett.* **2003**, *1*, 19–21. [CrossRef]
120. Skipp, B. Ground vibration—Codes and standards. In *Ground Dynamics and Man-Made Processes*; The Institution of Civil Engineers: Glasgow, UK, 1998; pp. 29–41.
121. Liu, H.; Motoda, H. *Feature Extraction, Construction and Selection: A Data Mining Perspective*; Springer Science & Business Media: Berlin, Germany, 1998; Volume 453.
122. Holcsik, P.; Pálfi, J.; Čonka, Z.; Avornicului, M. A Theoretical Approach to The Implementation of Low-Voltage Smart Switch Boards. *Acta Polytech. Hung.* **2019**, *16*, 123–149.
123. Pálfi, J.; Takács, M.; Holcsik, P.; Mitrik, Z. Determination of the fault identification accuracy in LV networks using the Fuzzy method. In Proceedings of the 2016 IEEE International Conference on Systems, Man, and Cybernetics, Budapest, Hungary, 9–12 October 2016; pp. 2756–2761.
124. Glowacz, A.; Głowacz, Z. Recognition of rotor damages in a DC motor using acoustic signals. *Bull. Pol. Acad. Sci. Tech. Sci.* **2017**, *65*, 187–194. [CrossRef]
125. Glowacz, A.; Glowacz, W. Vibration-Based Fault Diagnosis of Commutator Motor. *Shock Vib.* **2018**, *2018*. [CrossRef]
126. Glowacz, A. Acoustic fault analysis of three commutator motors. *Mech. Syst. Signal Process.* **2019**, *133*, 106226. [CrossRef]

2

Review of Soft Computing Models in Design and Control of Rotating Electrical Machines

Adrienn Dineva [1], Amir Mosavi [1,2,3,4], Sina Faizollahzadeh Ardabili [5], Istvan Vajda [1], Shahaboddin Shamshirband [6,7,*], Timon Rabczuk [8] and Kwok-Wing Chau [9]

1 Institute of Automation, Kando Kalman Faculty of Electrical Engineering, Obuda University, 1034 Budapest, Hungary; dineva.adrienn@kvk.uni-obuda.hu (A.D.); amir.mosavi@kvk.uni-obuda.hu or amirhosein.mosavi@qut.edu.au (A.M.); vajda@uni-obuda.hu (I.V.)
2 School of the Built Environment, Oxford Brookes University, Oxford OX3 0BP, UK
3 Queensland University of Technology (QUT), Centre for Accident Research Road Safety–Queensland (CARRS-Q), 130 Victoria Park Road, Queensland 4059, Australia
4 Institute of Structural Mechanics, Bauhaus University Weimar, D-99423 Weimar, Germany
5 Biosystem Engineering Department, University of Mohaghegh Ardabili, Ardabil 5619911367, Iran; sina_fa1990@yahoo.com
6 Department for Management of Science and Technology Development, Ton Duc Thang University, Ho Chi Minh City, Vietnam
7 Faculty of Information Technology, Ton Duc Thang University, Ho Chi Minh City, Vietnam
8 Department of Computer Engineering, College of Computer and Information Sciences, King Saud University, Riyadh 12372, Saudi Arabia; timon.rabczuk@uni-weimar.de
9 Department of Civil and Environmental Engineering, Hong Kong Polytechnic University, Hung Hom, Hong Kong, China; dr.kwok-wing.chau@polyu.edu.hk
* Correspondence: shahaboddin.shamshirband@tdtu.edu.vn

Abstract: Rotating electrical machines are electromechanical energy converters with a fundamental impact on the production and conversion of energy. Novelty and advancement in the control and high-performance design of these machines are of interest in energy management. Soft computing methods are known as the essential tools that significantly improve the performance of rotating electrical machines in both aspects of control and design. From this perspective, a wide range of energy conversion systems such as generators, high-performance electric engines, and electric vehicles, are highly reliant on the advancement of soft computing techniques used in rotating electrical machines. This article presents the-state-of-the-art of soft computing techniques and their applications, which have greatly influenced the progression of this significant realm of energy. Through a novel taxonomy of systems and applications, the most critical advancements in the field are reviewed for providing an insight into the future of control and design of rotating electrical machines.

Keywords: soft computing; artificial intelligence; machine learning; rotating electrical machines; energy systems; deep learning; electric vehicles; big data; hybrid models; ensemble models; energy informatics; electrical engineering; computational intelligence; data science; energy management; control; electric motor drives

---

1. Introduction

In 1831, Michael Faraday invented the disk machine which can be considered as the earliest form of the DC machine. Until 1870 when Thomas Edison started the commercial development of the DC generator, electrical machines were applied and investigated only in laboratories. Edison's pioneering concept of electric power distribution from central generation stations allowed the introduction of the electric grid infrastructure concept, which was the primary condition for widespread the application of electric motors [1]. The patent of the three-phase induction motor by Nikola Tesla in 1888 was an utmost important milestone in the history of the electric machines. After that, the construction of electrification began, which lasted until 1930 in the USA, and then conquered worldwide [2]. Nowadays, it is estimated that more than sixty-five percent of the energy demand of present-day industrialized countries used by electrical drives [3]. Constant or variable speed or servo-motor drives are employed almost everywhere: in households, industry, electric traction, transportation, aerospace, military equipment, medical equipment, agriculture, etc. These sectors are mainly focused on an alternative source for the existing power transfer technologies, where major subsystems would also be controlled and driven electronically [4].

Recently, electromechanical drives for position and speed control play a key role in process control, EVs, factory automation, robotics, autonomous vehicle, mobility and energy conservation [5–15]. Due to the introduction of vector control methods in the 1970s and low price and the high reliability of cage induction motors made induction motor as the most popular rotating electrical machine. The recent advancements of PM materials, solid-state devices and microelectronics have contributed to new energy efficient, high-performance electric drives which use modern PM brushless motors [16–19]. It is entirely possible that these permanent magnet brushless motor drives will become predominant in the next century [20]. Nevertheless, the control, design, and optimization of rotating electrical machines require advanced techniques and modern mathematical models to keep up with the ever-increasing demand for energy [21–25]. In fact, practical solutions to engineering problems involve model-integrated computing [26,27]. Model-based computing approaches or so-called soft computing (SC) provide a challenging way to replace the procedure with a knowledge-based method [28–33]. SC techniques denote a set of computational approaches that are used to approximate mathematical problems which are hard or unable to solve by the traditional time-consuming classical mathematical tools [34–36]. Figure 1 presents the principal SC algorithms widely used in engineering applications.

Furthermore, combining SC, non-conventional and novel data representation techniques is a possible way to overcome this difficulty [37–40]. Over the last decades, the benefits of SC techniques have been widely recognized and have brought several new solutions in the design and control of electrical machines [41,42]. However, there is a gap in the effectiveness of the SC models [43,44]. Thus, identification and evaluation of the SC models would be a practical approach to evaluate the progress and provide an insight into the future application of SC methods [45,46]. Consequently, the intention of the current work is to give a comprehensive overview of the recent state-of-the-art solutions using soft computing techniques in the design and control of rotating electrical machines. The organization of the rest of this article is as follow. In Section 2, the research methodology of the review is presented. Section 3 presents a review of the SC models used in electrical rotating machines. Section 4 presents the discussion followed by Section 5 with a conclusion to the review.

2. Methodology

The primary goal of this survey is to present the state of the art of SC techniques in the design and control of rotating electrical machines. The purpose of the research methodology is to identify, classify, and review the notable peer-reviewed articles concerned in top-level subject fields. In our comprehensive review using the Thomson Reuters Web-of-Science and Elsevier Scopus for implementation of the search queries would assure that any paper in the database would meet the essential quality measures, originality, high impact, and high h-index.

To identify the application of soft computing in design, control, and development of rotating electrical machines, the search queries were carefully chosen to build the initial database. The taxonomy of SC influenced the search queries. Figure 1 presents the principal SC tools and the subsections that we have identified by a slight modification of the classifications given in [27,28]. Accordingly, the keywords of the search queries to identify the SC tools were selected to be "fuzzy" or "neural network" or "evolutionary computation" or "genetic" or "meta-heuristics" or "ant colony" or "particle swarm" or "tabu search" or "cuckoo search" or "simulated annealing" or "Bayesian network" or "Markov logic network" or "rough set".

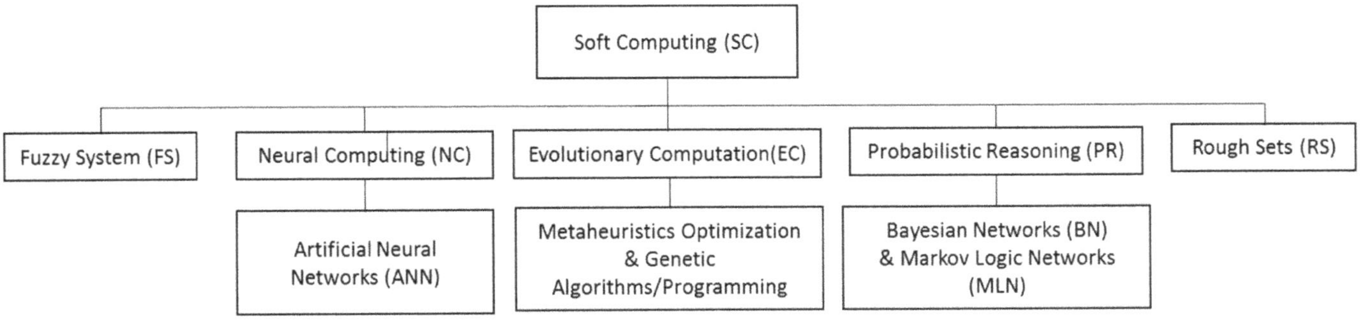

Figure 1. Taxonomy of soft computing methods used in making the initial database [47,48].

On the other hand, the search queries for the applications to rotating electrical machines may include various aspects of design and control, i.e., electric machine, rotating electrical machine, electric motor, electric generator, electromechanical generator, transformer and electrostatic, homopolar, permanent, brushed, magnet, reluctance and induction machines. Consequently, the entire search query is presented as: (TITLE-ABSTRACT-KEYWORDS ("electric machine" or "rotating electrical machine" or "electric motor" or "electric generator" or "electromechanical energy converter" or "generator" or "transformer" or "brushed machine" or "permanent magnet machine" or "induction machine") or TITLE-ABSTRACT-KEYWORDS ("reluctance machine" or "electrostatic machine" or "homopolar machine") and TITLE-ABSTRACT-KEYWORDS ("fuzzy" or "neural network" or "evolutionary computation" or "genetic" or "metaheuristics" or "ant colony" or "particle swarm" or "tabu search" or "cuckoo search" or "simulated annealing" or "Bayesian network" or "Markov logic network" or "rough set")) and PUBLICATION-YEAR > 1986 and PUBLICATION-YEAR < 2018. This query resulted in a total of 24,382 documents. It should be noted that each section contained a brief result and conclusion about the subject to help the authors and researchers make a sustainable and suitable decision about applications of soft-computing techniques.

Furthermore, to present an in-depth review and understanding of each modeling technique and its progress, we aimed at having different categories for the SC models used i.e., simple fuzzy systems, adaptive neuro-fuzzy inference system (ANFIS): neuro-fuzzy, advanced ANFIS/neuro-fuzzy: hybrids, neural computing: artificial neural network (ANN), evolutionary computation (EC) and metaheuristics, ant colony, tabu and cuckoo search, simulated annealing, probabilistic reasoning and Bayesian networks, and hybrid soft computing methods. Furthermore, the research methodology follows a comprehensive and structured workflow based on a systematic database search and cross-reference snowballing. The flowchart of the research methodology is presented in Figure 2.

In the step 1 of the methodology, the initial database of the relevant articles was identified based on the search queries of SC models through exploring the Thomson Reuters Web-of-Science and Elsevier Scopus databases. In step 2 the database was created with the relevant literature. For every SC model, we applied a new search query to suit that search well.

Nevertheless, some articles in the initial database might not be highly suitable for the review. For that matter, step 3 investigated an indent consideration of the literature to pass the irrelevant papers to step 4 to be excluded from the database. Note that the search queries will identify the relevant articles, yet the queries were uncertain as to whether the SC model belonged to a hybrid category. For instance, a hybrid model of SC may include single SC model. Therefore, step 5 is designed to reclassify the literature into one of the categories of SC models.

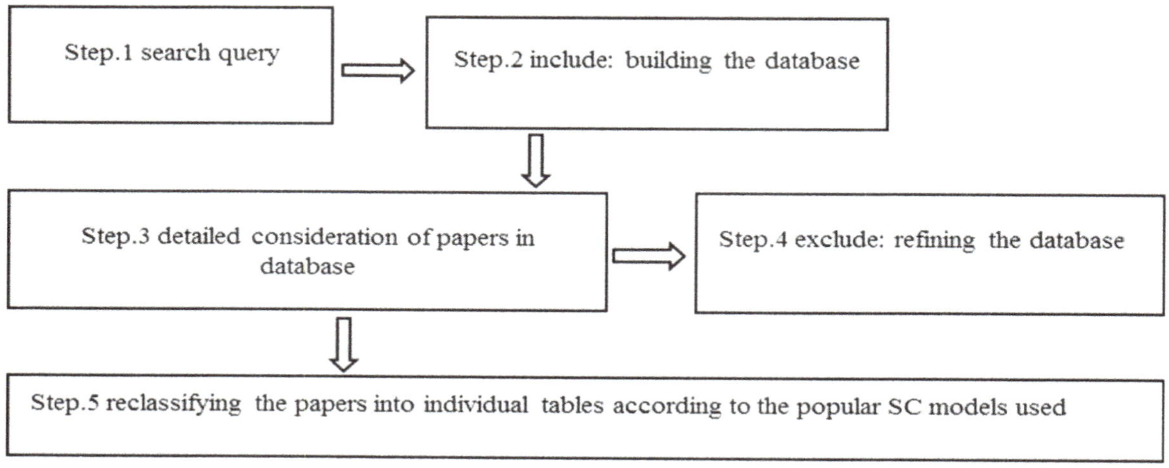

Figure 2. Methodology of research.

3. State of the Art of Soft Computing Techniques and Applications

Exploring the literature on rotating electrical machines shows continued progress on the design and advancement of rotating electrical machines. Almost half a million documents in this particular realm show the importance of this topic and the dependent technologies. Figure 3 shows the progress in some literature. It is also apparent that the use of SC models has been started from the late 80s and early 90s and boosted the design advancement.

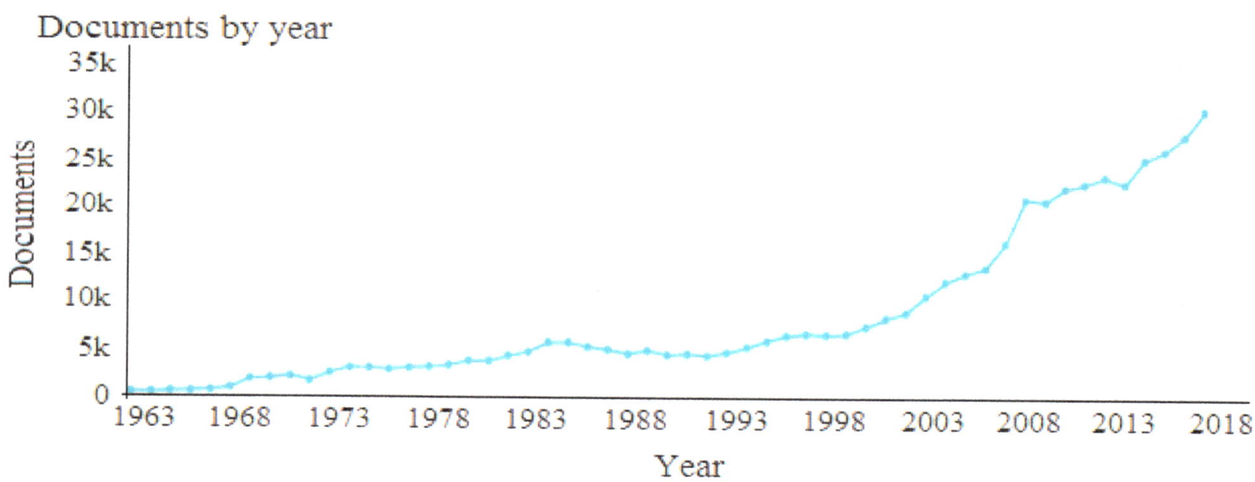

Figure 3. The number of published articles on rotating electrical machines from 1963–2017: 483,320 document results. (Source: Scopus & Web of Science).

Exploring the various SC models in wide applications to design and control the rotating electrical machines results in an extended database. The literature has been progressing since the early 1980s, with a fast-pace of involvement of SC methods in design and control purposes. Figure 4 shows the increasing popularity of SC, especially during the past two decades. It is apparent that statistics on the number of articles in literature change when using soft computing for advancement. Furthermore, the primary analysis of the database of the search shows that some SC methods have been more popular than the others and the usage of many SC methods has been very limited and not worth mentioning in the review. The most popular SC methods can be classified into five categories of fuzzy systems, neural computing: ANN, EC and metaheuristics, probabilistic reasoning and Bayesian Networks, and further hybrid soft computing methods.

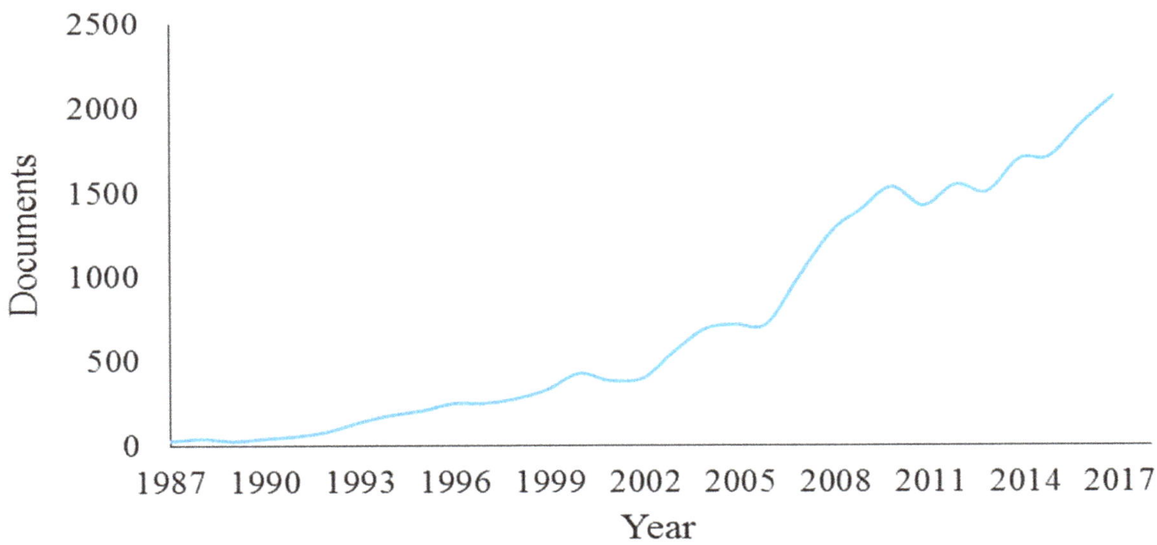

Figure 4. Analysis of the literature on soft computing in design, control, and development of rotating electrical machines from 1987–2017. Total of 30,382 document results. (Source: Scopus & Web of Science).

The analysis of literature type shows a significant number of articles were original papers written on the advancement and development of rotating electrical machines and only a tiny fraction of literature was devoted to reviewing papers (see, Figures 2–5).

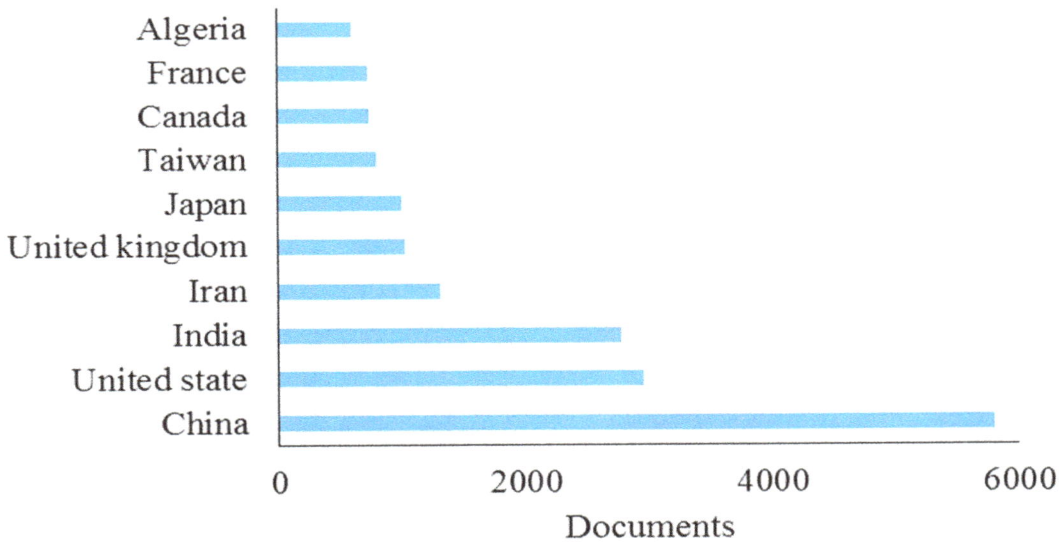

Figure 5. Analysis of the literature by based on countries (Source: Scopus & Web of Science).

Analysis of the literature based on countries showed that China, USA, India, Iran, and the UK are among the top five active regions on the advancement of rotating electrical machines using SC methods (see, Figure 6).

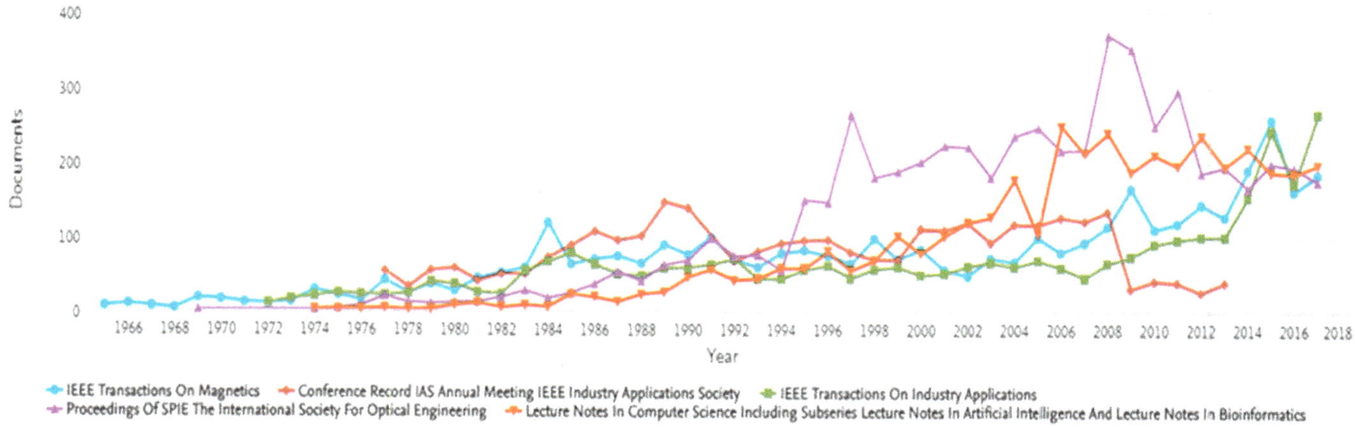

Figure 6. Analysis of the literature based on publishing sources (Source: Scopus & Web of Science).

The top five sources included: proceedings of the International Society for Optical Engineering (SPIE) (with 5841 papers), lecture notes in computer science, including subseries lecture notes in artificial intelligence and lecture notes in bioinformatics (with 4064 papers), Institute of Electrical and Electronics Engineers (IEEE) Transactions on Magnetics (with 3989 papers), and IEEE Transactions on Industry Applications (with 3306 papers), conference record Industry Applications Society IEEE Annual (with 2943 papers), for a total of 24,382 document results (see, Figure 6).

The analysis of literature type shows a significant number of articles are original papers written on the advancement and development of rotating electrical machine and only a tiny fracture of literature is devoted to reviewing papers (see, Figures 7 and 8).

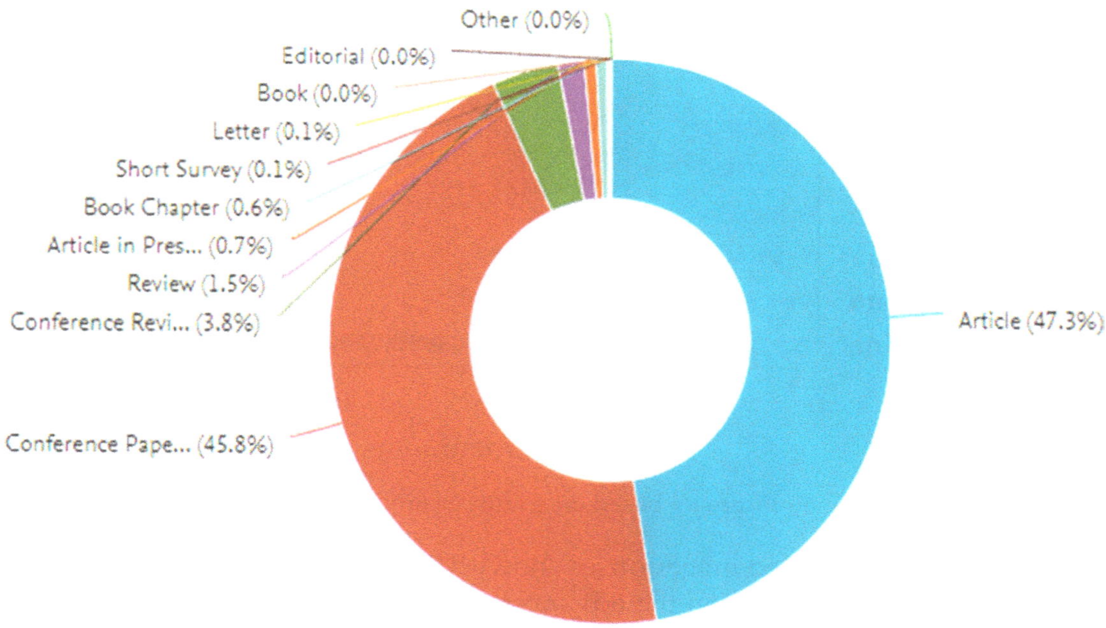

Figure 7. Literature types on the on rotating electrical machines using soft computing, from 1987–2017. (Source: Scopus & Web of Science).

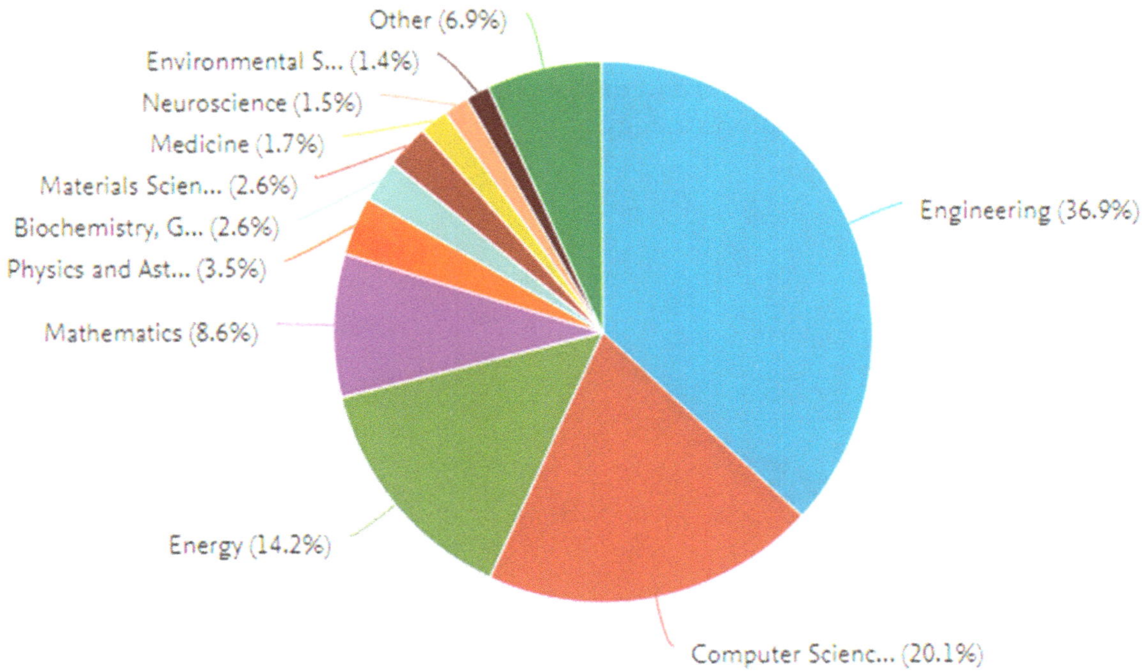

Figure 8. The subject area of rotating electrical machines using soft computing, from 1987–2017. The math, computer science, and energy subject areas had a rise (Source: Scopus & Web of Science).

3.1. *Fuzzy Systems*

Fuzzy logic is widely used for modeling complex and ill-defined systems. The core concept relies on the application of linguistic variables which are transformed into graded membership functions. Therefore, fuzzy logic is an extension of classical Boolean logic where logical statements are not only true or false but can also range from almost certain to very unlikely. A large number of practical applications apply fuzzy logic due to its excellent approximation properties.

3.1.1. Simple Fuzzy Systems

Fuzzy logic is widely used in nonlinear controllers due to their easy applicability and high performance. These are capable of replacing the most common Lyapunov's second method of nonlinear control, which is a complicated technique from a mathematical point of view and needs very skilled designers. The two most widely known systems are the Mamdani- and Sugeno-type inference systems. The main difference between them is that the latter allows functions as outputs instead of membership functions. Fuzzy logic serves as a useful tool for engineering practice; numerous examples found for the utilization of fuzzy logic in the control and design of electrical machines. Major works from the recent past collected in Table 1 illustrate that this trend is still unbroken.

Table 1. Simple fuzzy systems in rotating the electrical machine.

Reference	Year	SC Method	Application
Aguilar et al. [49]	2017	Fuzzy logic	Brake control
Rao Amulya [50]	2016	Mamdani type inference	Fault detection and vibration
Ben Smida, M.; Sakly, A. [51]	2016	Fuzzy controller	Power regulation
Z. Husain [52]	2018	Fuzzy expert system	Condition monitoring of power transformers via dissolved gas analysis test
Kahla, S.; et al. [53]	2018	Combination of standard on-off strategy with fuzzy logic	Maximize the power point tracking of wind energy

Aguilar [49] used a simple fuzzy logic algorithm for the intelligent brake control of an electric motorcycles engine in order to recover the maximum energy in braking processes while maintaining the vehicle's stability. The success of the employed method is very bold, such that the results of comparing the regenerative system with the ABS in a low adhesion condition have been presented in Figure 9 in terms of mean deceleration (a) and distance (b). Based on the results, the proposed systems could optimize the energy regeneration strategy.

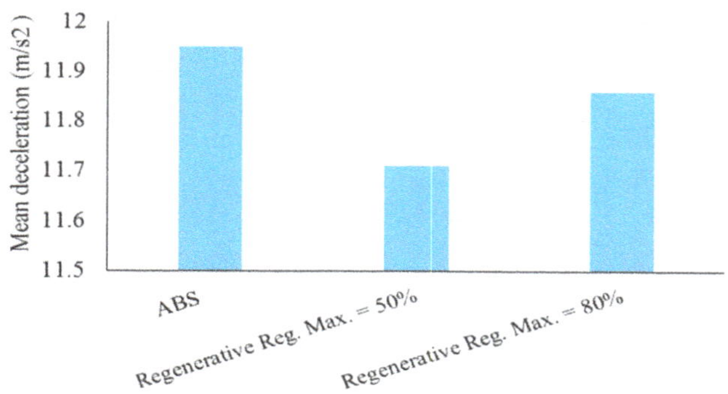

(a)

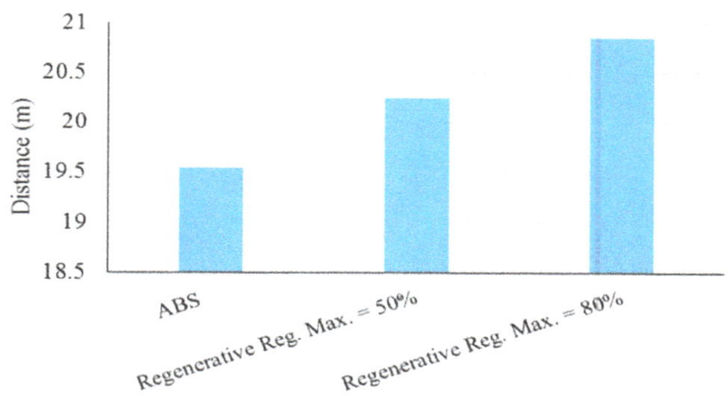

(b)

Figure 9. Results of the study by Aguilar [49] in terms of (**a**) mean deceleration and (**b**) distance.

In the paper by Amulya, in 2016 [50] a steam turbo-generator condition assessment problem was considered in which the condition is determined by fuzzy logic-based vibration analysis. Recommendations for improving the performance of the conventional pitch angle control strategy for power regulation in wind turbines are given in Ben Smida (2016) [51] in which the pitch angle is based on fuzzy logic. The study was developed using fuzzy logic and proportional integral derivative PID controllers. The comparison of the performance of the two systems was performed using mean absolute error (MAE) and mean percentage error (MPE). Such that, a fuzzy logic controller with low MAE (29.02%) and MPE (0.0088%) values provided a best-controlling ability compared that for PID controller which can make a smooth controlling process.

Husain (2018) [52] applies a fuzzy expert system for fault diagnosis of power systems which incorporates the information obtained from dissolved gas analysis test. The proposed approach could successfully reduce the issues raised by the conventional fault diagnosis methods. Also, the proposed method by Arumugam 2017 improves dynamic response and the energy saving and by the fast detection ability and removes the errors raised by the dynamic voltage restorer (DVR) response.

Kahla (2018) [53] introduces a fuzzy logic system in the on-off control strategy in case of wind energy conversion. The proposed method allows maximizing the power point tracking of wind energy and reducing the mechanical loads in comparison to the standard on-off control strategy.

3.1.2. ANFIS: Neuro-Fuzzy

Artificial neural systems represent systems acting as parallel distributed computing networks. Neural systems can find new associations, new functional dependencies and new patterns through learning. Therefore their primary advantage is their additivity. Consequently, combining neural networks with fuzzy logic techniques is evidently useful [54]. Thus, the excellent approximation properties are complemented with higher additivity and parallelism [40,55,56]. Such composed system is called fuzzy neural, neural fuzzy, neuro-fuzzy or fuzzy-neuro network, or ANFIS in which for instance neural networks can be used to tune membership functions of fuzzy systems. A vast amount of published papers supports its applicability [57–59]. However, in electrical machine design issues, only a few examples are found (see Table 2).

Table 2. Adaptive neuro-fuzzy systems in electrical machines.

Reference	Year	Method	Application
Arumugom [60]	2017	ANFIS based MIC controller	Dynamic Voltage Restorer Controller
Aruna [61]	2016	MWT and Accommodative ANFIS with ABC algorithmic rule.	IPMSM control
Bentouhami [62]	2016	Neuro-fuzzy combined method	Control of DSIM
Haritha [63]	2017	Q learning and ANFIS	Optimal relationship of rotor speed of the PMSG
Thankachan and Singh [64]	2016	Neuro-fuzzy	Speed and torque control of IM drive

An interesting application of neuro-fuzzy systems is described in Arumugom (2017) [60] where the authors attempt to design a DVR with energy conservation using a MIC and vibration energy harvester (VEH). Aruna [61] proposed in 2016 a hybrid method using ANFIS for the speed control of interior permanent magnet synchronous motor (IPMSM). Their results demonstrate that their solution enables better disturbance rejection compared to the classic PID controller. The settling, arising and overshooting times of the proposed controlling method has been compared with PID, MR-PID, and MWNN-particle swarm optimization (PSO) controlling algorithms. Figure 10 indicates the time values for each term. The related results indicate that the proposed controlling method is the best technique to eliminate the nonlinearity with high reliability, and sustainable performance compared with that for the other techniques.

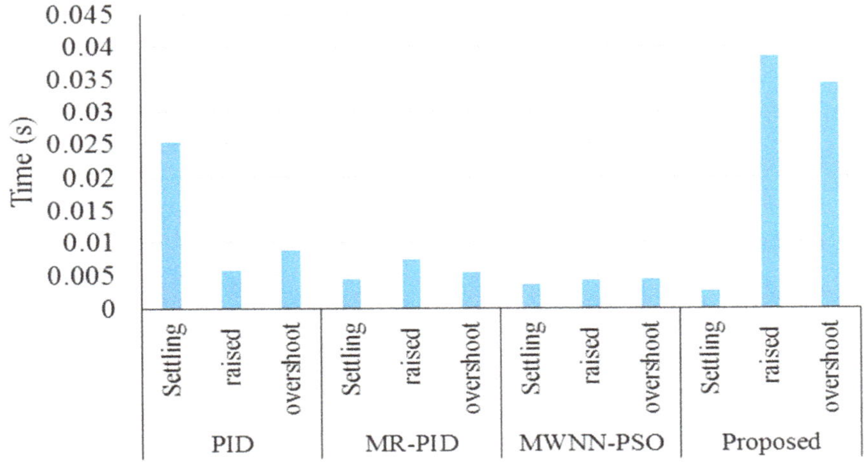

Figure 10. Comparing the time values for the study by Aruna (2016) [61].

Advanced method for the control of dual star induction machine (DSIM) supplied by Five-Level Inverter is shown by Bentouhami (2016) [62] who apply an ANFIS controller composed of an online learning algorithm with a neuro-fuzzy network. Haritha (2017) [63] introduced a neuro-fuzzy and Q learning-based technique for the maximum power point tracking control for improving learning efficiency of permanent magnet synchronous generator (PMSG) wind energy conversion system (WECS).

3.1.3. Advanced ANFIS/Neuro-Fuzzy: Hybrids

Adaptive neuro-fuzzy inference systems can be supported or combined further mathematical tools if the problem requires. The resulted systems are called hybrid fuzzy-neuro systems. In advanced ANFIS systems, the capabilities of the neural network are higher due to better learning algorithms (e.g., trust region reflective), so they can learn various parameters. These techniques are especially useful in highly nonlinear control problems of electrical machines or condition assessment issues (see Table 3).

Table 3. Hybrid techniques applied in electrical machine design and control.

Reference	Year	Technique	Application
Li et al. (2005) [65]	2005	Robust model reference fuzzy controller	DC servo motor system
Fatih Kececioglu [66]	2017	NFC	hybrid passive filter configuration proposed for PWM rectifiers
Gnanaprakasam [67]	2015	S-transform algorithm & ANFIS	detecting and classifying the vibration signal of IM
Hossain [68]	2015	FLC, SNC & ANFIS	Nonlinear controllers augment of a large-scale hybrid power system
Dehghani et al. [40]		ANFIS and GW	Hydropower generator performance

A vast amount of literature on nonlinear control solutions have been publishing in recent years. For instance, Li et al. (2005) [65] propose a control technique using a robust model reference controller with the combination of the hybrid fuzzy controller. The core of the control strategy is based on the classical Lyapunov technique which is a complicated technique, but the control signal includes fuzzy, classic and robust terms. The fuzzy logic controller was compared with the hybrid-fuzzy controller using root mean squared error (RMSE) values in terms of output tracking for DC servo motor and inverted pendulum. The results have been presented in Figure 11. Based on results, the hybrid controlling method had the best performance in both terms.

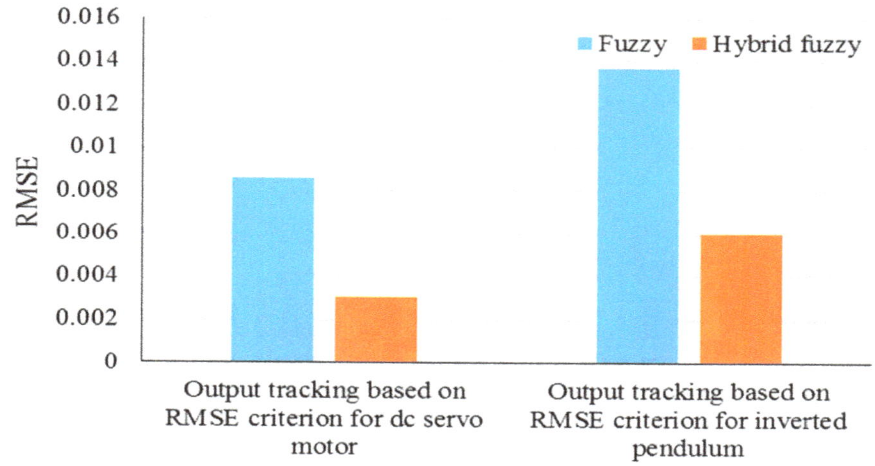

Figure 11. Results for the study by Li et al. (2005) [65].

In order to limit the current and voltage harmonics, and to improve total harmonic distortion and true power factor in Fatih Kececioglu, (2017) [66], a neuro-fuzzy controller is applied in a hybrid passive filter for PWM rectifiers. Gnanaprakasam (2015) [67] proposed a hybrid approach for the detection and classification of the vibration signal of an induction motor in which the fault detection process including the extraction of significant features of vibration signals is carried out by using the S-transformation algorithm. After, the ANFIS classification technique was employed to classify the signal into its faulty or the normal state. S-transform-ANFIS, S-transform- radial basis function neural network (RBFNN), S-transform-FFNN, and DWT-RBFNN were compared in terms of accuracy, sensitivity and specificity for IR faulty condition, centered or fault, opposite or fault, orthogonal or fault, and BB fault conditions. The general results have been presented in Figure 12. In all conditions, S-transform-ANFIS has the highest accuracy, sensitivity, and specificity compared with those for the other methods.

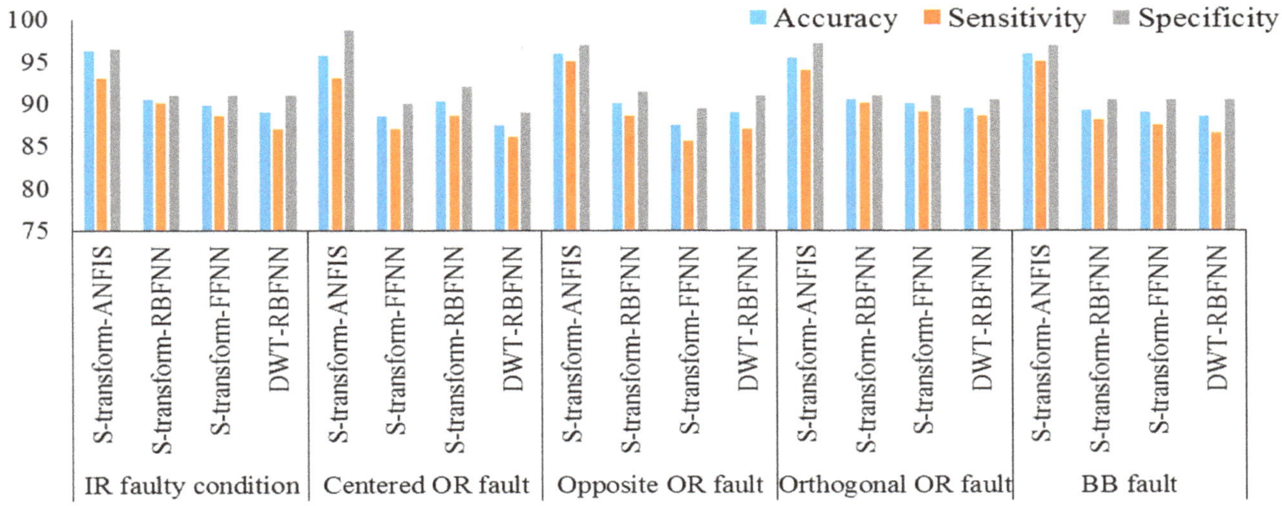

Figure 12. Results for the study by Gnanaprakasam (2015) [67].

In the paper by Hossain (2015) [68] three nonlinear control strategies are compared, namely the fuzzy logic controller, SNC, and ANFIS-based controller. Their results show that the proposed fuzzy logic controller-, SNC-, or ANFIS-based VR-FCL are effective in improving the transient stability of a large-scale hybrid power system consisting of a doubly-fed wind power generator (DFIG)-based wind farm, a PV plant, and SG.

In general, in all cases mentioning fuzzy systems, the fuzzy controller has a robust and smooth controlling process, which improves the control process and causes less damage to the hardware used and reduces the cost of servicing and maintaining of the systems. On the other hand, this would have a direct impact on the energy consumption of the control system and would lead to more sustainable control in the control system. This conclusion has also been confirmed in a study by Ardabili et al. (2016) [69] to design a fuzzy control system in a mushroom growing hall. Also, the operator's dominance and focus also increase with the use of fuzzy systems. Hyphenation hybrid systems can also optimize the system with the benefits of both fuzzy and neural systems in parallel with the creation of a fuzzy control sustainable system.

3.2. Neural Computing

The concept of neural networks is usually employed in itself. As the literature review reveals (see, Table 4). ANNs are a common tool of control tasks of electrical machines as other engineering applications [70,71]. However, it seems that the abilities of the nets have not yet been exploited in the field of machine design.

Table 4. Artificial neural networks.

Reference	Year	Technique	Application
Bouchiba [72]	2017	ANN-SMC	Control of multi-motor system coupled induction motors
Çelik et al. [73]	2017	multi-layer feedforward ANN	estimate the output power and efficiency of AFP SG
Zammit et al. [74]	2016	ANN controller	Direct torque control of DFIM
Zouggar et al. [75]	2018	ANN	Voltage and frequency control of a self-excited induction generator

Considering the ANNs–based solutions in the field of electrical machines the study of Bouchiba (2017) [72] provides a fresh example of the NN sliding mode controller for the control for multi-machine web winding systems. The advantage of this technique is that it can significantly reduce the chattering phenomenon and improve the error performance. In this study, the simulation and evaluation processes were performed by employing MATLAB/SIMULINK software. Based on the results, ANN-SMC controller has ignored the effect of disturbance. On the other hand, it is clear that the application of a hybrid PI-SMC controller is easy, but the performance of ANN-SMC controller is better than that for the PI-SMC. Detailed results indicate that the net performance improvement and the robustness of ANN-SMC controller are also strong superiority points compared with PI controller. Çelik et al. (2017) [73] designed a feed-forward multilayer NN in order to predict the efficiency and output power of an axial flux permanent magnet SG. Their tests have shown that the NN is highly suitable for this purpose according to the obtained 3% and 4% error percentages. The efficiency and power of a magnet generator were estimated using a neural network method. Based on results, the best structure was 2–18–12–2 which generated a maximum error of 3.587% for power and −3.909 for efficiencies value. Also, NN controller could estimate the power values higher than that for the experimental limits, and it can be a suitable technique. Some modification and improvements proposed for the classical direct torque controller, e.g., Zemmit et al. (2016) [75] designed an ANN- Direct Torque Control in which the IP and switching table have been replaced by a new artificial neural network. This strategy can reduce the high torque and flux ripples. The paper of Zidani (2018) presents a voltage and frequency control technique of self-excited induction generators that applies a NN-based inverse dynamic model of the system. Results indicate that the proposed technique can significantly increase the performance and stabilize the terminal voltage. This can be an economical and efficient way for stabilizing the voltage, and also it can be used by field programmable gate arrays controller cards on a real-time benchmark system.

In general, according to the results of the papers developed by NN techniques, it can be seen that this method has the most predictive effect on simulations and can, therefore, affect the performance of the system. Therefore, this system can be modeled from empirical data and based on the behavior of the predictive functional system, and the user's tastes can have the least interference, unlike the fuzzy system. Also, all studies have been conducted in order to improve existing conditions and offer the proposed system the best performance compared to previous systems.

3.3. EC and Metaheuristics

Metaheuristics are computational methods which optimize problems by iteratively making efforts to refine the function of a possible solution(s) to achieve a measurement value while guaranteeing polynomial time despite brute force optimization methods. The set of metaheuristics contain various algorithms, such as ant colony, evolutionary optimization, genetic algorithms, etc. EC is a subset of metaheuristics which involve algorithms or global optimization inspired by biological evolution. EC methods are stochastic optimization methods in which the initial set of candidate solutions are generated randomly. After, the new generations (new sets of possible solutions) are iteratively updated while mitigation the natural selection, mutation, etc. The fitting of a candidate solution is determined by a measure defined for the task under examination (see Table 5).

Table 5. Evolutionary computing and metaheuristics.

Reference	Year	Technique	Application
Mamede [76]	2018	GA and DE	Optimum design of SPSRM
Tamilselvi [77]	2018	Adaptive DE with FE	Design of PMMs for EV
Vanchinathan [78]	2018	BA	PID controller
Virtic [79]	2016	GA and analytical evaluation	Axial flux permanent magnet design

Optimization problems, especially ill-posed or multi-objective optimization tasks are common in control and design problems in most of the engineering problems, including electrical machine and drive systems. Most of the applications found in the latest paper employ the evolutionary computing techniques in nonlinear control tasks. There is only a limited number discussing the application or novelty of EC in rotating electrical machine design.

Design issues of electrical machines are usually raising complicated or multi-objective, ill-defined optimization problems. The application of Evolutionary-based solution or stochastic optimization methods can handle these problems by reducing the computational costs and simultaneously providing near-optimal solutions. For instance, Mamede (2018) [76] presents the combination of genetic algorithm (GA) and differential evolution (DE) techniques that allows maximizing the average torque and torque density and minimizing copper loss of single-phase switched reluctance machine (SPSRM). S. B. Tamilselvi, S. (2018) [77] proposes an evolutionary optimization procedure for the design of two, namely the surface-mounted and interior mounted permanent-magnet motor (PMMs) configurations using DE algorithm. Optimization tasks are the core of most control problems also. The metaheuristic optimization methods are effective alternatives in such cases, especially in sensorless drives. Vanchinathan (2018) [78] introduce the bat algorithm (BA) in the parameter tuning issue of a fractional–order PID controller for speed control of sensorless control of brushless DC electric motors. Modified genetic algorithm (MGA), ABC, and BA methods were employed in FOPID controller in three loads (0, 50 and 100% of full load) and the simulation was performed using MATLAB/Simulink software. Based on results, the proposed BA method, have the lowest steady-state error (%), peak time (s), rise time (s) and settling time(s) compared with those for the other methods in each load, separately (Figure 13). It seems that BA method has the best performance at 100% in case of settling, rise and peak times but in case of steady-state error, the condition of 0% provides the best performance for BA technique. The detailed results have been presented in Figure 13.

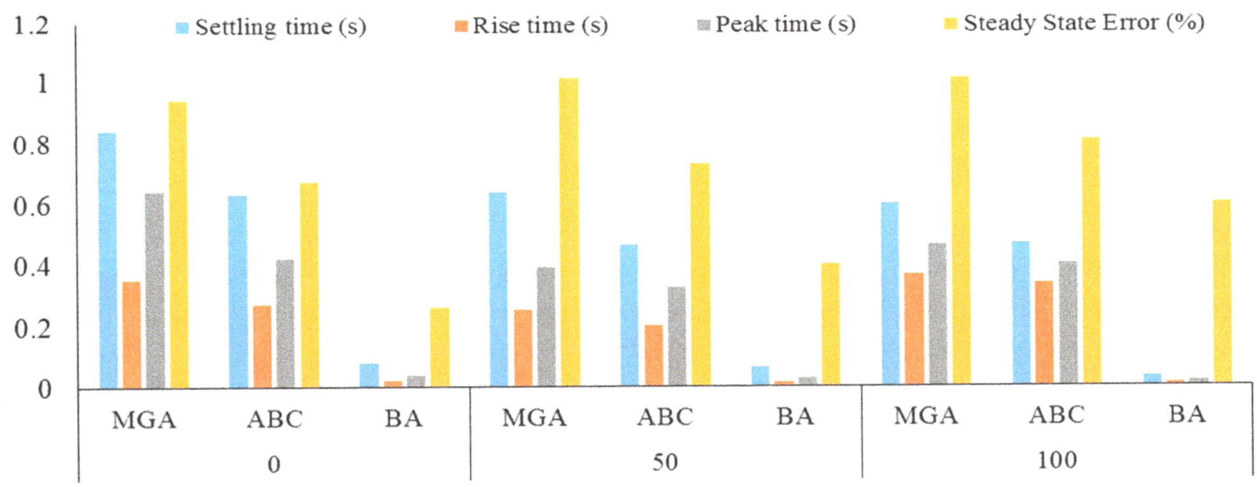

Figure 13. Results for the study by Vanchinathan (2018) [78].

The finding of Virtic (2016) [79] also supports that the evolutionary optimization with the analytical evaluation of objective functions significantly shortens the computational time required for design optimization in comparison with the finite element methods (FEM).

Considering the reasonable and suitable capabilities of the proposed method, it can be concluded that this method is considered as a process optimizer and according to its performance, it is suggested that the hybrid methods in this field can also be developed by researchers.

3.3.1. Ant colony

The ant colony system is a search metaheuristic which concept is based on the imitation of ants seeking nutrition. Ants can communicate with each other by a sophisticated way since they mark the different pathways from the anthill to the food sources and back by pheromone hormone respectively. The pheromone paths are perceived by another ant, and most likely followed. Pathways to foods can be very different, as well as obstacles between them. Ants tend to collect as much food as possible.

The characteristics of some individuals are different, but the unity of the colony is very effective. Each ant moves in an isolated and random way, but recognizing its pathways marked with pheromone is also likely to follow them. Meanwhile, an increase in their pheromone concentration by pheromone emissions, thereby increases the concentration of the pheromone attractiveness. Thus, the frequencies of frequently used routes increase the pheromone's level while being neglected paths. The pheromone emitted by the ants continuously vaporizes, that is, confirmation on the given route (releasing a new pheromone dose) without decreasing the pheromone level (until contains pheromone). If there are two paths leading to food, the shorter route is more likely to turn the ants (as they arrive sooner and later back). Thus, they often reinforce the pheromone level, thereby providing higher "attractiveness" for the route. However, this will attract more ants to the shorter journey. After a while, the shorter way will be taken by most ants, while the longer path for pheromone levels will decrease to a minimum level. The ant colony metaheuristic shares the search task between the ant agents. These agents have very basic subtleties and, to a certain extent, simulate the behavior of true ants. "Artificial ants" (agents) build on solutions based on appropriate problem-specific constructive heuristics. A pheromone matrix determines the order of the building blocks of the solution used to construct good solutions. The values stored in the pheromone matrix are taken into account by the other agents on a probability basis when constructing his solutions. The first ant colony algorithm was applied for the traveling salesman problem (TSP). Since then, it has been widespread in engineering practice. Table 6 includes application of ant colony in rotating electrical machines.

Table 6. Application of ant colony in electrical machines.

Reference	Year	Technique	Application
Ameli et al. [80]	2017	Ant colony	Simultaneous dynamic scheduling of feeder reconfiguration of DG units having uncertain and variant generations
Batista et al. [81]	2014	AS and ant colony algorithm on the graph	Interior permanent magnet motor design

Ameli (2017) [80] applies the ant colony algorithm in distribution systems. The purpose of their study is to minimize the total operational cost of the grid and to reduce the costs and losses. Based on Figure 14, the proposed method strongly reduces the network power losses for both total voltage profile index (TVPI) and total power loss index (TPLI) of the DG units.

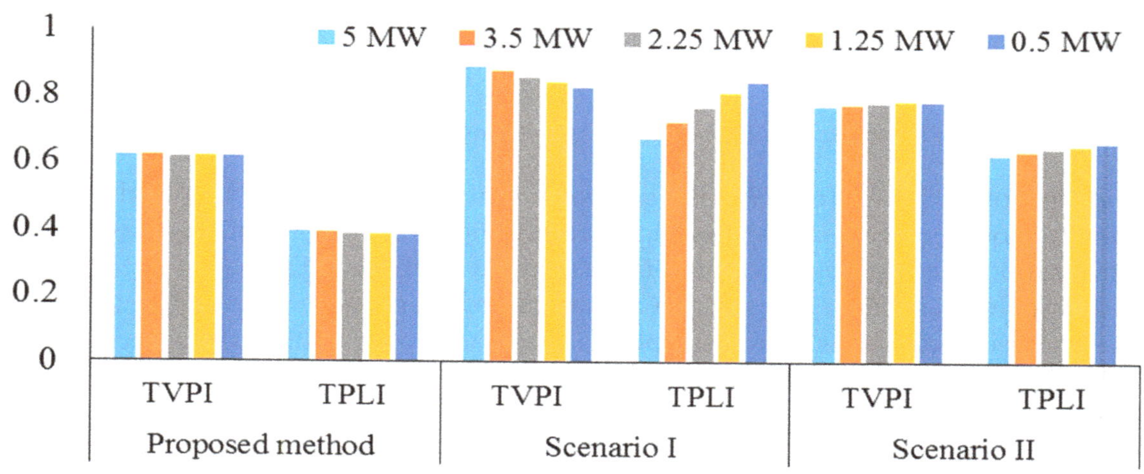

Figure 14. Results for the study by Ameli (2017) [80].

Batista (2014) [81] gives a novel solution where the permanent interior magnet motor design domain is discretized into a suitable equivalent graph representation and an ant system (AS) algorithm is employed to achieve an efficient distribution of materials into this graph. Ant systems have successfully applied also to control problems. The proposed method provided sustainable torque and suitable shapes for designed problems. The results show that the mechanism of optimization of the multi-dimensional topology of electromagnetic devices was successfully performed by the proposed method.

3.3.2. Tabu and Cuckoo Search

The tabu search proposed by Glover (1986) [82] is based entirely on the methods described above and their direct development. In the immediate environment of our current solution, we are investigating new, better solutions. In the tabu search method we are always trying to keep moving, so we always look for the best solution in the environment, even if it is worse than the solution being tested. Since this method has a lot to explore, we have to keep in mind that we need to select the direction in which the search was going and in which direction it was not yet. This so-called taboo lists that are marked by taboo tabs. Limitations on taboo tabs helping to get through the roads already gone by, increasing the search effectiveness. The cuckoo method [83] is a similar optimization method inspired by the breeding behavior of cuckoo species. In this approach, the eggs represent the possible solutions and the search is performed by the quality of the eggs, which leads to finding the best solution. The main differences and advantages of the discussed approaches can be found in some of the most interesting publications collected in Table 7.

Table 7. Tabu and cuckoo search techniques.

Reference	Year	Technique	Application
Chen et al. [84]	2017	The best-guided cuckoo search algorithm	Optimization of power systems
Wang et al. [85]	2018	Improved tabu search algorithm	Optimize a hybrid excited generator
Yang et al. [86]	2015	Improved tabu search algorithm	Performance evaluation of PM motors

The gbest-guided cuckoo search algorithm with the feedback control strategy and constraint domination rule for the security and economic operation of the power system is discussed in Chen et al. (2017) [84]. Results of the proposed method were compared with the results of another method developed by researchers in this field. The comparison results indicate that the proposed approach could provide high-quality, feasible solutions for different problems.

The paper by Wang et al. (2018) [85] proposes a hybrid excited generator for variable-speed wind power generation systems. The excitation sources are PMs on the rotor and the field windings on the stator. The proposed generator can achieve constant voltage control and maximum power point tracking control by controlling the field current while the optimization is carried out by a modified tabu search method. In Yang et al. (2015) [86] a new topological design of PM motor is investigated via five different types of permanent magnet arrangements. An improved tabu search algorithm and finite element method are proposed for the evaluation of the performance of these motors.

3.3.3. Simulated Annealing (SA)

The simulated annealing (SA) method was proposed by Kirkpatrick, Gelatt, and Vecchi in 1983 [87]. The SA searches for a new solution in the vicinity of a randomly chosen initial solution. However, if certain probability conditions are met, it also allows the solution to deteriorating. Therefore, it can escape from a local optimum that a gradient-based or a hill climbing type algorithm cannot. In the vertically improving methods, a series of random points is generated while in the target function improvement can be observed. To apply the SA strategy, it is necessary to define the following four main elements for each optimization problem; (a) solution space, (b) transition (i.e., how to generate a new random point), (c) fitness function, d) annealing schedule (i.e., determination of the number of iterations to be performed in the internal cycle and the method used to reduce the control parameter in the outer cycle). One of the advantages of the algorithm is that it can be easily implemented.

Furthermore, the SA algorithm with proper parameterization does not stick in a local optimum (see Table 8).

Table 8. Simulated annealing.

Reference	Year	Technique	Application
Chiu et al. [88]	2016	SA	Optimization space-constrained base-vibration system excited with a specific frequency
Farhani et al. [89]	2017	SA	Real-time efficiency optimization of the direct vector-controlled induction motor drives
Torrent–Fontbonne and López [90]	2016	SA	DGs

SA is an efficient tool for various optimization problems. It is shown by Chiu et al. (2016) [88]; that SA can be applied in the optimization task of space-constrained base-vibration systems. Based on the results, in the case of increasing the base-vibrating frequency, the extracted electrical energy will increase. Consequently, based on the buckling and fatigue analysis, the employed approach for an optimal designed one-mass vibration-based electromagnetic energy harvester is quite efficient in maximizing the energy.

Simulated annealing algorithm is used for finding a global optimal rotor flux while the suboptimal operating point is determined by a fast, analytical method using the induction machine's model in work of Farhani et al. (2017) [89]. The paper by Torrent–Fontbona and López (2016) [90] reviews the problem and provides a new solution for supporting grid planning with an optimized number DGs. A particle swarm optimization method is found to be the tool for finding the optimal number of DGs which allows for maximizing the profit of the generators, minimizing the system losses, and improving the voltage profile.

3.4. Probabilistic Reasoning and Bayesian Networks

The core of the theory of Bayesian interpretation of probability is based on a concept that probability is a measure of a rational agent's degree of belief in a proposition Ramsey (1926) [91], and de Finetti (1937) [92]. Probabilistic reasoning with graphical models, known as Bayesian networks or belief networks, has become an active field of research and practice in artificial intelligence, operations research, and statistics in the last two decades. In order to reduce the computational needs, an asymptotic approximation could be applied, but it reduces the precision level. Therefore, Bayesian-type methods are mostly applied in power system and network analysis as can be seen from Table 9. Few examples can be found for diagnostics problems or parameter estimation problems of power systems using Bayesian techniques.

Table 9. Application of probabilistic reasoning in the field of electrical machines and power systems.

Reference	Year	Technique	Application
Kari et al. [93]	2018	ANFIS and Dempster–Schafer	Fault detection of power transformers
Kazemdehdashti et al. [94]	2018	Density estimator based on generalized cross-entropy method	Optimal power flow problem in networks containing renewable energy resources, nonstationary loads
Torrent–Fontbona, F. and B. López [70]	2017	Probability theory methods and neuro-fuzzy modeling	Networks containing renewable energy sources

The dissolved gas analysis (DGA) approach is the basic tool for fault detection of power transformers. Kari (2018) [93] introduces an ANFIS system for these purposes. The outputs of the ANFIS model are evaluated by the Dempster–Schafer method which is originally developed for medical reasoning problems. The results of consistency, accuracy, and reliability have been presented in Figure 15. As is clear, the proposed method has the highest average consistency, accuracy, and reliability.

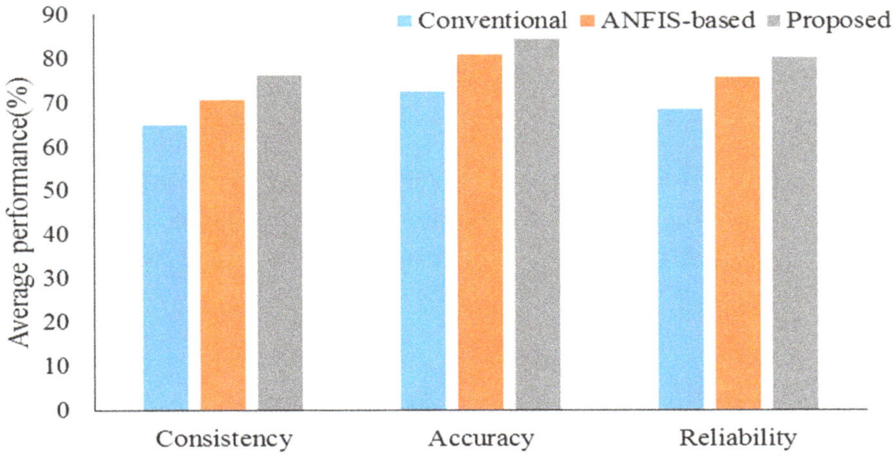

Figure 15. Results for the study by Kari (2018) [93].

Kazemdehdashti (2018) [94] introduces a new density estimator for optimal power flow problem of networks containing renewable energy sources, etc., using generalized cross-entropy method. Figure 16 presents the results of the study in terms of RMSE (a) and time (b) for comparing the proposed method for 14-bus. As is it shown, the RMSE of the proposed method is lower than that for the other methods, but this has increased the processing time. The two-point estimation method (TPEM) has a lower processing time and lowers accuracy.

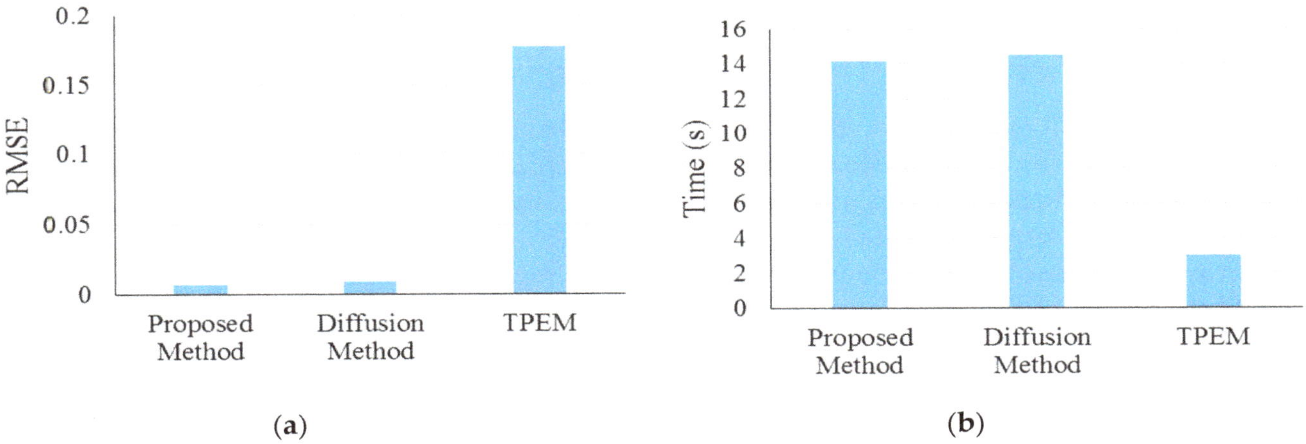

Figure 16. Results for the study by Kazemdehdashti (2018) [94] in terms of (**a**) root mean squared error (RMSE) and (**b**) time.

The paper by Jiang et al. (2015) [95] presents an integrated Bayesian probabilistic method for precise power splitting and the degradation diagnostics of a single-shaft combined cycle plant, accounting for uncertainties in the measured data. Lakehal et al. (2017) [96] introduces a Bayesian network based on the Duval triangle method in the dissolved gas analysis which is the primary technique for identifying faults in transformers. The proposed model makes a quantitative estimation analysis of transformer faults, which supported the IEEE C57.104 standard (see Table 10).

Table 10. Application of Bayesian methods in the field of electrical machines and power systems.

Reference	Year	Technique	Application
Jiang et al. [95]	2015	Bayesian inference method	Diagnostics of a CCGT power plant
Lakehal et al. [96]	2017	Bayesian network based on the Duval triangle method	Transformer condition monitoring via Dissolved gas analysis
Mansouri et al. [97]	2015	Bayesian methods	State and parameter estimation of IMs

Mansouri et al. (2015) [97] applies the Bayes method in a nonlinear time-varying state and parameter estimation task of induction machines (Ims). Based on results, the Bayes method has higher accuracy. Based on the results of the second comparative study, for all methods, estimation accuracy depends on estimating the model parameters as well as the convergence of the estimated parameters and states.

3.5. Hybrid Soft Computing Methods

Single SC techniques are found to be efficient due to their robustness and easy interpretability. At the same time, though their usage can be so advantageous, it is still limited by their exponentially increasing computational complexity. The combination of different techniques offers the synergy of their beneficial properties. Consequently, hybrid soft computing techniques have gained popularity (see Table 11).

Table 11. Application of hybrid methods in the field of electrical machines and power systems.

Reference	Year	Technique	Application
Dai et al. [98]	2016	Quantum-behaved PSO for parameter identification with SA	Ensure the accuracy of the DFIG for control performance of the generator
McDonald [99]	2017	Hybrid GA and pattern search process	Magnet SGs for offshore direct drive wind turbines
Meo et al. [100]	2016	Combining a multi-objective PWO and ANN	Design optimization of a direct-drive permanent magnet flux switching generators

Dai et al (2016) [98] propose a new hybrid quantum-behaved particle swarm optimization-based solution for improving the precision of parameter identification (five parameters including stator resistance, stator inductance, rotor resistance, rotor inductance, and mutual inductance of stator and rotor) of the DFIG. Results indicated that the proposed algorithm accelerated the computing process by reducing the processing time. For the optimization of direct-drive permanent magnet synchronous generators, McDonald, (2017) [99] used a hybrid genetic algorithm and pattern search process and has found that the surface-mounted permanent magnet generator produces the lower cost of energy.

With the purpose of reducing the costs and weight of the machine while maximizing the amplitude of the induced voltage as well as minimizing its total harmonic distortion Meo et al. (2016) [100] has presented a new hybrid approach for the design optimization of a direct-drive permanent magnet flux switching generators. Figure 17 presents the results of hybrid model, non-dominated sorting genetic algorithm (NSGA-II), abyss and proposed ANN-multi-objective particle swarm optimization (SMPSO) in terms of cost (a), total harmonic distortions of voltage (THD) (b), weight (c) and rated line voltage; em (d). As is clear from Figure 17, the proposed method reduces the cost ($) by about 23.85%, reduces the THD by about 17%, reduces the weight by about 44.73% and increases em by about 2.6%. These values claim that the proposed hybrid method improves the condition compared with that for the other techniques.

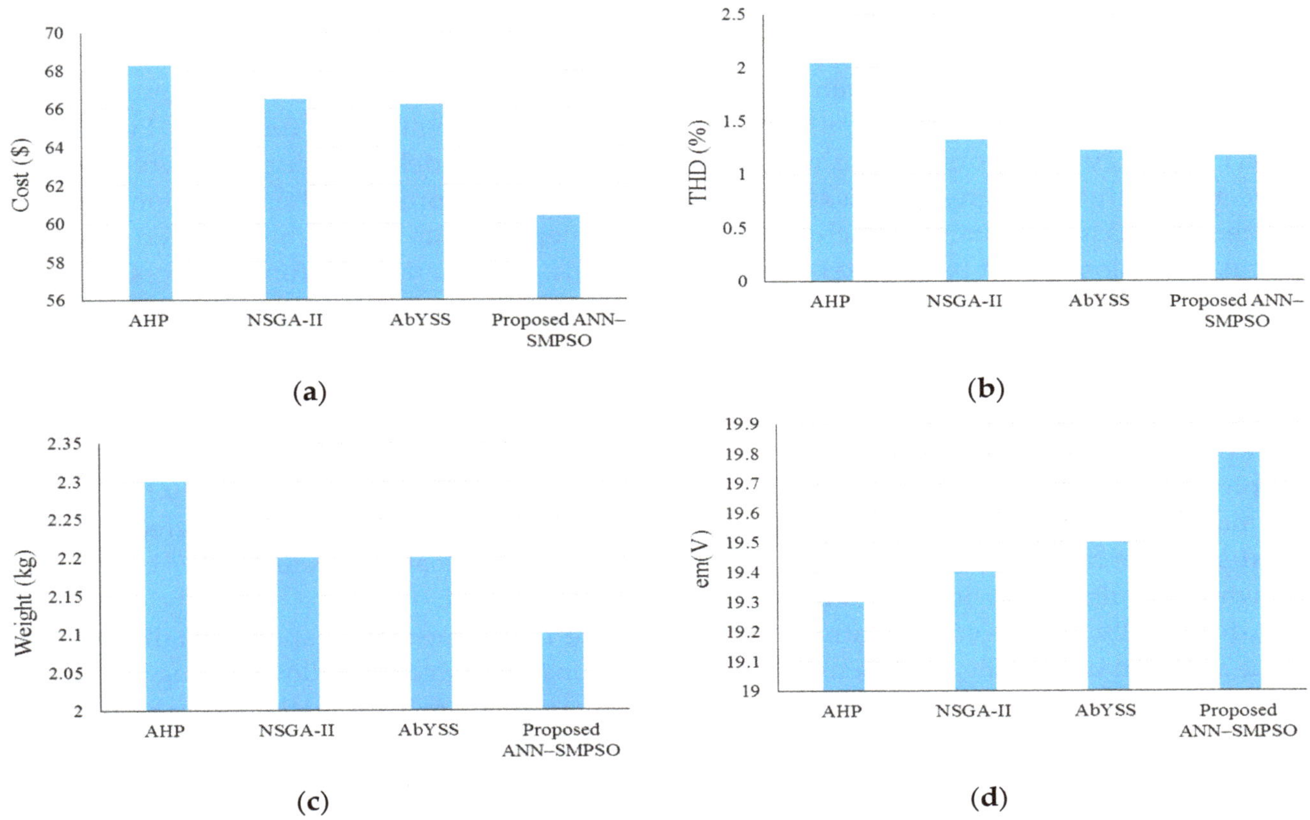

Figure 17. Results for the study by Meo et al. (2016) [100] in terms of (**a**) cost; (**b**) THD; (**c**) weight and (**d**) em.

In general, it can be claimed that hybrid methods can be the most successful methods than single methods because these methods can take advantage of several methods simultaneously and reduce the defects in each method. With the advancement of artificial intelligence and soft computing methods, the trend towards hybrid methods increases for the reasons given. There are still many hybrid methods that have not been used, which suggests that exploration of these methods be increased which can clarify the dark points in this and increase the orientation and usage by the researchers.

4. Discussions

The trend of recent years suggests that the greatest flexibility of design may be ensured by soft computing tools. In this paper, we explore the latest techniques of intelligent algorithmic methods for design and control of electrical machines and investigate how the reviewed methods affect the solution.

The great majority of automated control tasks needs the existence of the system model on some level. General practice of the simple feedback control applies the a priori knowledge, i.e., the system is regulated by the variation of the measured output. However, in practice, the system under control is highly nonlinear, only a few variables are measurable external disturbances degrade the performance, and also the technological processes generate further constraints in the desired control of the system dynamics [101]. Also, adaptive and advanced control methods usually apply the mathematically complicated Lyapunov method [102,103]. It is clear that the system modeling and mathematically easy manageability provided by advanced methods are the key factors in a successful control method design [23]. The choice of the proper control law in case of a specific electric machine and drive system raises many questions. Below we attempt to highlight how the replacement of conventional techniques with the various "soft" approaches collected in this paper may allow the greatest benefits.

The algorithmic approaches rely on the physical system equations of motion described by mathematical equations. Such techniques calculate analytically, for example, the controller's parameters. In such cases when information may be incomplete, or the system model is not fully available, fuzzy logic can be applied as a universal approximator due to the linguistic terms. It is suitable for modeling the system behavior, deducing the control law by fuzzy reasoning built on implications or tuning the parameters by applying the suitable membership functions. Furthermore, the various fuzzy operators allow the flexible fitting of the model behavior to the real scenario. Engineers usually face regression tasks during the control design process. The regression in practice is related to measured data. For each measurement, an error value should be assigned so that the classical regression cannot be handled in common models. Despite this, the fuzzy regression assigns the error values to the measurement error and if the regression line is within this error range that line is accepted. However, an infinite number of lines with such properties may exist within the error range. Therefore, by applying a fuzzy membership function which follows the error probability's distribution and setting a rule basis could easily solve this problem without large computational burdens.

A potential advantage of fuzzy logic-based approximation is that they allow us to optimize multiple variants and configurations and reduce complexity so that the problem may have limited to a smaller number of parameters. Furthermore, the linguistic variables ensure great flexibility context-sensitive machine modeling. This will allow the system to learn by example. We can construct rules whose results are as close as possible for the expected results.

Additionally, the classical control methods require the immediate response of the system in various situation depending on the environment's reactions. Fuzzy control allows taking into account not just the physical properties but also the rules that cannot be exactly mathematically described. A real situation would require a large number of parameters that cannot be totally measured under the operation conditions, for instance, if we wish to teach an electric car how to behave in specific traffic situations. Such cases demand immediate responses, and we cannot wait minutes while the onboard computer solves complicated differential equations since the application of fuzzy logic and other soft computing tools allow the system to learn from examples. After the proper rule system is set up, the system can give the nearly ideal responses.

It is obvious when the only the approximate system model is available, heuristic or cognitive techniques can be applied which describe the context- or situation-dependent behavior of the system. The neural networks emerging from statistical methods are capable of approximating a regular function [104]. This property is highly advantageous in nonlinear control problems where analytically unknow nonlinear functions are required to be found using only a few (measurement) data. Today a significant number of the NN variants are known, therefore specific and highly nonlinear cases can be efficiently handled. We should emphasize that both NN and fuzzy methods can be adapted in most of the conventional control and system modeling techniques.

The estimation of the physical parameters which characterize the model of the machine is also essential for the efficient control of electrical machines. Parameter estimation is particularly fundamental in condition monitoring tasks. Here, the parameter estimation techniques are similar to the methods applied in the control law synthetization and state observations. Such tasks mainly result in nonlinear programming problems where the previous knowledge may become crucial. Since the Bayesian statistic may enhance the estimation performance in contrast to classical probabilistic solutions, examples can be found, e.g., in [105] and in the publications mentioned in Section 3.4. Furthermore, an important step of the estimation is its convergence. Resulted optimization problems may have many constraints. At this point, the evolutionary computing techniques, for instance the heuristic search methods, are applicable, such as genetic algorithms, taboo search, cuckoo method, etc.

Optimization issues arise in most engineering disciplines, and we encounter them in our everyday life. The optimization task aims to achieve the best possible solution according to some objective function by setting the appropriate values of the decision variables. Finding the optimal solution is typically a difficult task because, in practice, the target function is nonlinear, different constraints may occur, etc., and often the algorithm stuck into local extrema of the search space. In case of electrical machine design, although the general practice gives the main directions [4], all of the modeling assumptions are necessary to be satisfied from the beginning of the design process [106] since design optimization may dispute further questions. Constrained multivariate optimization problems usually arise during machine design issues from which it follows that the mathematical models become more complicated.

The classical design practice can describe the problems only with complicated differential equations. Of course, the parameters must be known to solve the equations. Usually, it would take many parameters to describe a real system, or it would take a too long time to solve the problem. The parameters in most cases otherwise cannot be measured. It is, therefore, necessary to estimate them. Another solution is soft computing, especially fuzzy logic.

The general feature of designing technical systems is that the same task can be solved in several ways [107]. Comparison of individual solutions is rather subjective, weighing the advantages and disadvantages relative to each other basically influences the assessment. The solution that is considered to be optimal can only be selected by a pre-formulated criteria system. The subjectivity of judgment is carried by the accepted criteria system. Since its adoption, the optimum system selection has become a specific task. We should be aware that one solution that is considered to be optimal by one of the criteria systems, using another set of criteria, is no longer optimal. Fuzzy logic is proven to be suitable in such assessments. The optimal solution may depend on the applied model parameters. Therefore, once the optimal model has been set up a post-optimality analysis should be performed, usually referred to as parametric study. The role of sensitivity analysis is revealing the sensitive parameters. Classical techniques of sensitivity analyses are observing the influence of the perturbation of the parameters. Soft computing tools could also support this step of the process by providing soft boundaries of the sets.

We see that finding a single optimum solution can be a major challenge. However, finding several optimal solutions at once may result in a more complex task. The so-called multi-modal optimization (MMO) approach serves as a solution to these latter optimization problems [108] during the machine design process. Multimodal optimization could benefit from the various evolutionary-based methods, for instance in founding the basin of attraction of the optimum, or reducing the search domain. Also, the above chapters give insight into modern heuristics which may speed up the design process with a combination of classical optimization because they allow easy parallelization and system-level optimization. Similarly, the fulfillment of the various conditions of the model configurations at the same time is problematic and requires skillful designers. MMO performs the design process by optimizing multiple models having some shared variables. The multi-model optimization approach uses different representations, including coarser and more detailed models, and allows multiple configurations [108]. Therefore, in these cases, stochastic searching methods play a prominent role. Many studies suggest various evolutionary algorithms to multimodal problems (see, e.g., [109]).

However, the expected improvement depends on many conditions, of which the most prominent may be the performance. Most of the suggested algorithmic techniques are only partially effective or applicable to a specific class of problems without the capability of simultaneously satisfying all the different requirements. Further investigations are needed to achieve more reliable and general methods. The theory of fuzzy logic has been elaborated and well established by today, but its introduction into new areas is still ongoing. From the above presented general panorama of the literature, it is obvious that there are only a few results on the global optimization solutions of evolutionary techniques and there is still limited contribution for its fundamental theory. Altogether, these observations suggest that by transferring the approaches between intelligent computational methods in the direction of MMO approaches could be an excellent step toward new and effective algorithmic techniques and plausible theory.

5. Conclusions

Electrical machines and drive systems account for more than half of the global electricity used, and most of it is consumed by electric motors. Thus, there is a great demand for efficient designs to satisfy the competitive requirements. Furthermore, due to the trend of automation, demonstrated in the review, the control and drive systems of electrical rotating machines are expected to continue to grow. Having mentioned that, the control systems and design methodologies and techniques will be evolved to achieve higher performance.

Several SC models reviewed that dramatically improved the efficiency of rotating electrical machines through improvement of parameters for finite element analysis, and nonlinear time-domain finite element analysis. It was observed that most of the issues of each design and control steps are covered by advanced methods. A wide range of intelligent computational techniques serves for the designers, and all of them has advantages and disadvantages. Choosing the optimal method of SC shown to be not an easy task. Such a selection task requires the user to understand the differences among the computation algorithms. This brief literature survey indicates that the fundamental issue is finding the trade-off among model complexity, accuracy, and computing time. Designers could use a combination of various soft computing techniques as the optimal solution to obtain the required machine performances as hybrid techniques.

Recent studies have provided, from a theoretical point of view, how modern heuristics and soft computing techniques may improve the efficiency of engineering practice in the field of electrical machine design and control. Due to the problem-specific nature of SC methods, new effective algorithmic techniques are necessary to be designed for implementing them in any stage of the process. Potential advantages of SC relying on the capabilities to deal with context-sensitivity, high flexibility despite classical methods based on sharp borders. Additionally, without complex mathematical operations, sorting can be automatically performed on alternatives. Also, the relationships between categories can be easily examined.

At present, there is still no general theory or methodology for designing such heuristics, e.g., ANN, GA, etc. The primary intention of the designer should be the specification of the structure of the neural network or the fuzzy reasoning system, etc. After, the key parameters must be identified. Experimental results may help in training these systems.

Our review supports that the hybrid methods outperform the standard ones. Considerable progress in the field could be achieved by the synergetic utilization of the heuristic algorithmic and MMO methods. Our research could support designers in finding enhanced methods.

Author Contributions: Review, writing, data curation and analyzing the results, A.D., A.M., and S.F.A.; electrical machine expertise, I.V., and A.D.; soft computing, and machine learning expertise, A.M., A.D., S.S., and S.F.; management, database, conceptualization, administration and methodology, A.M.; visualization, S.F.; supervision, resources, software, resources, expertise, and controlling the results, S.S., K.-W.C., I.V., S.S., and T.R.

Acknowledgments: Mosavi contributed to the preparation of this paper during his Endeavour Research Leadership Award of Australian Education International at Queensland University of Technology, CARRS-Q, under supervision of Andry Rakotonirainy.

Acronyms

ABC	Artificial bee colony
AS	Ant system
ANFIS	Adaptive neuro-fuzzy inference system
ANN	Artificial neural network
ANN-SMC	Artificial neural networks - sliding mode control
ANN-SMPSO	ANN-multi-objective particle swarm optimization
BA	Bat algorithm
CCGT	Combined cycle gas turbine
DC	Direct current
DFIG	Doubly fed induction generator
DL	Deep learning

DGA	Dissolved gas analysis
DSIM	Dual star induction machine
DG	Distributed generation
DE	Differential evolution
EC	Evolutionary computation
FE	Finite-element
EV	Electric vehicle
FR	Feeder reconfiguration
IM	Induction machine
MLP	Multi-layered perceptron
SA	Simulated annealing
IM	Induction machine
SPSRM	Single-phase switched reluctance machine
PMSG	Permanent magnet synchronous generator
PMMs	Permanent-magnet motors
PID	Proportional integral derivative
TPLI	Total power loss index
NSGA-II	Non-dominated sorting genetic algorithm
PV	Photovoltaic
DFIM	Double fed induction machine
GA	Genetic algorithm
NFC	Neuro-fuzzy controller
MAE	Mean absolute error
MPE	Mean percentage error
DWT	Discrete wavelet transform
SVM	Support vector machine
PM	Permanent magnet
MIC	Multi-functional intelligent controller
IPMSM	Interior permanent magnet synchronous motor
MMO	Multi-model or multi-modal optimization
MWT	Multi ripple remodel
PSO	Particle swarm optimization
MGA	Modified genetic algorithm
SC	Soft computing
RBF	Radial basis function
TPEM	Two-point estimation method
ML	Machine learning
VEH	Vibration energy harvester
SG	Synchronous generator
RMSE	Root mean squared error
SNC	Static nonlinear controller
GRNN	Generalized regression neural networks
RBFNN	Radial basis function neural network
TVPI	Total voltage profile index
GW	Grey wolf
AFPMSG	Axial flux permanent magnet synchronous generator

References

1. Chapman, S. *Electric Machinery Fundamentals*; Tata McGraw-Hill Education: New York, NY, USA, 2005.
2. Daniels, A.R. *Introduction to Electrical Machines*; Macmillan: London, UK, 1976.
3. Tavner, P.; Ran, L.; Penman, J.; Sedding, H. *Condition Monitoring of Rotating Electrical Machines*; IET: Stevenage, UK, 2008; Volume 56.

4. Pyrhonen, J.; Jokinen, T.; Hrabovcova, V. *Design of Rotating Electrical Machines*; John Wiley & Sons: Hoboken, NJ, USA, 2013.
5. Dehkordi, S.G.; Larue, G.S.; Cholette, M.E.; Rakotonirainy, A.; Rakha, H.A. Ecological and safe driving: A model predictive control approach considering spatial and temporal constraints. *Trans. Res. Part D Transp. Environ.* **2019**, *67*, 208–222. [CrossRef]
6. Liberto, C.; Valenti, G.; Orchi, S.; Lelli, M.; Nigro, M.; Ferrara, M. The impact of electric mobility scenarios in large urban areas: The rome case study. *IEEE Trans. Intell. Transp. Syst.* **2018**, *99*, 1–10. [CrossRef]
7. Mosavi, A.; Varkonyi-Koczy, A.R. Integration of Machine Learning and Optimization for Robot Learning. In *Recent Global Research and Education: Technological Challenges*; Springer: Berlin/Heidelberg, Germany, 2017; pp. 349–355.
8. Mosavi, A.; Varkonyi, A. Learning in Robotics. *Int. J. Comput. Appl.* **2017**, *157*, 0975–8887. [CrossRef]
9. Merkelbach, S.; Murrenhoff, H.; Brecher, C.; Fey, M.; Eßer, B. Pneumatic or electromechanical drives–a comparison regarding their exergy efficiency. In Proceedings of the 10th International Fluid Power Conference, Dresden, Germany, 8–10 March 2016.
10. Boldea, I.; Nasar, S.A. *Electric Drives*; CRC Press: Boca Raton, FL, USA, 2016.
11. Lyshevski, S.E. *Electromechanical Systems, Electric Machines, and Applied Mechatronics*; CRC Press: Boca Raton, FL, USA, 2018.
12. Chau, K. *Electric Vehicle Machines and Drives: Design, Analysis and Application*; John Wiley & Sons: Hoboken, NJ, USA, 2015.
13. Gan, C.; Wu, J.; Hu, Y.; Yang, S.; Cao, W.; Guerrero, J.M. New integrated multilevel converter for switched reluctance motor drives in plug-in hybrid electric vehicles with flexible energy conversion. *IEEE Trans. Power Electron.* **2017**, *32*, 3754–3766. [CrossRef]
14. Tah, J.H.; Abanda, F.H. Internet of Things for Urban Sustainability. In *Integrating Information in Built Environments*; Routledge: Abington, UK, 2017; pp. 52–69.
15. Abanda, H.; Tah, J. Linked data in renewable energy domain. In Proceedings of the 6th International Congress on Environmental Modelling and Software, Leipzig, Germany, 1–5 July 2012.
16. Zhu, Z.; Howe, D. Halbach permanent magnet machines and applications: A review. *IEE Proce. Electr. Power Appl.* **2001**, *148*, 299–308. [CrossRef]
17. Mosavi, A.; Rabczuk, T.; Varkonyi-Koczy, A.R. Reviewing the Novel Machine Learning Tools for Materials Design. In *Proceedings of the International Conference on Global Research and Education*; Springer: Berlin/Heidelberg, Germany, 2017.
18. Mosavi, A.; Rituraj, R.; Varkonyi-Koczy, A.R. Review on the Usage of the Multiobjective Optimization Package of modeFrontier in the Energy Sector. In *Proceedings of the International Conference on Global Research and Education*; Springer: Berlin/Heidelberg, Germany, 2017.
19. Mosavi, A.; Rabczuk, T. Learning and intelligent optimization for material design innovation. In *Proceedings of the International Conference on Learning and Intelligent Optimization*; Springer: Berlin/Heidelberg, Germany, 2017.
20. Gieras, J.F. *Permanent Magnet Motor Technology: Design and Applications*; CRC Press: Boca Raton, FL, USA, 2009.
21. Salon, S.J. *Finite Element Analysis of Electrical Machines*; Kluwer Academic Publishers: Boston, MA, USA, 1995; Volume 101.
22. Wilkinson, M.D. The FAIR Guiding Principles for Scientific Data Management and Stewardship. *Sci. Data* **2016**, *3*, 160018. [CrossRef] [PubMed]
23. Husson, R. *Control Methods for Electrical Machines*; John Wiley & Sons: Hoboken, NJ, USA, 2013.
24. Torabi, M. A Hybrid clustering and classification technique for forecasting short-term energy consumption. *Environ. Prog. Sustain. Energy* **2018**, *38*, 66–76. [CrossRef]
25. Hosseini Imani, M.; Zalzar, S.; Mosavi, A.; Shamshirband, S. Strategic behavior of retailers for risk reduction and profit increment via distributed generators and demand response programs. *Energies* **2018**, *11*, 1602. [CrossRef]
26. Baranyai, M.; Mosavi, A.; Vajda, I.; Varkonyi-Koczy, A.R. Optimal Design of Electrical Machines: State of the Art Survey. In *Proceedings of the International Conference on Global Research and Education*; Springer: Berlin/Heidelberg, Germany, 2017.
27. Krasopoulos, C.T.; Beniakar, M.E.; Kladas, A.G. Multicriteria PM Motor Design Based on ANFIS Evaluation of EV Driving Cycle Efficiency. *IEEE Trans. Transp. Electrif.* **2018**, *4*, 525–535. [CrossRef]

28. Fotovatikhah, F. Survey of computational intelligence as basis to big flood management: Challenges, research directions and future work. *Eng. Appl. Comput. Fluid Mech.* **2018**, *12*, 411–437. [CrossRef]
29. Chaturvedi, D. Soft Computing Techniques and Their Applications. In *Mathematical Models, Methods and Applications*; Springer: Berlin/Heidelberg, Germany, 2015; pp. 31–40.
30. Sheng, H.; Xiao, J. Electric vehicle state of charge estimation: Nonlinear correlation and fuzzy support vector machine. *J. Power Sources* **2015**, *281*, 131–137. [CrossRef]
31. Jang, J.-S.R.; Sun, C.-T.; Mizutani, E. Neuro-fuzzy and soft computing-a computational approach to learning and machine intelligence [Book Review]. *IEEE Trans. Autom. Control* **1997**, *42*, 1482–1484. [CrossRef]
32. Zadeh, L.A. Soft computing and fuzzy logic. In *Fuzzy Sets, Fuzzy Logic, and Fuzzy Systems: Selected Papers by Lotfi a Zadeh*; World Scientific: Singapore, 1996; pp. 796–804.
33. Mitra, S.; Hayashi, Y. Neuro-Fuzzy Rule Generation: Survey in Soft Computing Framework. *IEEE Trans. Neural Netw.* **2000**, *11*, 748–768. [CrossRef] [PubMed]
34. Jurado, S.; Nebot, À.; Mugica, F.; Avellana, N. Hybrid methodologies for electricity load forecasting: Entropy-based feature selection with machine learning and soft computing techniques. *Energy* **2015**, *86*, 276–291. [CrossRef]
35. Zhu, Q.; Azar, A.T. *Complex System Modelling and Control Through Intelligent Soft Computations*; Springer: Berlin/Heidelberg, Germany, 2015.
36. Liu, R.; Yang, B.; Zio, E.; Chen, X. Artificial intelligence for fault diagnosis of rotating machinery: A review. *Mech. Syst. Signal Process.* **2018**, *108*, 33–47. [CrossRef]
37. Moeini, I.; Ahmadpour, M.; Mosavi, A.; Alharbi, N.; Gorji, N.E. Modeling the time-dependent characteristics of perovskite solar cells. *Solar Energy* **2018**, *170*, 969–973. [CrossRef]
38. Mosavi, A.; Lopez, A.; Varkonyi-Koczy, A.R. Industrial Applications of Big Data: State of the Art Survey. In *Proceedings of the International Conference on Global Research and Education*; Springer: Berlin/Heidelberg, Germany, 2017.
39. Ijadi Maghsoodi, A.; Ijadi Maghsoodi, A.; Mosavi, A.; Rabczuk, T.; Zavadskas, E. Renewable Energy Technology Selection Problem Using Integrated H-SWARA-MULTIMOORA Approach. *Sustainability* **2018**, *10*, 4481. [CrossRef]
40. Dehghani, M.; Riahi-Madvar, H.; Hooshyaripor, F.; Mosavi, A.; Shamshirband, S.; Zavadskas, E.K.; Chau, K.W. Prediction of Hydropower Generation Using Grey Wolf Optimization Adaptive Neuro-Fuzzy Inference System. *Energies* **2019**, *12*, 289. [CrossRef]
41. Krasopoulos, C.T.; Beniakar, M.E.; Kladas, A.G. Robust optimization of high-speed PM motor design. *IEEE Trans. Magn.* **2017**, *53*, 1–4. [CrossRef]
42. Meghni, B.; Dib, D.; Azar, A.T. A second-order sliding mode and fuzzy logic control to optimal energy management in wind turbine with battery storage. *Neural Comput. Appl.* **2017**, *28*, 1417–1434. [CrossRef]
43. Bayram, D.; Şeker, S. Anfis model for vibration signals based on aging process in electric motors. *Soft Comput.* **2015**, *19*, 1107–1114. [CrossRef]
44. Van, T.L.; Nguyen, T.H.; Lee, D.-C. Advanced pitch angle control based on fuzzy logic for variable-speed wind turbine systems. *IEEE Trans. Energy Convers.* **2015**, *30*, 578–587. [CrossRef]
45. Zhang, S.; Mishra, Y.; Shahidehpour, M. Fuzzy-logic based frequency controller for wind farms augmented with energy storage systems. *IEEE Trans. Power Syst.* **2016**, *31*, 1595–1603. [CrossRef]
46. Salimi, M. State of the art of machine learning in energy systems. *Energies* **2019**, *12*.
47. Mitra, S.; Pal, S.K.; Mitra, P.J.I. Data mining in soft computing framework: A survey. *IEEE Trans. Neural Netw.* **2002**, *13*, 3–14. [CrossRef] [PubMed]
48. Jang, J.-S.R.; Sun, C.-T.; Mizutani, E. *Neuro-Fuzzy and Soft Computing; a Computational Approach to Learning and Machine Intelligence*; Pearson: London, UK, 1997.
49. Aguilar, J.J.C.F.; García, J.M.V.; Carrillo, J.A.C. Regenerative intelligent brake control for electric motorcycles. *Energies* **2017**, *10*, 1648. [CrossRef]
50. Amulya, C.S.; Rao, M.S. Condition Monitoring of Turbogenerators of a Thermal Power Plant Using Fuzzy Logic. *Int. J. Mech. Prod. Eng. Res. Dev.* **2016**, *6*, 25–34.
51. Ben Smida, M.; Sakly, A. Pitch angle control for grid-connected variable-speed wind turbine system using fuzzy logic: A comparative study. *Wind Eng.* **2016**, *40*, 528–539. [CrossRef]
52. Husain, Z. Fuzzy Logic Expert System for Incipient Fault Diagnosis of Power Transformers. *Int. J. Electr. Eng. Inform.* **2018**, *10*, 300–317. [CrossRef]

53. Kahla, S.; Sedraoui, M.; Bechouat, M.; Soufi, Y. Robust Fuzzy On–Off Synthesis Controller for Maximum Power Point Tracking of Wind Energy Conversion. *Trans. Electr. Electron. Mater.* **2018**, *19*, 146–156. [CrossRef]
54. Jang, J.-S. ANFIS: Adaptive-network-based fuzzy inference system. *IEEE Trans. Syst. Man Cybern.* **1993**, *23*, 665–685. [CrossRef]
55. Rezakazemi, M.; Mosavi, A.; Shirazian, S. ANFIS pattern for molecular membranes separation optimization. *J. Mol. Liquids* **2019**, *274*, 470–476. [CrossRef]
56. Mosavi, A.; Edalatifar, M. A Hybrid Neuro-Fuzzy Algorithm for Prediction of Reference Evapotranspiration. In *Proceedings of the International Conference on Global Research and Education*; Springer: Berlin/Heidelberg, Germany, 2018.
57. Gu, Z.Q.; Oyadiji, S.O. Application of MR damper in structural control using ANFIS method. *Comput. Struct.* **2008**, *86*, 427–436. [CrossRef]
58. Mosavi, A.; Ozturk, P.; Chau, K.-W. Flood prediction using machine learning models: Literature review. *Water* **2018**, *10*, 1536. [CrossRef]
59. Najafi, B.; Ardabili, S.F. Application of ANFIS, ANN, and logistic methods in estimating biogas production from spent mushroom compost (SMC). *Resour. Conserv. Recycl.* **2018**, *133*, 169–178. [CrossRef]
60. Arumugom, S.R. Vibration energy harvester based dynamic voltage restorer with energy conservation using multi-functional intelligent controller. *J. Comput. Theor. Nanosci.* **2017**, *14*, 3624–3635. [CrossRef]
61. Aruna, R.S. An intelligent technique based MRPID controller for controlling the speed of an IPMSM drive system. *Indian J. Sci. Technol.* **2016**, *9*. [CrossRef]
62. Bentouhami, L.A.; Bendjeddou, Y.; Merabet, E. Neuro-fuzzy control of a dual star induction machine. *J. Electr. Eng.* **2016**, *16*, 139–147.
63. Haritha, D.R. MPPT control for improving learning efficiency of PMSG wecs based on anfis method. *J. Adv. Res. Dyn. Control Syst.* **2017**, *9*, 1509–1524.
64. Naik, N.V.T.; Singh, S.P. *A Neuro-Fuzzy Direct Torque Control Using Bus-Clamped Space Vector Modulation*; IETE Technical Review; Institution of Electronics and Telecommunication Engineers: Delhi, India, 2016; Volume 33, pp. 205–217.
65. Li, H.X.; Zhang, L.; Cai, K.Y.; Chen, G. An improved robust fuzzy-PID controller with optimal fuzzy reasoning. *IEEE Trans. Syst. Man Cybern. Part B (Cybern.)* **2005**, *35*, 1283–1294. [CrossRef]
66. Fatih Kececioglu, O.A.; Yildiz, C.; Gani, A.; Sekkeli, M. Power quality improvement using hybrid passive filter configuration for wind energy systems. *J. Electr. Eng. Technol.* **2017**, *12*, 207–216. [CrossRef]
67. Gnanaprakasam, C.; Chitra, K. S-transform and ANFIS for detecting and classifying the vibration signals of induction motor. *J. Intell. Fuzzy Syst.* **2015**, *29*, 2073–2085. [CrossRef]
68. Hossain, M.K.; Ali, M.H. Transient stability augmentation of PV/DFIG/SG-based hybrid power system by nonlinear control-based variable resistive FCL. *IEEE Trans. Sustain. Energy* **2015**, *6*, 1638–1649. [CrossRef]
69. Ardabili, F. Modeling and comparison of fuzzy and on/off controller in a mushroom growing hall. *Measurement* **2016**, *90*, 127–134. [CrossRef]
70. Najafi, B.; Faizollahzadeh Ardabili, S.; Mosavi, A.; Shamshirband, S.; Rabczuk, T. An Intelligent Artificial Neural Network-Response Surface Methodology Method for Accessing the Optimum Biodiesel and Diesel Fuel Blending Conditions in a Diesel Engine from the Viewpoint of Exergy and Energy Analysis. *Energies* **2018**, *11*, 860. [CrossRef]
71. Wlas, M.; Krzeminski, Z.; Guzinski, J.; Abu-Rub, H.; Toliyat, H.A. Artificial-neural-network-based sensorless nonlinear control of induction motors. *IEEE Trans. Energy Convers.* **2005**, *20*, 520–528. [CrossRef]
72. Bouchiba, B.; Bousserhane, I.K.; Fellah, M.K.; Hazzab, A. Artificial neural network sliding mode control for multi-machine web winding system. *Revue Roumaine des Sciences Techniques-Serie Electrotechnique et Energetique* **2017**, *62*, 109–113.
73. Celik, E.; Gör, H.; Öztürk, N.; Kurt, E. Application of artificial neural network to estimate power generation and efficiency of a new axial flux permanent magnet synchronous generator. *Int. J. Hydrogen Energy* **2017**, *42*, 17692–17699. [CrossRef]
74. Zemmit, A.; Messalti, S.; Harrag, A. Innovative improved Direct Torque Control of Doubly Fed Induction Machine (DFIM) using Artificial Neural Network (ANN-DTC). *Int. J. Appl. Eng. Res.* **2016**, *11*, 9099–9105.
75. Zidani, Y.; Zouggar, S.; Elbacha, A. Steady-State Analysis and Voltage Control of the Self-Excited Induction Generator Using Artificial Neural Network and an Active Filter. *Iran. J. Sci. Technol. Trans. Electr. Eng.* **2018**, *42*, 41–48. [CrossRef]

76. Mamede, A.C.F.; Camacho, J.R. Evolutionary Algorithms for Optimization of 4/4 Single Phase Switched Reluctance Machine. *IEEE Latin Am. Trans.* **2018**, *16*, 1684–1691. [CrossRef]
77. Tamilselvi, S. Multi objective evolutionary algorithm for designing energy efficient distribution transformers. *Swarm Evol. Comput.* **2018**, *42*, 109–124. [CrossRef]
78. Vanchinathan, K.; Valluvan, K. A metaheuristic optimization approach for tuning of fractional-order PID controller for speed control of sensorless BLDC motor. *J. Circuits Syst. Comput.* **2018**, *27*, 1850123. [CrossRef]
79. Virtič, P.; Vražić, M.; Papa, G. Design of an axial flux permanent magnet synchronous machine using analytical method and evolutionary optimization. *IEEE Trans. Energy Convers.* **2016**, *31*, 150–158. [CrossRef]
80. Ameli, A. A dynamic method for feeder reconfiguration and capacitor switching in smart distribution systems. *Int. J. Electr. Power Energy Syst.* **2017**, *85*, 200–211. [CrossRef]
81. Batista, L.S.; Campelo, F.; Guimarães, F.G.; Ramírez, J.A.; Li, M.; Lowther, D.A. Ant colony optimization for the topological design of interior permanent magnet (IPM) machines. *COMPEL Int. J. Comput. Math. Electr. Electron. Eng.* **2014**, *33*, 927–940. [CrossRef]
82. Glover, F. Future paths for integer programming and links to artificial intelligence. *Comput. Oper. Res.* **1986**, *13*, 533–549. [CrossRef]
83. Yang, X.-S.; Deb, S. Cuckoo search via Lévy flights. In Proceedings of the NaBIC 2009. World Congress on Nature & Biologically Inspired Computing, Coimbatore, India, 9–11 December 2009.
84. Chen, G.; Qiu, S.; Zhang, Z.; Sun, Z.; Liao, H. Optimal Power Flow Using Gbest-Guided Cuckoo Search Algorithm with Feedback Control Strategy and Constraint Domination Rule. *Math. Prob. Eng.* **2017**, *2017*, 9067520. [CrossRef]
85. Wang, Q.; Niu, S. Design, Modeling, and Control of a Novel Hybrid-Excited Flux-Bidirectional-Modulated Generator-Based Wind Power Generation System. *IEEE Trans. Power Electron.* **2018**, *33*, 3086–3096. [CrossRef]
86. Yang, L. Design optimization of a permanent magnet motor derived from a general magnetization pattern. *IEEE Trans. Magn.* **2015**, *51*, 1–4. [CrossRef]
87. Kirkpatrick, S.; Gelatt, C.D.; Vecchi, M.P. Optimization by simulated annealing. *Science* **1983**, *220*, 671–680. [CrossRef]
88. Chiu, M.C.; Chang, Y.C.; Yeh, L.J.; Chung, C.H. Numerical Assessment of a One-Mass Spring-Based Electromagnetic Energy Harvester on a Vibrating Object. *Arch. Acoust.* **2016**, *41*, 119–131. [CrossRef]
89. Farhani, F.; Zaafouri, A.; Chaari, A. Real time induction motor efficiency optimization. *J. Frank. Inst.* **2017**, *354*, 3289–3304. [CrossRef]
90. Torrent-Fontbona, F.; López, B. Decision support for grid-connected renewable energy generators planning. *Energy* **2016**, *115*, 577–590. [CrossRef]
91. Myerson, R.B. *Game Theory*; Harvard University Press: Cambridge, MA, USA, 2013.
92. De Finetti, B. Cambridge probability theorists. *Rivista di Matematica per le Scienze Economiche e Sociali* **1985**, *8*, 79–91.
93. Kari, T.; Gao, W.; Zhao, D.; Zhang, Z.; Mo, W.; Wang, Y.; Luan, L. An integrated method of ANFIS and Dempster-Shafer theory for fault diagnosis of power transformer. *IEEE Trans. Dielectr. Electr. Insul.* **2018**, *25*, 360–371. [CrossRef]
94. Kazemdehdashti, A.; Mohammadi, M.; Seifi, A.R. The Generalized Cross-Entropy Method in Probabilistic Optimal Power Flow. *IEEE Trans. Power Syst.* **2018**, *33*, 5738–5748. [CrossRef]
95. Jiang, X.; Lin, T.; Mendoza, E. Bayesian Calibration for Power Splitting in Single-Shaft Combined Cycle Plant Diagnostics. *J. Eng. Gas Turbines Power* **2016**, *138*, 051702. [CrossRef]
96. Lakehal, A.; Tachi, F. Bayesian Duval Triangle Method for Fault Prediction and Assessment of Oil Immersed Transformers. *Meas. Control* **2017**, *50*, 103–109. [CrossRef]
97. Mansouri, M.M.; Mohamed-Seghir, M.M.; Nounou, H.N.; Nounou, M.N.; Abu-Rub, H.A. Bayesian methods for time-varying state and parameter estimation in induction machines. *Int. J. Adapt. Control Signal Process.* **2015**, *29*, 905–924. [CrossRef]
98. Dai, S.; Liwang, M.; Liu, Y.; Gao, Z.; Huang, L.; Du, X. Hybrid Quantum-Behaved Particle Swarm Optimization for Mobile-Edge Computation Offloading in Internet of Things. In *Proceedings of the International Conference on Mobile Ad-Hoc and Sensor Networks*; Springer: Berlin/Heidelberg, Germany, 2017.
99. McDonald, A.; Bhuiyan, N.A. On the optimization of generators for offshore direct drive wind turbines. *IEEE Trans. Energy Convers.* **2017**, *32*, 348–358. [CrossRef]

100. Meo, S.; Zohoori, A.; Vahedi, A. Optimal design of permanent magnet flux switching generator for wind applications via artificial neural network and multi-objective particle swarm optimization hybrid approach. *Energy Convers. Manag.* **2016**, *110*, 230–239. [CrossRef]
101. Bupp, R.T.; Bernstein, D.S.; Coppola, V.T. Vibration suppression of multi-modal translational motion using a rotational actuator. In Proceedings of the 1994 33rd IEEE Conference on Decision and Control, Lake Buena Vista, FL, USA, 14–16 December 1994.
102. Levant, A. Sliding order and sliding accuracy in sliding mode control. *Int. J. Control* **1993**, *58*, 1247–1263. [CrossRef]
103. Gutman, S. Uncertain dynamical systems—A Lyapunov min-max approach. *IEEE Trans. Autom. Control* **1979**, *24*, 437–443. [CrossRef]
104. Urbani, D.; Roussel-Ragot, P.; Personnaz, L.; Dreyfus, G. The selection of neural models of nonlinear dynamical systems by statistical tests. In Proceedings of the IEEE Workshop on Neural Networks for Signal Processing, Ermioni, Greece, 6–8 September 1994.
105. Trigeassou, J.-C.; Poinot, T.; Bachir, S. *Parameter Estimation for Knowledge and Diagnosis of Electrical Machines*; ISTE Ltd.: London, UK; John Wiley & Sons Inc.: Hoboken, NJ, USA, 2009.
106. Papalambros, P.Y.; Wilde, D.J. *Principles of Optimal Design: Modeling and Computation*; Cambridge University Press: Cambridge, UK, 2000.
107. Filizadeh, S. *Electric Machines and Drives: Principles, Control, Modeling, and Simulation*; CRC Press: Boca Raton, FL, USA, 2016.
108. Preuss, M. *Multimodal Optimization by Means of Evolutionary Algorithms*; Springer: Berlin/Heidelberg, Germany, 2015.
109. Yoo, C.-H. A New Multi-Modal Optimization Approach and Its Application to the Design of Electric Machines. *IEEE Trans. Magn.* **2018**, *54*, 1–4. [CrossRef]

3

High Frequency Square-Wave Voltage Injection Scheme-Based Position Sensorless Control of IPMSM in the Low- and Zero- Speed Range

Shuang Wang, Jianfei Zhao * and Kang Yang

School of Mechatronic Engineering and Automation, Shanghai University, Baoshan District, Shanghai 200444, China; wang-shuang@shu.edu.cn (S.W.); yangk@shu.edu.cn (K.Y.)

* Correspondence: jfzhao@shu.edu.cn.

Abstract: In this paper, a new sensorless control scheme with the injection of a high-frequency square-wave voltage of an interior permanent-magnet synchronous motor (IPMSM) at low- and zero-speed operation is proposed. Conventional schemes may face the problems of obvious current sampling noise and slow identification in the process of magnetic polarity detection at zero speed operation, and the effects of inverter voltage error on the rotor position estimation accuracy at low speed operation. Based on the principle analysis of d-axis magnetic circuit characteristics, a method for determining the direction of magnetic polarity of d-axis two-opposite DC voltage offset by uninterruptible square-wave injection is proposed, which is fast in convergence rate of magnetic polarity detection and more distinct. In addition, the strategy injects a two-opposite high-frequency square-wave voltage vectors other than the one voltage vector into the estimated synchronous reference frame (SRF), which can reduce the effects of inverter voltage error on the rotor position estimation accuracy. With this approach, low-pass filter (LPF) and band-pass filter (BPF), which are used to obtain the fundamental current component and high-frequency current response with rotor position information respectively in the conventional sensorless control, are removed to simplify the signal process for estimating the rotor position and further improve control bandwidth. Finally, the experimental results on an IPMSM drive platform indicate that the rotor position with good steady state and dynamic performance can be obtained accurately at low-and zero-speed operation with the sensorless control strategy.

Keywords: sensorless control; high frequency square-wave voltage; interior permanent-magnet synchronous motor (IPMSM); magnetic polarity detection; rotor position estimation

1. Introduction

Permanent-magnet synchronous motors (PMSMs) have been widely applied in industrial fields [1–4] for the excellent features of high reliability, high efficiency, high torque density, good dynamic performance, etc. According to the structure of the permanent magnet of the rotor, the PMSM can be divided into interior permanent-magnet synchronous motor (IPMSM) and surface-mounted permanent-magnet synchronous motor (SPMSM). Compared with the SPMSM, the IPMSMs have attracted much attention in industrial fields recently due to their greatly improved overload capacity, power density and speed regulation range, which ascribes to the permanent magnet of the IPMSM is located inside the rotor.

To take advantage of these features, the information of rotor position acquired generally by the mechanical sensors, i.e., resolver or encoder, is necessarily required when the field-oriented control (FOC) scheme [5,6] is adopted. However, these mechanical sensors [7,8] mounted on the shaft of a PMSM bring several disadvantages such as extra cost, extra volume, low reliability, etc. In order to overcome these disadvantages, various kinds of position-sensorless control strategies that estimate the rotor position information without a mechanical sensor have been proposed in past decades [9–12]. Conventional sensorless control methods are mainly divided into two groups: (1) For medium- to high-speed operation, the schemes based on the estimation of back electromotive force (EMF), which contains the position information, (2) for low-speed operation, usually under 5% of the rated speed, the schemes based on the machine saliency injects an additional high-frequency (HF) signal to the PMSM which generates a response containing information of the rotor position.

This paper investigates the position-sensorless control strategy using the HF signal injection method in the low- and zero-speed operation, and according to the types of the injected HF signals, the signal can be mainly divided into rotating sinusoidal voltage injection (RSVI) [13], pulsating sinusoidal voltage injection (PSVI) [14], and pulsating square-wave voltage injection (PUVI) [15]. The first HF rotating sinusoidal voltage injection method was proposed by Lorenz R.D in the early time, and in this method, the balanced rotating voltage signals were injected into the stationary reference frame (SRF) and then the induced currents were extracted to obtain the rotor position information. However, this method injects an additional HF voltage into the SRF system, which will lead to the torque ripple caused by the fluctuation of q-axis current. Moreover, the saliency will be reduced under heavy load and the detection accuracy will be worse due to the magnetic saturation effect. For the improvement of the RSVI, pulsating sinusoidal voltage injection and pulsating square-wave voltage injection are proposed, which inject HF voltages in the estimated d-axis of the rotational reference frame (RRF) other than the SRF. To overcome these problems, a PSVI method [16] injects HF voltage into the estimated d-axis reference frame. Similar to RSVI method, the injection voltage frequency of PSVI method is usually about 1/10 of the carrier frequency. Therefore, this method should use low-pass filter (LPF) and band-pass filter (BPF) to separate the fundamental current component and HF currents, respectively, which decreases the control bandwidth and the dynamic performance. To improve the dynamic performance, reference [17] proposed the PUVI method, in which the HF square-wave injected into the estimated d-axis and increased the frequency of the injected signal as high as possible to the pulse width modulation (PWM) carrier frequency. Since the frequency of the injected HF square-wave voltage signal is much higher than the cut-off frequency of current loop, the LPFs in the current feedback loop can be omitted [18], which improves the response speed to some extent. Seung et al. [19] proposed a method that the frequency of the injected HF square-wave voltage signal is the same as the PWM switching frequency, in which two current samples were taken in a square-wave period and three arithmetic operations were performed to obtain the HF induced current signal. Besides, the dynamics of the position-sensorless control can be improved and the acoustic noise can be remarkably reduced. However, this method still needs to use a BPF for obtaining the feedback current signal. In [20], the fundamental current signal of the stator winding is obtained by arithmetic operation, without using a filter, but it is still necessary to sample the current twice in one square-wave period.

The conventional square-wave injection method has many advantages, such as low current noise, high bandwidth, and high steady state performance, etc. However, the voltage error caused by the nonlinearity of the inverter still needs to be taken seriously and solved. In addition, the magnetic polarity detection, which is one of the main problems to be improved, is essential for smooth startup and robust control of IPMSM. The conventional method [21] proposes to detect the initial position of the rotor by pulse voltage vector method, which injects a series of pulse voltage vectors with the same amplitude and different direction into the motor stator winding, and estimates the rotor pole position by comparing the magnitude of the response current based on the nonlinear saturation characteristics of the motor stator core. However, the current amplitude does not change much as the voltage vector approaches the rotor pole position. At the same time, the phase current measurement error and the inverter nonlinearity will affect the measurement accuracy [16].

In order to solve the problem of initial position detection in conventional sensorless control of IPMSM, a rotating voltage injection sensorless method of estimating the initial rotor position of a direct torque controlled IPMSM drive is proposed in [22], which injects a HF voltage to the windings and extracts the amplitude of the corresponding stator current components based on motor salient effect. However, this method still has some problems, such as obvious current sampling noise, influence by motor parameter variations, and special position interference. The improvement of the RSVI, PUVI, which has been proven to be the best injection type for the low- and zero-speed position sensorless control [23], can rarely be found for the initial rotor position detection in the sensorless control system. Xie et al. [24] proposed a PUVI method in which the two opposite voltage vectors were injected to reduce the effects of inverter voltage error on the position estimation accuracy. However, this method only studies the low-speed range, and does not conduct research and analysis on the zero-speed range, especially the lack of research on magnetic polarity detection, which is essential for smooth startup and robust control of IPMSM. In [25], the positive and negative test pulse voltages are injected into the estimated d-axis, and the positive direction of the magnetic polarity is judged by the time when the currents at different magnetic polarity are attenuated from the steady state value to 0. However, the position estimation is stopped in the process of polarity identification, which makes the structure of the code more complicated. This problem also exists in [26] and [27], and the interval of signal injection must be existed, resulting in long execution time and poor stability. Then, [28] uses the difference in current measured at zero time of each PWM period to determine the magnetic polarity. The advantage is that no additional injection voltage is required, and the convergence speed is fast. The disadvantage of this method is that the position estimation depends on the accuracy of the current. Once the current signal-to-noise ratio is too small, the rotor position cannot be accurately obtained. Based on the above references and the improvement points of this article, the comparison of Table 1 is obtained.

Table 1. List of the research contents covered by the articles.

Literature	Remove LPF	Remove BPF/HPF	Reduce Audible Noise	Nonlinearity of the Inverter	Magnetic Polarity Judgment
Ref. [4]	×	√	×	×	×
Ref. [8]	√	×	×	×	√
Ref. [10]	×	×	√	×	×
Ref. [11]	×	×	√	×	×
Ref. [12]	√	√	√	×	×
Ref. [13]	√	×	×	×	√
Ref. [15]	√	√	×	×	×
Ref. [16]	×	√	×	×	√
Ref. [18]	√	√	×	√	×
Ref. [23]	√	√	×	×	√
Ref. [24]	√	√	×	√	×
This paper	√	√	×	√	√

Therefore, in view of the existing sensorless estimation methods for rotor position, a HF square-wave voltage injection scheme-based rotor position estimation of IPMSM in low-speed range and a new method of magnetic polarity detection in zero-speed range are proposed. Based on the principle analysis of d-axis magnetic circuit characteristics, a method for determining the direction of magnetic polarity of d-axis two-opposite DC voltage offset by uninterruptible square-wave injection is proposed, which is simple and can quickly converge by comparing the absolute value of the peak-to-peak value in the d-axis high-frequency current response. Therefore, the method of magnetic polarity identification proposed in this paper is not only suitable for the case where the motor is stationary, but also suitable for the free running condition of the motor. At the same time, the strategy injects a two-opposite HF square-wave voltage vectors into the estimated SRF, which considers the effects of inverter voltage error on the rotor position estimation accuracy. With this approach, the LPF and the BPF are removed to simplify the signal process for estimating the rotor position and further improve control bandwidth.

This paper is organized as follows. First, the rotor position estimation strategy based on conventional HF square-wave voltage injection is analyzed in Section 2. In Section 3, the rotor position estimation strategy based on improved HF square-wave voltage injection and proposed magnetic polarity detection are investigated. Then, in Section 4, comprehensive simulation and experimental setup are introduced and experiments are provided to prove the effectiveness of the improved sensorless control strategy and proposed magnetic polarity detection method. Finally, Section 5 concludes this paper.

2. Analysis of Rotor Position Estimation Strategy Based on Conventional Square-Wave Voltage Injection

2.1. Mathematical Model of IPMSM

Assuming that the IPMSM operates in an unsaturated state with negligible hysteresis loss and eddy current loss, the mathematical model of the IPMSM in the α-β stationary reference frame is given as:

$$\begin{bmatrix} u_{\alpha s} \\ u_{\beta s} \end{bmatrix} = R_s \begin{bmatrix} i_{\alpha s} \\ i_{\beta s} \end{bmatrix} + \begin{bmatrix} L^r_{sum} + L^r_{dif}\cos 2\theta_r & L^r_{dif}\sin 2\theta_r \\ L^r_{dif}\sin 2\theta_r & L^r_{sum} - L^r_{dif}\cos\theta_r \end{bmatrix} \frac{d}{dt} \begin{bmatrix} i_{\alpha s} \\ i_{\beta s} \end{bmatrix} + \\ 2L^r_{dif}\omega_r \begin{bmatrix} -\sin 2\theta_r & \cos 2\theta_r \\ \cos 2\theta_r & \sin\theta_r \end{bmatrix} \begin{bmatrix} i_{\alpha s} \\ i_{\beta s} \end{bmatrix} + \omega_r\psi_f \begin{bmatrix} -\sin\theta_r \\ \cos\theta_r \end{bmatrix} \tag{1}$$

where $u_{\alpha,\beta s}$, $i_{\alpha,\beta s}$ are the α- and β-axes stator voltage and current, respectively. R_s is the stator resistance; ω_r is the rotor speed; ψ_r is the linkage magnetic flux, and d/dt represents derivative operator. $L^r{}_{sum}$, $L^r{}_{dif}$ are average inductance and differential inductance, respectively, and are defined as $L^r{}_{sum} = (L^r_d + L^r_q)/2$ and $L^r{}_{dif} = (L^r_d - L^r_q)/2$, respectively, where $L^r{}_d$ and $L^r{}_q$ are d-and q-axis inductance, respectively. θ_r is the actual rotor position, and the physical model of actual and estimated rotor reference frames is shown in Figure 1.

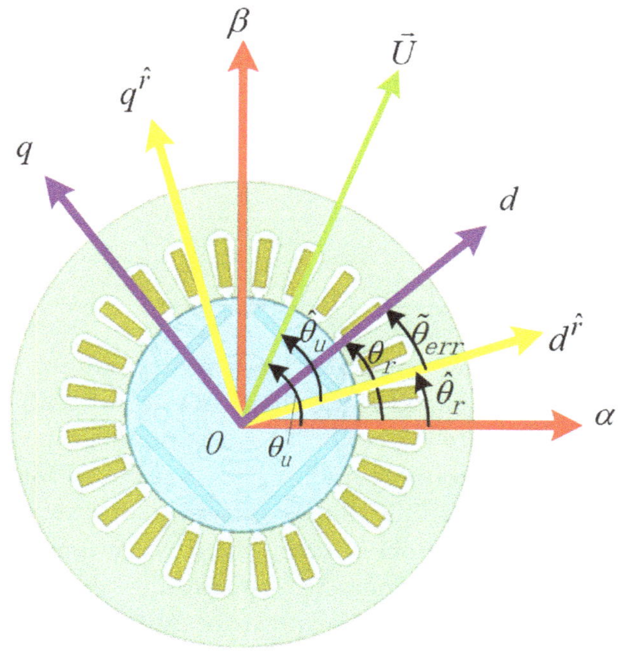

Figure 1. Physical model of the actual and estimated rotor reference frames.

According to Euler's formula, the voltage and current vector can be described as

$$\begin{cases} \vec{U}_{\alpha s} = U_{\alpha s} e^{j\theta_r} \\ \vec{U}_{\beta s} = U_{\beta s} j e^{j\theta_r} \\ \vec{I}_{\alpha s} = I_{\alpha s} e^{j\theta_r} \\ \vec{I}_{\beta s} = I_{\beta s} j e^{j\theta_r} \end{cases} \tag{2}$$

where $U_{\alpha s}$, $I_{\alpha s}$, $U_{\beta s}$, and $I_{\beta s}$ are the magnitudes of $\vec{U}_{\alpha s}$ and $\vec{I}_{\beta s}$ on the α- and β-axes, respectively; e is the natural constant; j is the imaginary unit.

Under the premise that IPMSM operates at zero- and low-speed range, the product term related to ω_r can be omitted because the speed is close to 0. Therefore, (1) can be simplified as

$$\begin{cases} \vec{U}_{\alpha s} = R_s \vec{I}_{\alpha s} + \left(L^r_{sum} + L^r_{dif} \cos 2\theta_r\right)\frac{d\vec{I}_{\alpha s}}{dt} - jL^r_{dif} \sin 2\theta_r \frac{d\vec{I}_{\beta s}}{dt} \\ \vec{U}_{\beta s} = R_s \vec{I}_{\beta s} + jL^r_{dif} \sin 2\theta_r \frac{d\vec{I}_{\alpha s}}{dt} + \left(L^r_{sum} - L^r_{dif} \cos 2\theta_r\right)\frac{d\vec{I}_{\beta s}}{dt} \end{cases} \tag{3}$$

By adding the two equations in (3), (3) is rewritten as

$$\begin{aligned} \vec{U}_{\alpha s} + \vec{U}_{\beta s} = {} & R_s\left(\vec{I}_{\alpha s} + \vec{I}_{\beta s}\right) + \frac{(L^r_d + L^r_q)}{2}\left(\frac{d\vec{I}_{\alpha s}}{dt} + \frac{d\vec{I}_{\beta s}}{dt}\right) \\ & + \frac{(L^r_d - L^r_q)}{2} \cos 2\theta_r \left(\frac{d\vec{I}_{\alpha s}}{dt} - \frac{d\vec{I}_{\beta s}}{dt}\right) + j\frac{(L^r_d - L^r_q)}{2} \sin 2\theta_r \left(\frac{d\vec{I}_{\alpha s}}{dt} - \frac{d\vec{I}_{\beta s}}{dt}\right) \end{aligned} \tag{4}$$

2.2. *Signal-Process Method in the Estimated Rotor Reference Frame*

Figure 2 shows the control system scheme for obtaining position information with the conventional square-wave injection method. By reasonably selecting the injection voltage, the voltage generated on the stator resistance in (4) can be neglected. After Euler transform, (4) is derived as

$$d\vec{I}_{\alpha\beta s} = \frac{L^r_d(\vec{U}_{\alpha\beta s} - \vec{U}^*_{\alpha\beta s} e^{j2\theta_r}) + L^r_q(\vec{U}_{\alpha\beta s} + \vec{U}^*_{\alpha\beta s} e^{j2\theta_r})}{2L^r_d L^r_q} dt \tag{5}$$

where $\vec{U}^*_{\alpha\beta s}$ is the conjugate vector of $\vec{U}_{\alpha\beta s}$. Since $d\vec{I}_{\alpha\beta s}/dt$ can be regarded as $\Delta\vec{I}_{\alpha\beta s}/\Delta t$ during one PWM switching period, where $\Delta\vec{I}_{\alpha\beta s}$ is the α- and β-axes current variation in the stationary reference frame. Therefore, the current variation can be expressed as

$$\Delta\vec{I}_{\alpha\beta s} = \left(\frac{L^r_d + L^r_q}{2L^r_d L^r_q} U_{\alpha\beta s} e^{j\theta_u} - \frac{L^r_d - L^r_q}{2L^r_d L^r_q} U_{\alpha\beta s} e^{j(2\theta_r - \theta_u)}\right)\Delta t \tag{6}$$

where θ_u is the angle of the voltage vector in the stationary reference frame. By converting the current variation in the α-β stationary reference frame to the estimated $d^{\hat{r}} - q^{\hat{r}}$ rotating reference frame, the current variation can be rewritten as

$$\Delta\vec{I}_{d^{\hat{r}}q^{\hat{r}}s} = \left(\frac{L^r_d + L^r_q}{2L^r_d L^r_q} - \frac{L^r_d - L^r_q}{2L^r_d L^r_q} e^{j2(\theta_r - \hat{\theta}_u - \hat{\theta}_r)}\right)\Delta t \cdot \vec{U}_{d^{\hat{r}}q^{\hat{r}}s} \tag{7}$$

where $\hat{\theta}_r$ is the estimated position angle. As shown in Figure 1, the relationship between different angles are given as

$$\begin{cases} \theta_u = \hat{\theta}_u + \hat{\theta}_r \\ \theta_r = \widetilde{\theta}_{err} + \hat{\theta}_r \end{cases} \tag{8}$$

where $\widetilde{\theta}_{err}$ and $\hat{\theta}_u$ are the error angle of the actual rotating reference frame and the estimated rotating reference frame, the angle of the voltage vector in the estimated rotating reference frame, respectively. If $d^{\hat{r}}$-axis is selected as the injection axis, the form of square-wave voltage is shown as

$$U_{inj}(t) = \begin{cases} U_{inj}e^{j2\hat{\theta}_u}, 0 < t_m(T) < \frac{T}{2} \\ 0, \frac{T}{2} < t_m(T) < T \end{cases} \tag{9}$$

where T and $t_m(T)$ are the square-wave period, the remainder of t divided by T, respectively.

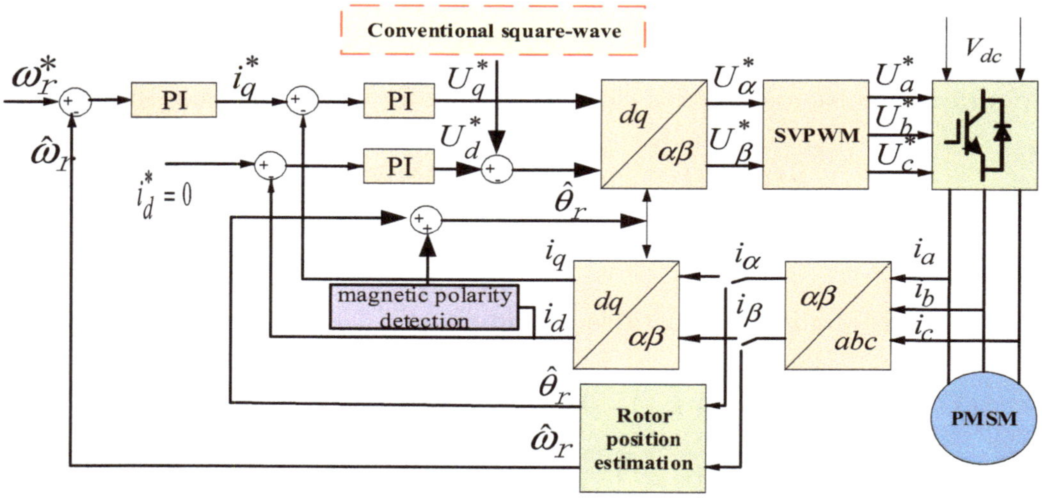

Figure 2. Control system scheme for obtaining position information with the conventional square-wave injection method.

The current variation in the estimated $d^{\hat{r}} - q^{\hat{r}}$ rotating reference frame after injecting the conventional square-wave voltage can be calculated as

$$\Delta \vec{I}_{d^{\hat{r}}q^{\hat{r}}s} = \frac{\Delta t(L_d^r + L_q^r)}{2L_d^r L_q^r} U_{inj} - \frac{\Delta t(L_d^r - L_q^r)}{2L_d^r L_q^r} U_{inj}\cos(2\widetilde{\theta}_{err}) - \frac{\Delta t(L_d^r - L_q^r)}{2L_d^r L_q^r} U_{inj}\sin(2\widetilde{\theta}_{err})j \tag{10}$$

It can be seen from (10) that the error angle can be directly extracted from the imaginary part of the current variation when the error angle is small, which is obtained as

$$\mathrm{Im}(\Delta \vec{I}_{d^{\hat{r}}q^{\hat{r}}s}) = \Delta \vec{I}_{q^{\hat{r}}s} = C \cdot \sin(2\widetilde{\theta}_{err}) \approx 2C \cdot \widetilde{\theta}_{err} \tag{11}$$

where $C = \frac{\Delta t(L_d^r - L_q^r)}{2L_d^r L_q^r} u_{inj}$, which is a constant. The estimated rotor position $\hat{\theta}_r$ is obtained by signal-processing in the sensorless control, as shown in Figure 3. The full name of PLL in Figure 3 is phase-locked loop.

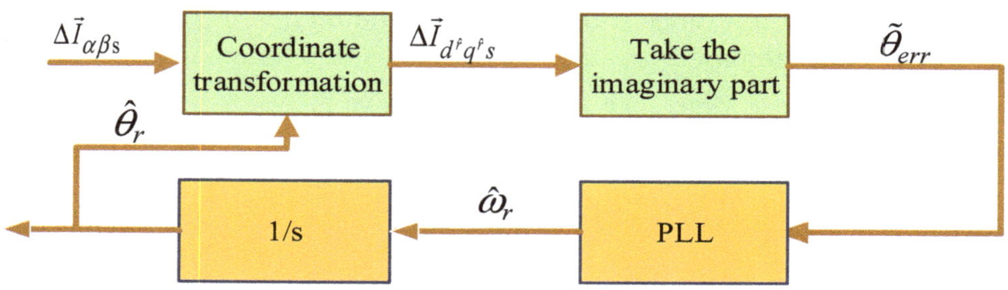

Figure 3. Signal processing diagram of the conventional square-wave voltage injection method for IPMSM.

2.3. *Voltage Vector Injection Scheme*

Since the motor speed is extremely low, i.e., less than 5% of rated speed and the motor state changes minimally during one PWM switching period, the performance of motor is basically the same when the PI controller integrates once every two PWM switching periods and integrates once for each PWM switching period. In the subsequent experiments, one switching frequency of 10 kHz was used, and the injection frequency was 1/2 of the switching frequency. Although the system of rotor position estimation can extract the position information of the rotor and obtain an estimated value, the estimated value may be consistent with the actual position, or may be offset by π rad, so that the positive direction of the actual d-axis cannot be determined. Therefore, the magnetic polarity must be judged before the motor runs at low speeds. The conventional method to solve the problem is usually to inject voltage pulses of equal width in both positive and negative directions into the d-axis in the estimated rotating reference frame. Considering that the duty period of the PWM calculated in the ARM controller will be updated in the next period, the conventional control sequence of injection voltage and sampling current under position-sensorless control at zero- and low-speed is shown in Figure 4.

As can be seen from Figure 4, current sampling was performed at the beginning of each current action period, and the current variation obtained by the difference between the measured currents in the two periods can be used for position estimation. The PI controller of the current loop only acts after the current sampling step in the FOC control period, so by separating the FOC control period and the voltage injection period, an additional low-pass filter for extracting the estimated rotor position can be omitted.

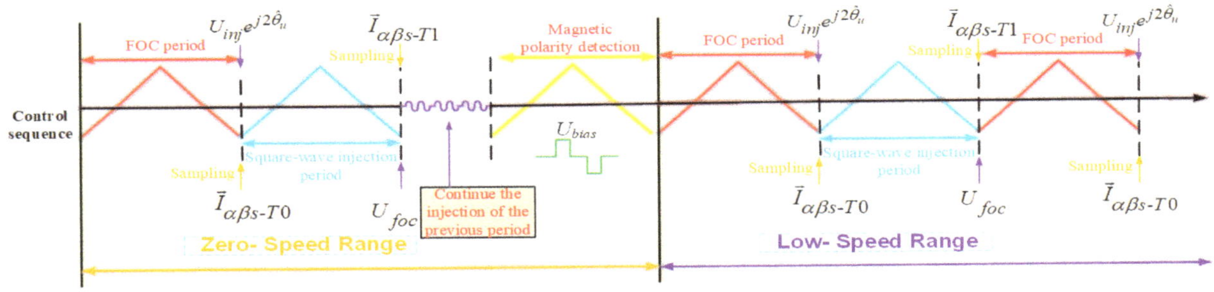

Figure 4. Conventional control sequence of injection voltage and sampling current at zero- and low-speed.

3. Analysis of Rotor Position Estimation Strategy Based on Improved Square-Wave Voltage Injection

3.1. *Improved Signal-Process Method in the Estimated Rotor Reference Frame*

At low speeds, various position-sensorless driving methods are generally affected by the nonlinear voltage error of the inverter. The nonlinear factors in the voltage-type inverter cause the HF response current to be distorted, which results in the rotor position estimation error and affects the stability of sensorless control. If the error is not compensated, the injected voltage vector cannot be injected into the target axis accurately, resulting in the observation position offset. In (6), $\left|\frac{L_d^r+L_q^r}{2L_d^rL_q^r}\right| = \left|-\frac{L_d^r-L_q^r}{2L_d^rL_q^r}\right|$, when square-wave voltage is injected to the estimated $d^{\hat{r}}$ -axis, the rotor position angle error signal is obtained from the term $\left[-\frac{\Delta t(L_d^r-L_q^r)}{2L_d^rL_q^r}u_{inj}\sin(2\widetilde{\theta}_{err})\right]$ in the imaginary part of $\Delta\vec{I}_{d^{\hat{r}}q^{\hat{r}}s}$ However, due to the nonlinearity of the voltage-source inverters, the square-wave voltage cannot be injected into the target position of $d^{\hat{r}}$ -axis accurately, so a small error will be enlarged from the term $\left[\frac{\Delta t(L_d^r-L_q^r)}{2L_d^rL_q^r}u_{inj}\right]$ in the real part of $\Delta\vec{I}_{d^{\hat{r}}q^{\hat{r}}s}$ Therefore, when extracting the rotor position angle error signal, the term $\left[\frac{\Delta t(L_d^r-L_q^r)}{2L_d^rL_q^r}u_{inj}\right]$ will not be completely eliminated, which will seriously affect the accuracy of conventional square-wave injection for estimating the rotor position angle. In the conventional method, the observation value of rotor position can be obtained by injecting only one voltage vector in the injection period, but the voltage error affected by the nonlinearity of the voltage-source inverters is not compensated. In addition, the conventional method first calculates the error and then compensates the error, which is not only complicated to operate, but also the digital control systems, e.g., DSP and dSPACE, have a delay of switching periods. Therefore, the voltage error calculated in each switching period will only be compensated in the next switching period, which leads to inaccurate voltage error compensation. Therefore, on the basis of the above injection method, it is important to study a simple and accurate inverter voltage error compensation method to improve the accuracy of rotor position estimation and the control performance of IPMSM.

In order to realize the compensation of the voltage error caused by the nonlinear factor of the inverter, e.g., dead-time of switches and turn-on and turn-off voltage drop of switches, another voltage vector with the same amplitude and opposite direction can be injected in the next switching period of injecting the positive voltage vector on the basis of separating the FOC control period and the square-wave injection period, at which time the PI controller of current loop acts and updates every three switching periods. Figure 5 shows the inverter single-phase bridge arm structure, and the improved physical model of actual and estimated rotor reference frames is shown in Figure 6.

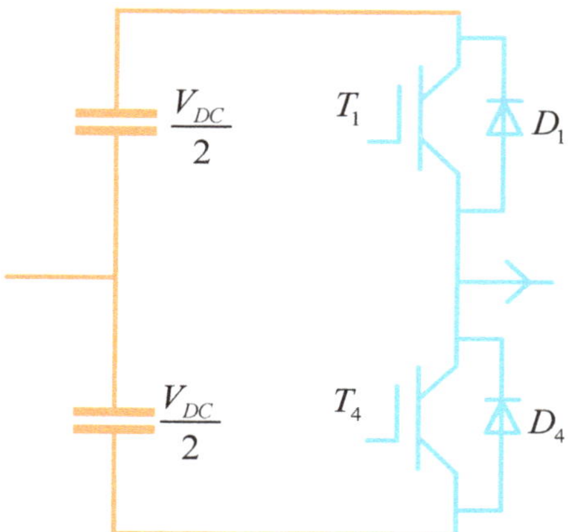

Figure 5. Inverter single-phase bridge arm structure.

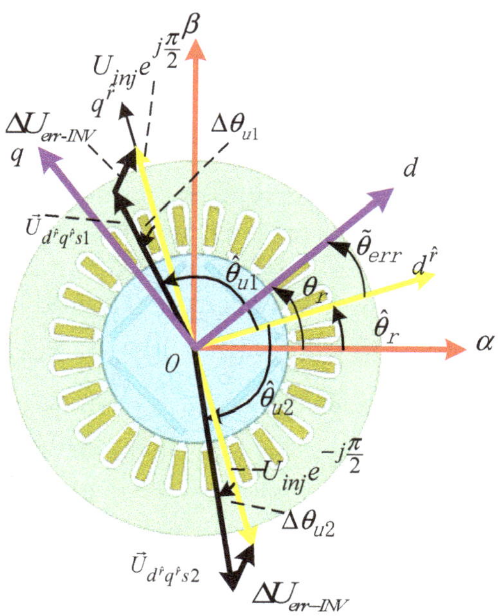

Figure 6. The improved physical model of actual and estimated rotor reference frames.

3.2. Improved Voltage Vector Injection Scheme

Similarly, the details of the ARM processor update in the next period after calculating the PWM duty period need to be considered. Figure 7 shows the improved control sequence of injection voltage and sampling current under position-sensorless control at zero- and low-speed. As shown in Figure 7, when the IPMSM operates at zero-speed, the initial position detection needs to be performed first, and then the magnetic polarity detection needs to be considered. When the IPMSM operates at low-speed, in the first PWM carrier period, the sensorless control system performs FOC control, without superimposing any high frequency vector. The current response generated by the FOC vector acts as the control current of three PWM periodic current loops. Then, before the beginning of the second PWM carrier period, a positive square-wave voltage is injected into the forward direction of $q^{\hat{r}}$, and then a negative square-wave voltage of the same magnitude is injected into the reverse direction of $q^{\hat{r}}$, before the beginning of the third PWM carrier period. Finally, the currents acquired by the second and third PWM periods are compared with the current acquired by the previous PWM period. By making a difference, two varying currents are obtained to calculate the rotor position angle.

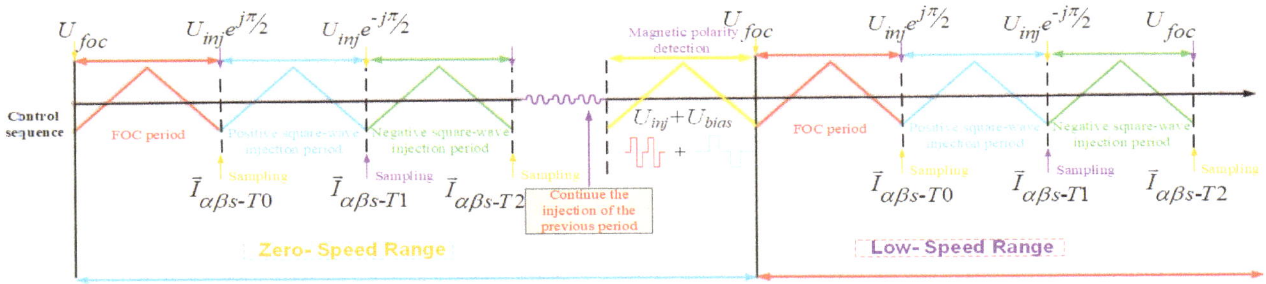

Figure 7. The improved control sequence of injection voltage and sampling current at zero- and low-speed.

Define the inverter voltage error as $\Delta U_{err-INV}$. Since the motor speed is very low and the switching period is very short, assuming that $\Delta U_{err-INV}$ does not change during the two switching periods, the voltage error caused by the nonlinearity of the inverter can be expressed as

$$\begin{cases} \vec{U}_{d^{\hat{r}}q^{\hat{r}}s1} = U_{inj}e^{j\hat{\theta}_{u1}} \\ \quad \approx U_{inj}e^{\ j\pi/2} - \Delta\vec{U}_{err-INV} \\ \vec{U}_{d^{\hat{r}}q^{\hat{r}}s2} = U_{inj}e^{j\hat{\theta}_{u2}} \\ \quad \approx U_{inj}e^{-j\pi/2} - \Delta\vec{U}_{err-INV} \end{cases} \tag{12}$$

By substituting (12) into (7), the simplified equation is rewritten as (13) and (14)

$$\begin{aligned} \Delta\vec{I}_{d^{\hat{r}}q^{\hat{r}}s1} &= \frac{\Delta t(L_d^r+L_q^r)}{2L_d^rL_q^r}(U_{inj}e^{\ j\pi/2} - \Delta\vec{U}_{err-INV}) - \frac{\Delta t(L_d^r-L_q^r)}{2L_d^rL_q^r}U_{inj}e^{j(2\tilde{\theta}_{err}-\hat{\theta}_{u1})} \\ &= \Delta t\left(\frac{L_d^r+L_q^r}{2L_d^rL_q^r}U_{inj}e^{\ j\pi/2} - \frac{L_d^r+L_q^r}{2L_d^rL_q^r}\Delta\vec{U}_{err-INV} - \frac{L_d^r-L_q^r}{2L_d^rL_q^r}U_{inj}e^{j(2\tilde{\theta}_{err}-\hat{\theta}_{u1})}\right) \end{aligned} \tag{13}$$

$$\begin{aligned} \Delta\vec{I}_{d^{\hat{r}}q^{\hat{r}}s2} &= \frac{\Delta t(L_d^r+L_q^r)}{2L_d^rL_q^r}(U_{inj}e^{\ -j\pi/2} - \Delta\vec{U}_{err-INV}) - \frac{\Delta t(L_d^r-L_q^r)}{2L_d^rL_q^r}U_{inj}e^{j(2\tilde{\theta}_{err}-\hat{\theta}_{u2})} \\ &= \Delta t\left(\frac{L_d^r+L_q^r}{2L_d^rL_q^r}U_{inj}e^{\ -j\pi/2} - \frac{L_d^r+L_q^r}{2L_d^rL_q^r}\Delta\vec{U}_{err-INV} - \frac{L_d^r-L_q^r}{2L_d^rL_q^r}U_{inj}e^{j(2\tilde{\theta}_{err}-\hat{\theta}_{u2})}\right) \end{aligned} \tag{14}$$

Thus, the rotor position estimation scheme can be realized by an improved physical model of actual and estimated rotor reference frames, as shown in Figure 6. In Figure 6, the relationship between angles can be described as

$$\begin{cases} \hat{\theta}_{u1} = \pi/2 + \Delta\theta_{u1} \\ \hat{\theta}_{u2} = -\pi/2 - \Delta\theta_{u2} \\ \Delta\theta_{u1} \approx \Delta\theta_{u2} \end{cases} \quad (15)$$

Since $\Delta\theta_{u1}$ is very close to 0, combining (13), (14), and (15), the obtained equation is calculated as (16).

$$\mathrm{Re}(\Delta\vec{I}_{d^{\hat{r}}q^{\hat{r}}s1} - \Delta\vec{I}_{d^{\hat{r}}q^{\hat{r}}s2}) = -\frac{\Delta t(L_d^r - L_q^r)}{L_d^r L_q^r} U_{inj}\cos\Delta\theta_{u1}\cdot\sin 2\widetilde{\theta}_{err} = 2C\cdot\sin 2\widetilde{\theta}_{err} \approx 4C\cdot\widetilde{\theta}_{err} \quad (16)$$

Similarly, if the injection voltage from the $d^{\hat{r}}$-axis is selected, the estimated rotor position angle can be expressed as

$$\mathrm{Im}(\Delta\vec{I}_{d^{\hat{r}}q^{\hat{r}}s1} - \Delta\vec{I}_{d^{\hat{r}}q^{\hat{r}}s2}) = 2C\cdot\sin 2\widetilde{\theta}_{err} \approx 4C\cdot\widetilde{\theta}_{err} \quad (17)$$

It is observed that the (16) and (17) can finally estimate the rotor position angle after considering the voltage error caused by the nonlinearity of the inverter. Thus, the rotor position-sensorless estimation scheme can be realized by an improved square-wave injection method, as shown in Figure 8.

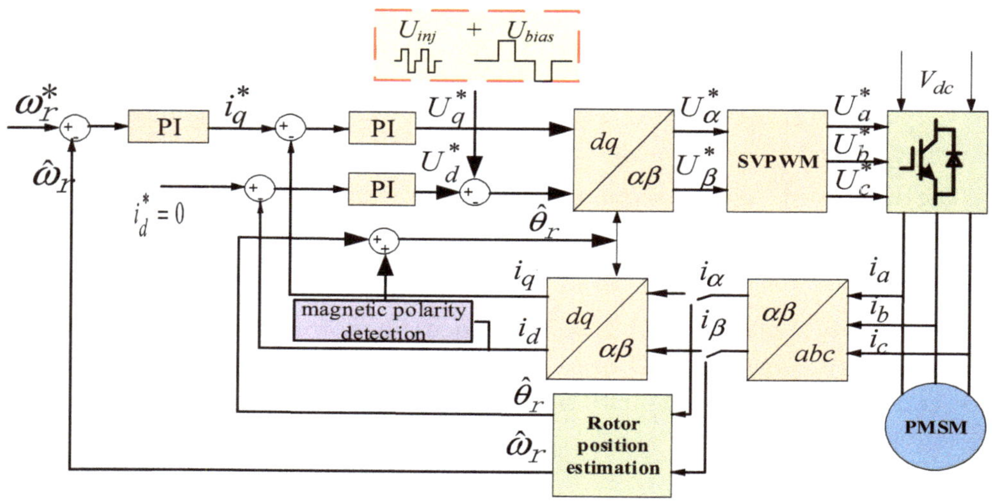

Figure 8. Control system scheme for obtaining position information with the improved square-wave injection method.

3.3. Determining the Direction of Magnetic Polarity

In the conventional method, after the square-wave voltage is injected, the magnetic polarity is determined by injecting two square-wave pulses of opposite directions and equal durations. However, in the process of software algorithm implementation, the switching task of the algorithm state machine is additionally increased, which makes the structure of the code more complicated, and the noise of the current sampling affects the accuracy of the judgment, which may lead to the magnetic polarity judgment error.

In this paper, a fast-initial position identification method is proposed. After the positive and negative square-wave injection, the given d-axis bias voltage U_{bias} is added and the direction of the bias voltage is changed on the basis of the uninterrupted square-wave injection. By comparing the absolute value of the peak-to-peak value of the d-axis HF current response, the magnetic polarity identification is completed, which is relatively simple.

At the same time, the bias current is not directly given in this paper, but the bias current is generated by a given bias voltage. Figure 9 shows the characteristics curve of the d-axis magnetic circuit and the high-frequency current response diagram. The incremental inductance at *X1* and *X2* can be defined as

$$\begin{cases} L_1 = \frac{\partial \psi}{\partial i}\big|_{X_1} \\ L_2 = \frac{\partial \psi}{\partial i}\big|_{X_2} \end{cases} \tag{18}$$

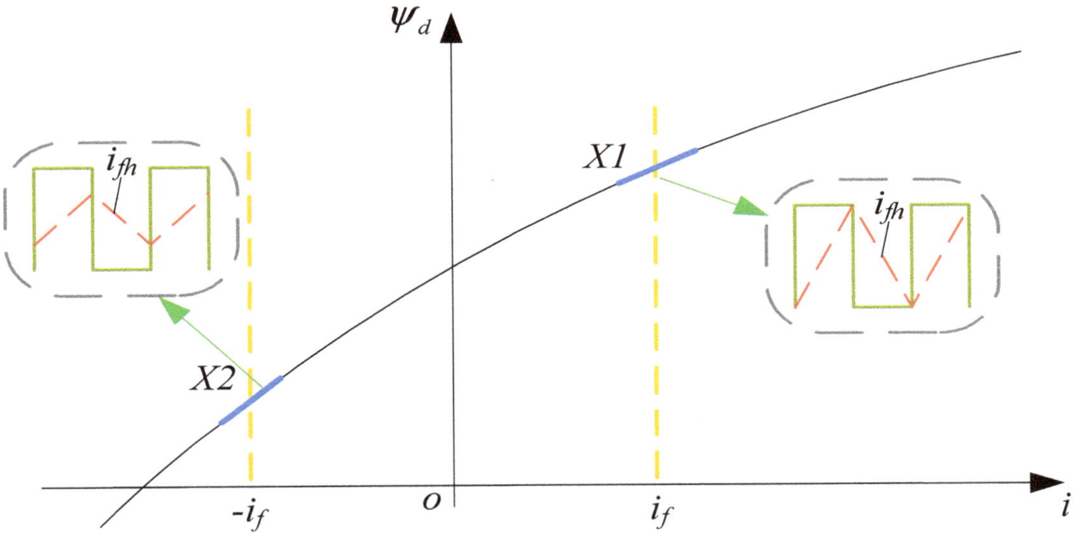

Figure 9. The characteristics curve of the d-axis magnetic circuit and the high-frequency current response diagram.

As shown in Figure 9, the incremental inductance *L2* is larger than *L1*. Therefore, when the same flux is changed, the current at *X1* changes greatly and the current at *X2* changes little. Since the square-wave is not interrupted during the initial position identification process, the convergence speed of magnetic polarity identification is fast. In addition, it is necessary to set U_{bias} to 0 at an intermediate time period of a given bias voltage $\pm U_{bias}$, so that the fundamental current returns to the initial state. When the d-axis DC bias voltage is the same as the magnetic polarity of the rotor (*X1*), the stator flux saturation is increased, the incremental inductance is decreased, and the absolute value of the d-axis HF response current peak-to-peak value is increased. When the voltage is opposite to the magnetic polarity of the rotor (*X2*), the saturation of the stator flux is weakened, the incremental inductance is increased, and the absolute value of the peak-to-peak value of the d-axis HF response current is decreased.

Therefore, the magnetic polarity identification of the rotor can be realized by comparing the absolute values of the peak-to-peak value of the HF current response generated by the HF voltage under the given bias of the positive and negative d-axis voltage. If the peak-to-peak value of forward HF current is greater than peak-to-peak value of the reverse HF current, the estimated position angle direction is directed to the N pole; i.e., the estimated position angle is the actual position angle of the rotor. In addition, if the peak-to-peak value of forward HF current is less than the peak-to-peak value of reverse HF current, the estimated position angle direction is directed to the S pole, which means the rotor position angle needs to be compensated for π. Figure 10 shows the flow charts of two methods for magnetic polarity identification of the rotor.

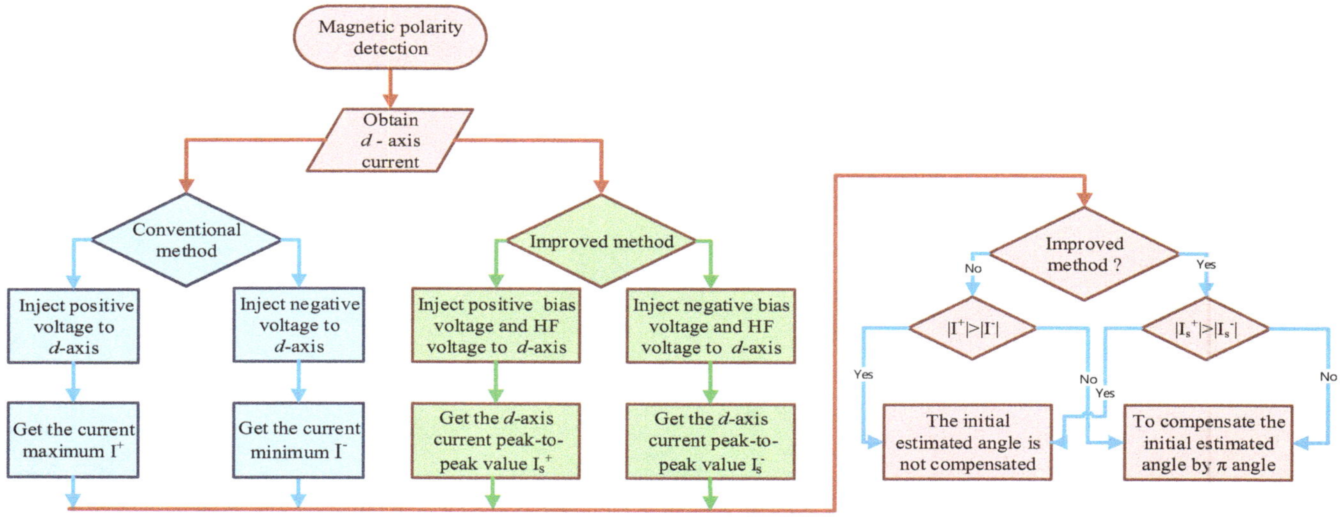

Figure 10. The flow charts of two methods for magnetic polarity identification of the rotor.

4. Simulation and Experimental Results

4.1. Simulationl Results

Because of the need to implement discrete simulation models in Matlab/Simulink, this paper uses the M function in Simulink and the data storage unit to save the discrete data for the next period of calculation. Before conducting the experiment, it is first determined by simulation that the square-wave injection is performed by injecting a HF voltage into the d-axis or a HF voltage into the q-axis. In the simulation, the bus voltage is 310 V, and the amplitude of the injected HF voltage vector is 70 V.

Figure 11 shows the current response when the q-axis is injected with positive and negative pulse voltages for position estimation. Figure 12 shows the current response when the d-axis is injected with positive and negative pulse voltages for position estimation. According to the torque equation of an IPMSM, the q-axis current has a predominant influence on the torque. As shown in Figure 11b, the HF current response with high amplitude is generated in the q-axis, which has a great influence on the torque ripple. However, as shown in Figure 12b, it can be seen that the amplitude of the HF current response in the q-axis is small, which is more conducive to the stability of the control system. Therefore, in subsequent experiments, the rotor position estimation is performed by injecting a HF voltage vector in the *d*-axis instead of the q-axis injection.

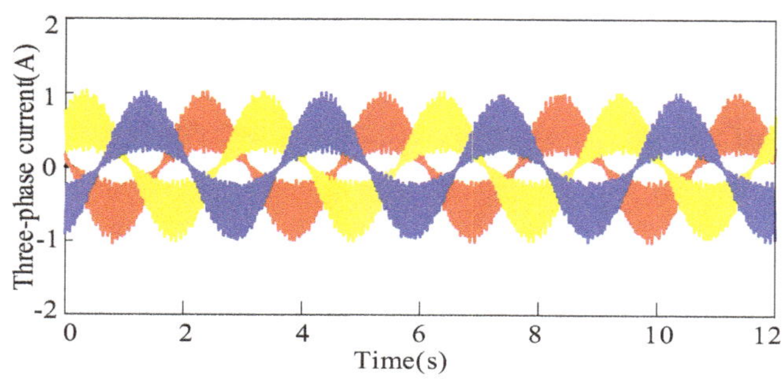

(a) Three-phase current waveform.

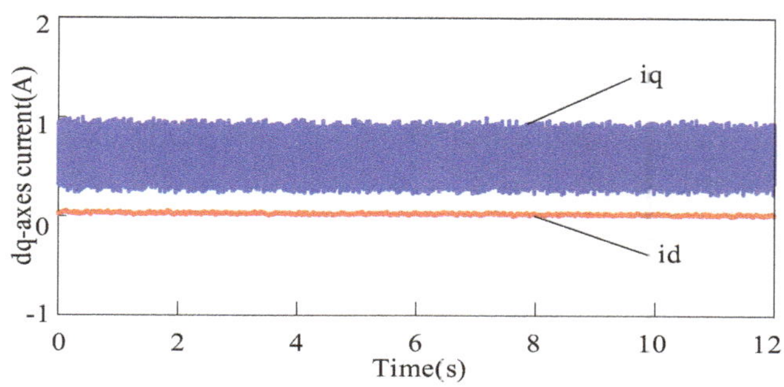

(b) *dq*-axes current waveform.

Figure 11. Current response when the q-axis is injected with positive and negative pulse voltages for position estimation.

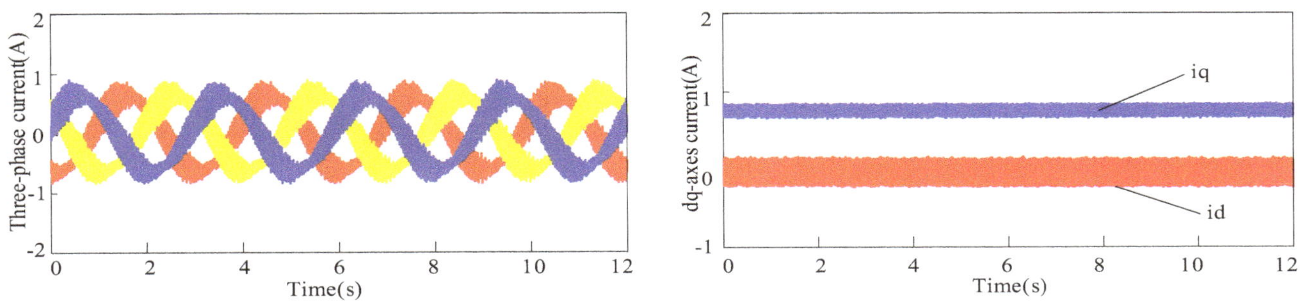

(**a**) Three-phase current waveform. (**b**) dq-axes current waveform.

Figure 12. Current response when the d-axis is injected with positive and negative pulse voltages for position estimation.

4.2. Experimental Platform

In order to verify the feasibility of this method in engineering and the accuracy of rotor position estimation, the proposed position-sensorless control scheme was verified on the platform with a 400W IPMSM, as shown in Figure 13. As shown in Figure 13, the experimental platform mainly includes two parts: the tested system and the load system.

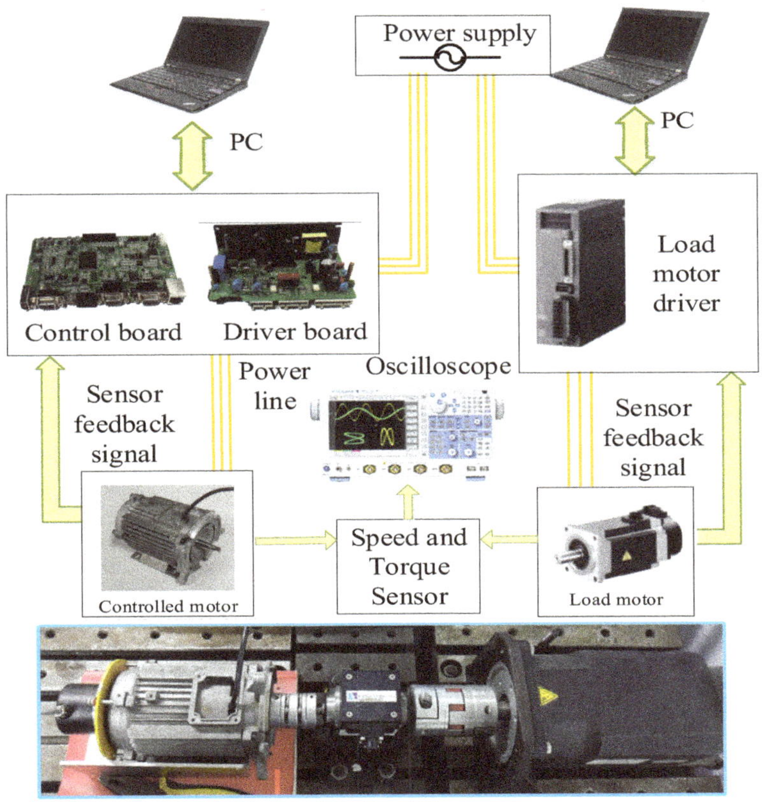

Figure 13. The experimental platform.

The tested system can be mainly divided into three groups: (1) the upper computer for status monitoring, (2) the built-in pump motor as the controlled object, and (3) the control and drive system with the Infineon XMC4500 chip as the core. The load system can be mainly divided into three groups: (1) the upper computer for command transmission and status monitoring, (2) the industrial servo drive, and (3) the high-performance servo motor as load. In addition, the controlled motor and the load motor are connected by speed and torque sensors. The IPMSM and system parameters are shown in Table 2.

Table 2. IPMSM and system parameters for experiment.

Parameter	Quantity	Unit
Pole pairs	2	poles
Resistance	1.6	[Ω]
d-axis inductance	15	[mH]
q-axis inductance	18.8	[mH]
Rated speed	3000	[rpm]
Rated power	400	[W]
Rated voltage	220	[V]
Rated current	2.28	[A]
Rated torque	1.27	[N·m]
PWM switching frequency	10	[kHz]
Injection voltage magnitude	70	[V]

4.3. Initial Rotor Position Estimation

In the experiment, the PWM switching frequency remains unchanged at 10 kHz. The rotor position estimated by the conventional HF square-wave voltage injection method and the improved HF square-wave voltage injection method are compared, and the obtained experimental waveforms are shown in Figures 14 and 15.

The experimental conditions are as follows: the conventional HF square-wave injection frequency is 10 kHz, and the rotor position is estimated every two PWM switching periods. The improved HF square-wave injection frequency is 10 kHz, and rotor position estimation is performed every three PWM switching periods. The initial position angle is artificially fixed at 30°, 60°, 120°, and 150° in advance by acquiring the angle of the encoder. Figure 16 shows the offset angles of the estimated and actual angles of the two methods. The initial position detection in Figure 14c,d and Figure 15c,d are both before the magnetic polarity judgment, and the magnetic polarity judgment are analyzed in the subsequent experiments.

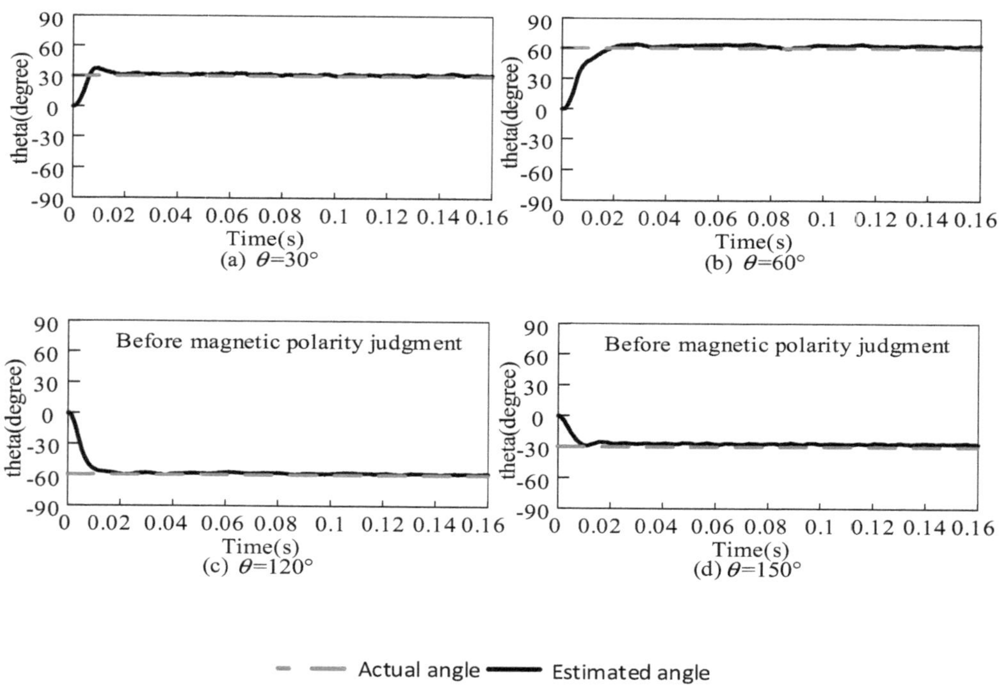

Figure 14. Initial position estimation response curve under conventional HF square-wave voltage injection.

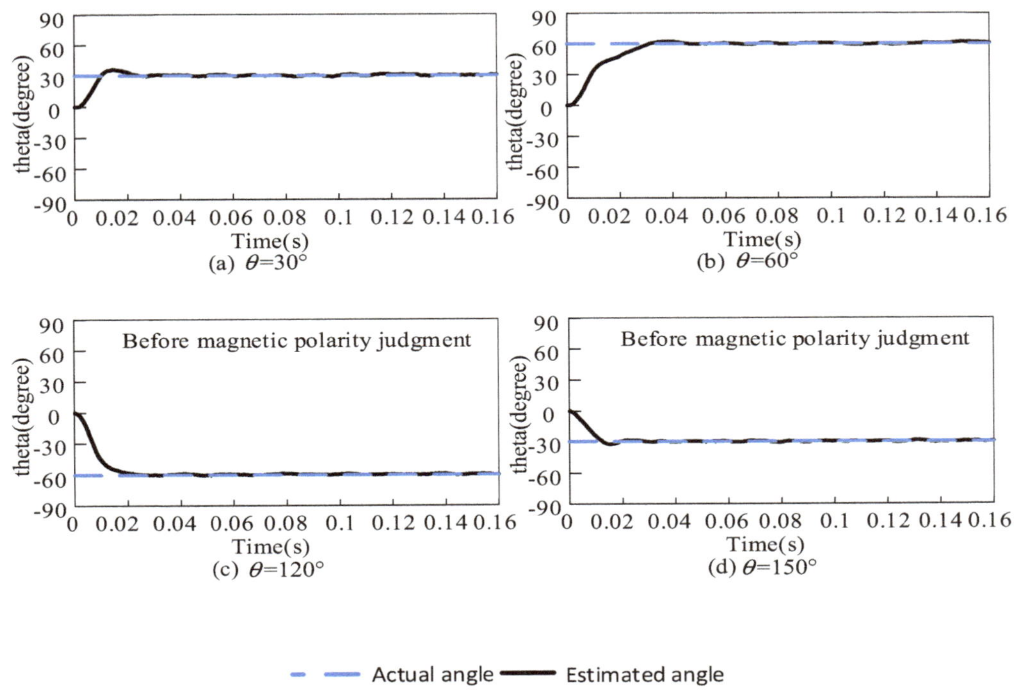

Figure 15. Initial position estimation response curve under improved HF square-wave voltage injection.

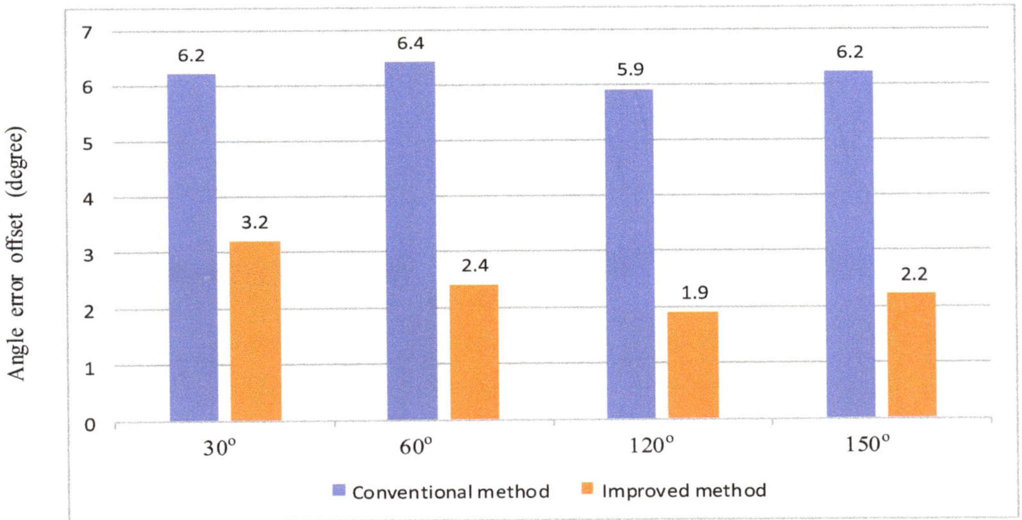

Figure 16. The angle error offset of the two methods.

By summarizing the eight initial position estimation response curves in Figures 14 and 15, the relevant data in Table 3 can be sorted out. In addition, the angle error offset is the size of the offset angle generated on the basis of the actual angle, and the fluctuation range of the angle error is generated on the basis of the angle error offset.

Table 3. Comparison of initial position estimation parameters for two injection methods.

Injection Method	30°		60°		120°		150°	
	Time (s)	error (°)	Time (s)	error (°)	Time (s)	error (°)	Time (s)	error (°)
Conventional method	0.018	±3.3 + 6.2	0.02	±3.4 + 6.4	0.016	±3.2 + 5.9	0.012	±3.6 + 6.2
Improved method	0.022	±3.4 + 3.2	0.032	±3.2 + 2.4	0.023	±2.9 + 1.9	0.017	±3.6 + 2.2

As shown in Figure 16, the estimated angle of the conventional HF square-wave voltage injection method is offset from the actual angle, but this offset is weakened in the improved HF square-wave voltage injection method, in the conventional method, the angle error offset keep within 6.4°, which is caused by the inverter voltage error, in the improved method, the angle error offset keep within 3.2°, it can be seen that the angle error offset is greatly reduced, which is reduced by about 50%, because the improved method reduces the voltage error caused by the nonlinearity of the inverter. It can effectively compensate the voltage error and reduce the influence of the error on the accuracy of rotor position estimation.

4.4. Estimation of Rotor Position at Low Speed

4.4.1. Comparison of Rotor Position Estimation Before and After Improvement

In the HF injection sensorless control period, the FOC control period time is 100 microseconds, and the positive voltage injection period and negative voltage injection period are also 100 microseconds. Figure 17 shows the comparison of rotor position estimation before and after improvement. As shown in Figure 17a, in the conventional method, the estimation error results in an offset error of around 5° and an estimated ripple error of ±7°. However, in the improved method of Figure 17b, the offset error is close to 0° and the estimated ripple error is kept within ±5°. The analysis diagram of steady-state estimation error of two methods under different low speed conditions is shown in Figure 18.

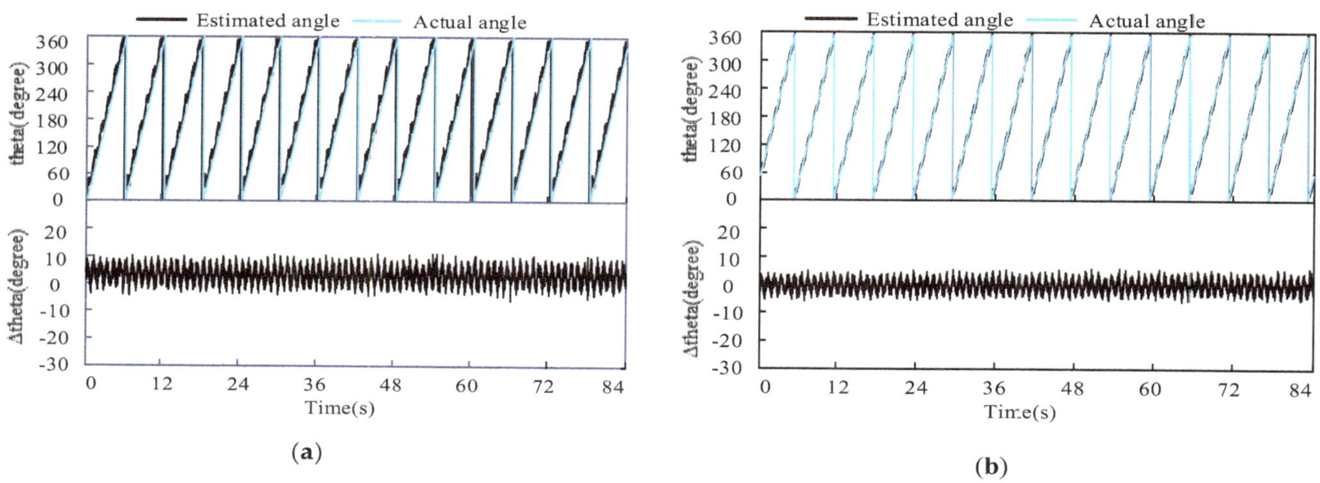

Figure 17. Comparison of rotor position estimation effects between two methods (**a**) Conventional method; (**b**) Improved method.

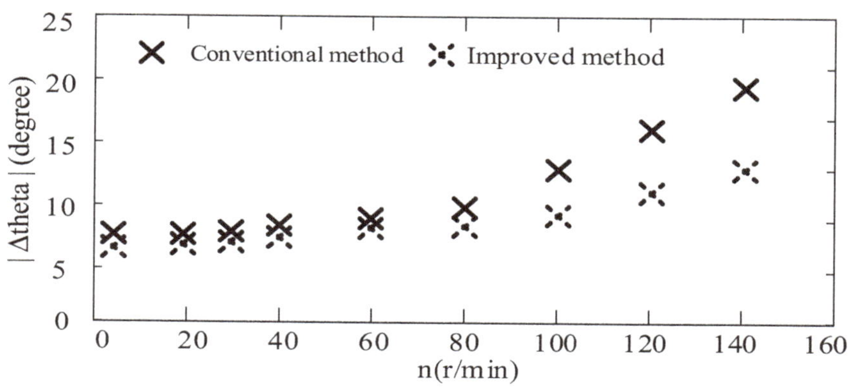

Figure 18. Steady-state estimation error of two injection methods at different speeds.

4.4.2. Rotor Position Observation Experiment Under Forward and Reverse

Figure 19a shows the rotor angle and estimation error waveform of conventional HF square-wave voltage injection method switching back and forth from 5 r/min to −5 r/min. It can be seen that the estimated ripple error can be stabilized within ±10° when the rotation speed is 5 r/min, as shown in Figure 19a. The rotor angle and estimation error waveform of conventional HF square-wave voltage injection method switches back and forth from 20 r/min to −20 r/min, as shown in Figure 19b. It can be seen that the estimated ripple error can be stabilized within ±10° when the rotation speed is 20 r/min, as shown in Figure 19b.

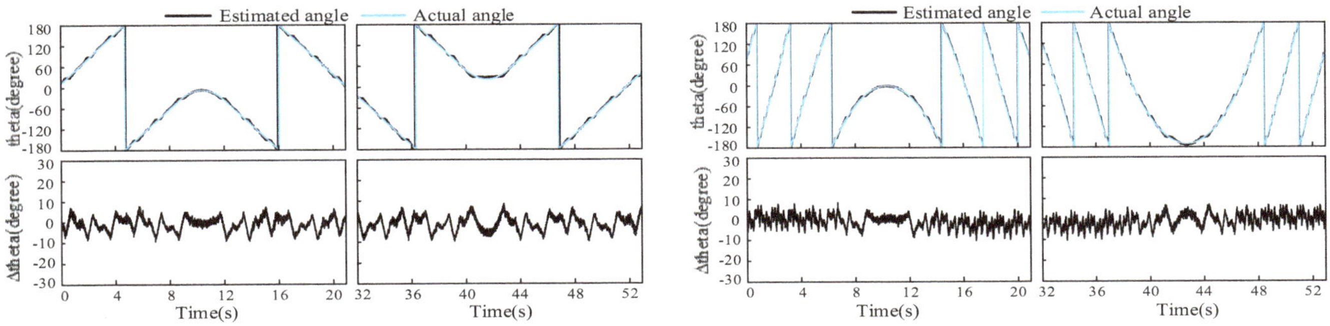

(**a**) 5 r/min (**b**) 20 r/min

Figure 19. The rotor angle and estimation error waveform of conventional HF square-wave voltage injection method at low-speed operation (**a**) at 5 r/min. (**b**) at 20 r/min.

The rotor angle and estimated error waveform switching back and forth from 5 r/min to −5 r/min and the rotor angle and estimated error waveform switching back and forth from 20 r/min to −20 r/min are shown in Figure 20a,b, respectively. As shown in Figure 20a, it can be seen that the estimated ripple error can be stabilized within ±6° when the rotation speed is 5 r/min. It can be seen that the estimated ripple error can be stabilized within ±8° when the rotation speed is 20 r/min, as shown in Figure 20b.

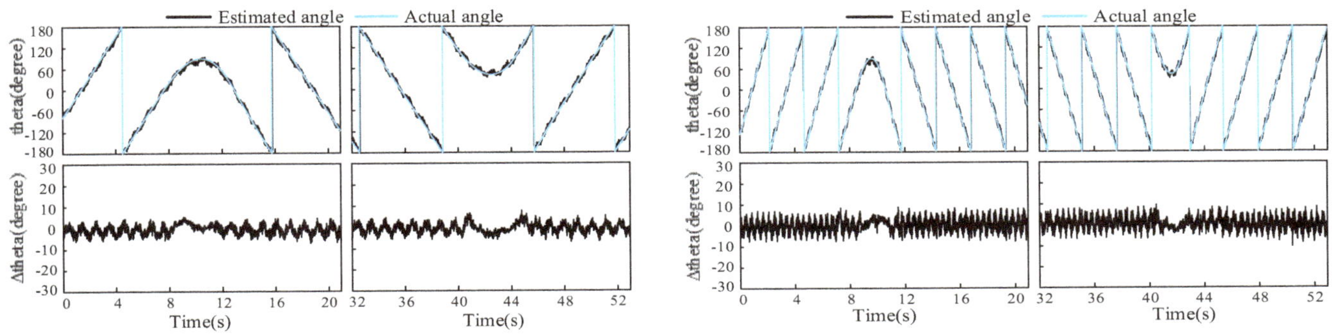

(**a**) 5 r/min (**b**) 20 r/min

Figure 20. The rotor angle and estimation error waveform of improved HF square-wave voltage injection method at low-speed operation (**a**) at 5 r/min. (**b**) at 20 r/min.

4.5. *Judging Compensation of Magnetic Polarity Direction*

In this section, the effects of the two magnetic polarity discrimination methods are compared at 30° and 120°, respectively. The magnetic polarity discrimination experiments of the two methods at 30° are shown in Figure 21a,b, respectively. As shown in Figure 21a, the conventional method uses a 20 V DC bias voltage to identify the magnetic polarity. The proposed method uses a 20 V DC bias voltage to superimpose a 16 V HF square-wave voltage signal, as shown in Figure 21b.

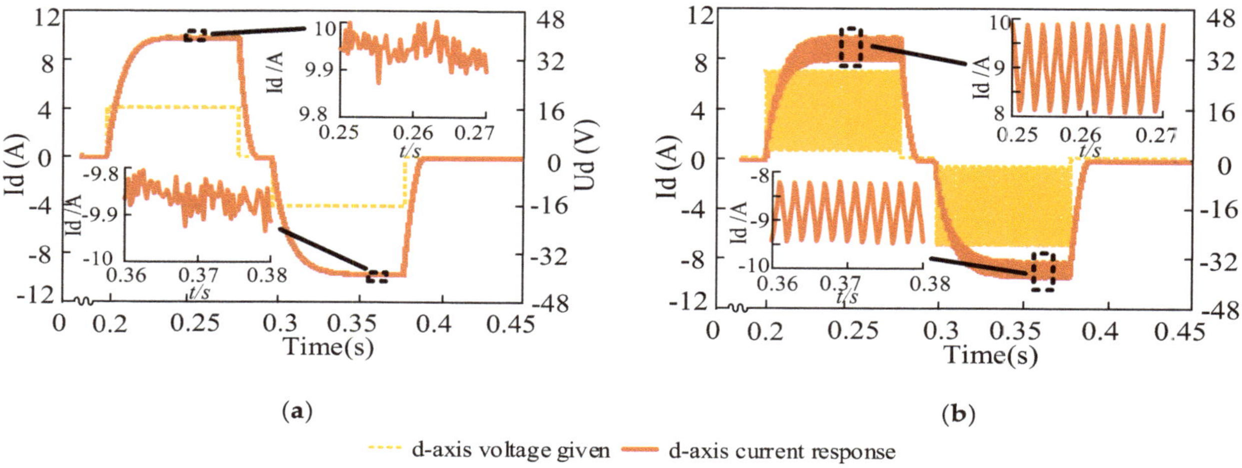

Figure 21. The d-axis current characteristic of the rotor at 30° for magnetic polarity judgment (**a**) conventional method. (**b**) proposed method.

Similarly, the magnetic polarity discrimination experiments of the two methods at 120° are shown in Figure 22a,b, respectively. The experimental data are compared in detail, as shown in Figures 23 and 24. We can divide the process into two stages. When t is between 0 and 0.18 s, the first stage period is the from standstill to startup, that is no procedure and no equipment to be performed at standstill, and the rotor is fixed initially in the startup period.

Then, during the second stage, when t is between 0.18 and 0.38 s, the magnetic polarity judgment is performed in order to ensure the accuracy of rotor position estimation. The magnetic polarity identification method proposed in this paper is a combination of HF square-wave voltage injection method and DC bias voltage. In the process of magnetic polarity identification, the HF square-wave voltage injection method has also been updating its angle.

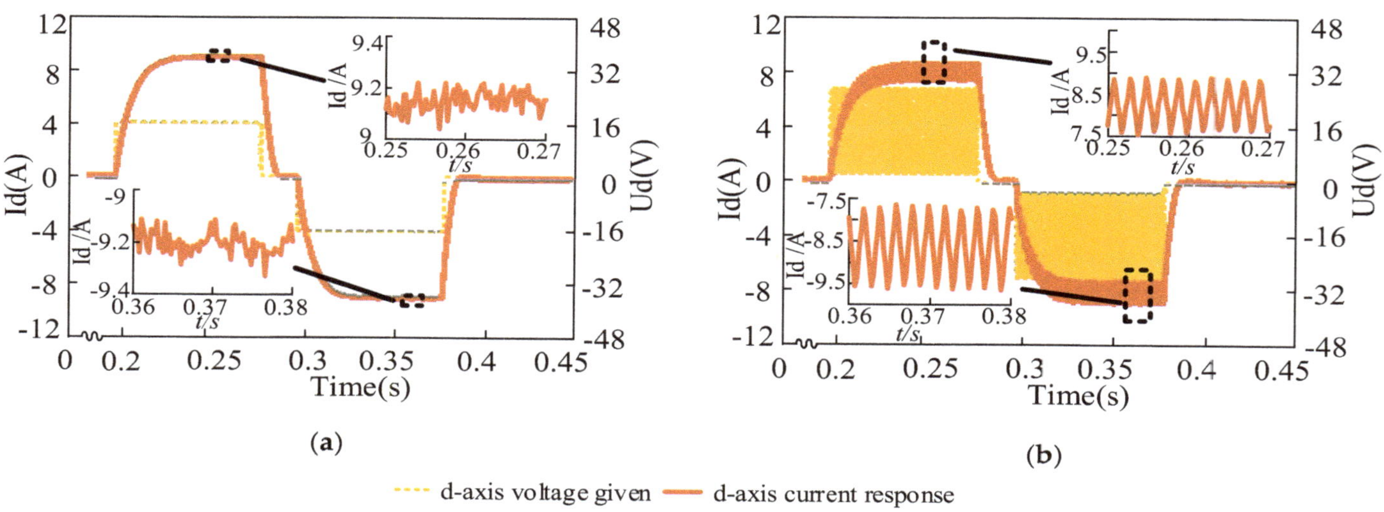

Figure 22. The d-axis current characteristic of the rotor at 120° for magnetic polarity judgment (**a**) conventional method. (**b**) proposed method.

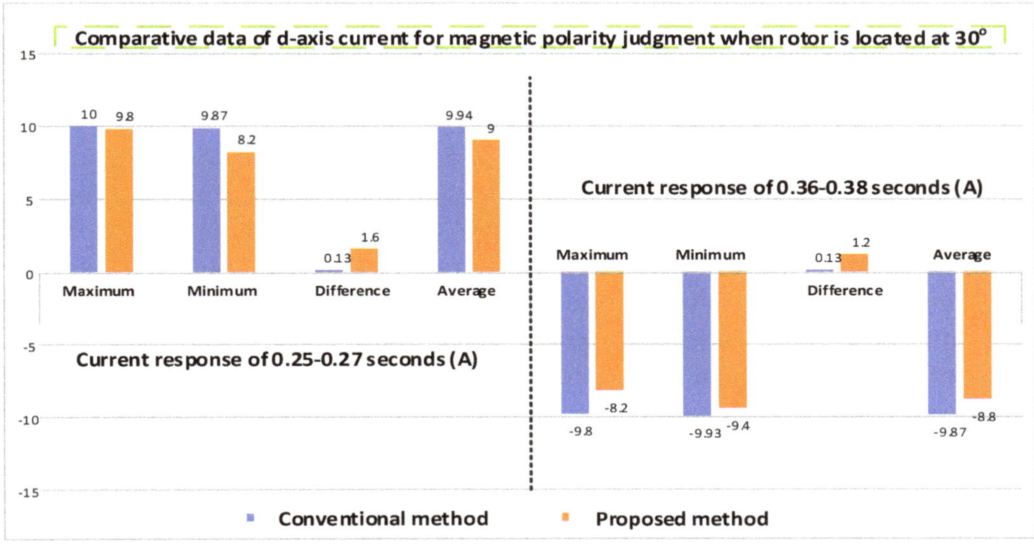

Figure 23. Comparative data of *d*-axis current characteristics for magnetic polarity judgment when rotor is located at 30°.

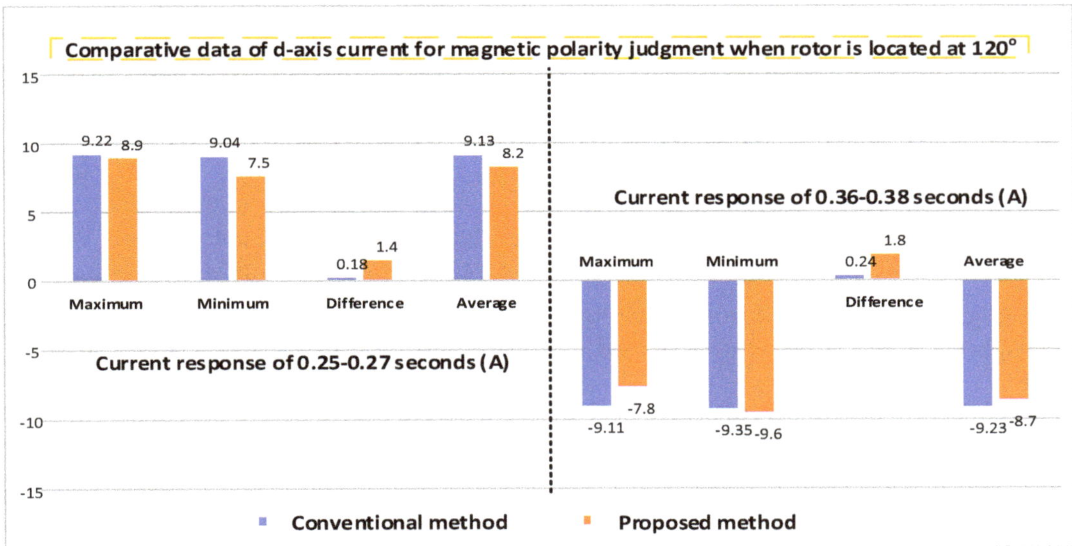

Figure 24. Comparative data of *d*-axis current characteristics for magnetic polarity judgment when rotor is located at 120°.

Therefore, the method of magnetic polarity identification proposed in this paper is not only suitable for the case where the motor is stationary, but also suitable for the free running condition of the motor. As can be seen from the Figures 21 and 22, the effect of only injecting DC bias voltage on the initial position observation is not very obvious either from the difference of the extreme value of the d-axis current response or from the average value of the steady-state current of the d-axis. If the burr of the current is large, it is easy to affect the judgment result. In 50 repetitive experiments, the probability of method (a) judging errors was about 10%. For method (b), the discrimination is more obvious from the difference of high frequency response of current. In 50 repetitive experiments, the probability of misjudgment was 0, which directly proved that the improved direction judgment of magnetic polarity method proposed in this paper is effective.

5. Conclusions

In this paper, a HF square-wave voltage injection scheme-based position sensorless control of IPMSM in the low-speed range and a new method of magnetic polarity detection in zero-speed range are proposed. The strategy realizes the switch between zero-speed algorithm and low-speed algorithm for sensorless control of IPMSM. Since the proposed method is an improvement on the conventional square-wave injection method, the method is applicable to AC motors with salient pole effect that can be applied in the conventional method. Furthermore, the voltage error caused by the nonlinearity of the inverter are compensated to improve the accuracy of rotor position estimation. In addition, based on the principle analysis of d-axis magnetic circuit characteristics, a method for determining the direction of magnetic polarity of d-axis two-opposite DC voltage offset by uninterruptible square-wave injection is proposed, which is fast in the convergence rate of magnetic polarity detection Compared with the conventional method, the new method is more distinct and the success rate is higher. Finally, comprehensive simulation and experimental results verify that the improved HF square-wave voltage injection method and the detection of magnetic polarity method perform faster and more accurately compared with the conventional method.

Author Contributions: Conceptualization, S.W. and J.Z.; methodology, S.W. and J.Z.; software, K.Y.; validation, S.W., J.Z. and K.Y.; formal analysis, W.S.; data curation, K.Y.; writing—original draft preparation, S.W. and J.Z.; writing—review and editing, S.W. and J.Z. All authors gave advice for the manuscript.

References

1. Onambele, C.; Elsied, M.; Mpanda Mabwe, A.; El Hajjaji, A. Multi-Phase Modular Drive System: A Case Study in Electrical Aircraft Applications. *Energies* **2018**, *11*, 5. [CrossRef]
2. Zhang, C.; Guo, Q.; Li, L. System Efficiency Improvement for Electric Vehicles Adopting a Permanent Magnet Synchronous Motor Direct Drive System. *Energies* **2017**, *10*, 2030. [CrossRef]
3. Hua, W.; Zhou, K.L. Investigation of a Co-Axial Dual-Mechanical Ports Flux-Switching Permanent Magnet Machine for Hybrid Electric Vehicles. *Energies* **2015**, *8*, 14361–14379. [CrossRef]
4. Wang, S.; Yang, K.; Chen, K. An Improved Position-Sensorless Control Method at Low Speed for PMSM Based on High-Frequency Signal Injection into a Rotating Reference Frame. *IEEE Access* **2019**, *7*, 86510–86521. [CrossRef]
5. Zhang, X.; Hou, B.; Mei, Y. Deadbeat Predictive Current Control of Permanent-Magnet Synchronous Motors with Stator Current and Disturbance Observer. *IEEE Trans. Power Electron.* **2017**, *32*, 3818–3834. [CrossRef]
6. Park, Y.; Sul, S. Sensorless Control Method for PMSM Based on Frequency-Adaptive Disturbance Observer. *IEEE J. Emerg. Sel. Top. Power Electron.* **2014**, *2*, 143–151. [CrossRef]
7. Zhang, G.; Wang, G.; Xu, D.; Zhao, N. ADALINE-Network-Based PLL for Position Sensorless Interior Permanent Magnet Synchronous Motor Drives. *IEEE Trans. Power Electron.* **2016**, *31*, 1450–1460. [CrossRef]
8. Nguyen, H.Q.; Yang, S. Rotor Position Sensorless Control of Wound-Field Flux-Switching Machine Based on High Frequency Square-Wave Voltage Injection. *IEEE Access* **2018**, *6*, 48776–48784. [CrossRef]
9. Wang, T.; Zhang, H.; Gao, Q.; Xu, Z.; Li, J.; Gerada, C. Enhanced Self-Sensing Capability of Permanent-Magnet Synchronous Machines: A Novel Saliency Modulation Rotor End Approach. *IEEE Trans. Ind. Electron.* **2017**, *64*, 3548–3556. [CrossRef]
10. Wang, G.; Yang, L.; Yuan, B.; Wang, B.; Zhang, G.; Xu, D. Pseudo-Random High-Frequency Square-Wave Voltage Injection Based Sensorless Control of IPMSM Drives for Audible Noise Reduction. *IEEE Trans. Ind. Electron.* **2016**, *63*, 7423–7433. [CrossRef]
11. Wang, G.; Yang, L.; Zhang, G.; Zhang, X.; Xu, D. Comparative Investigation of Pseudorandom High-Frequency Signal Injection Schemes for Sensorless IPMSM Drives. *IEEE Trans. Power Electron.* **2017**, *32*, 2123–2132. [CrossRef]
12. Wang, G.; Xiao, D.; Zhao, N.; Zhang, X.; Wang, W.; Xu, D. Low-Frequency Pulse Voltage Injection Scheme-Based Sensorless Control of IPMSM Drives for Audible Noise Reduction. *IEEE Trans. Ind. Electron.* **2017**, *64*, 8415–8426. [CrossRef]

13. Almarhoon, A.H.; Zhu, Z.Q.; Xu, P. Improved Rotor Position Estimation Accuracy by Rotating Carrier Signal Injection Utilizing Zero-Sequence Carrier Voltage for Dual Three-Phase PMSM. *IEEE Trans. Ind. Electron.* **2017**, *64*, 4454–4462. [CrossRef]
14. Seilmeier, M.; Piepenbreier, B. Sensorless Control of PMSM for the Whole Speed Range Using Two-Degree-of-Freedom Current Control and HF Test Current Injection for Low-Speed Range. *IEEE Trans. Power Electron.* **2015**, *30*, 4394–4403. [CrossRef]
15. Zhang, G.; Wang, G.; Yuan, B.; Liu, R.; Xu, D. Active Disturbance Rejection Control Strategy for Signal Injection-Based Sensorless IPMSM Drives. *IEEE Trans. Transp. Electrif.* **2018**, *4*, 330–339. [CrossRef]
16. Luo, X.; Tang, Q.; Shen, A.; Zhang, Q. PMSM Sensorless Control by Injecting HF Pulsating Carrier Signal into Estimated Fixed-Frequency Rotating Reference Frame. *IEEE Trans. Ind. Electron.* **2016**, *63*, 2294–2303. [CrossRef]
17. Yoon, Y.; Sul, S.; Morimoto, S.; Ide, K. High-Bandwidth Sensorless Algorithm for AC Machines Based on Square-Wave-Type Voltage Injection. *IEEE Trans. Ind. Appl.* **2011**, *47*, 1361–1370. [CrossRef]
18. Ni, R.; Xu, D.; Blaabjerg, F.; Lu, K.; Wang, G.; Zhang, G. Square-Wave Voltage Injection Algorithm for PMSM Position Sensorless Control with High Robustness to Voltage Errors. *IEEE Trans. Power Electron.* **2017**, *32*, 5425–5437. [CrossRef]
19. Kim, S.; Ha, J.; Sul, S. PWM Switching Frequency Signal Injection Sensorless Method in IPMSM. *IEEE Trans. Ind. Appl.* **2012**, *48*, 1576–1587. [CrossRef]
20. Park, N.; Kim, S. Simple sensorless algorithm for interior permanent magnet synchronous motors based on high-frequency voltage injection method. *IET Electr. Power Appl.* **2014**, *8*, 68–75. [CrossRef]
21. Chen, L.; Götting, G.; Dietrich, S.; Hahn, I. Self-Sensing Control of Permanent-Magnet Synchronous Machines with Multiple Saliencies Using Pulse-Voltage-Injection. *IEEE Trans. Ind. Appl.* **2016**, *52*, 3480–3491. [CrossRef]
22. Hu, J.; Liu, J.; Xu, L. Eddy Current Effects on Rotor Position Estimation and Magnetic Pole Identification of PMSM at Zero and Low Speeds. *IEEE Trans. Power Electron.* **2008**, *23*, 2565–2575. [CrossRef]
23. Jin, X.; Ni, R.; Chen, W.; Blaabjerg, F.; Xu, D. High-Frequency Voltage-Injection Methods and Observer Design for Initial Position Detection of Permanent Magnet Synchronous Machines. *IEEE Trans. Power Electron.* **2018**, *33*, 7971–7979. [CrossRef]
24. Xie, G.; Lu, K.; Dwivedi, S.K.; Rosholm, J.R.; Blaabjerg, F. Minimum-Voltage Vector Injection Method for Sensorless Control of PMSM for Low-Speed Operations. *IEEE Trans. Power Electron.* **2016**, *31*, 1785–1794. [CrossRef]
25. Liu, Y.; Zhou, B.; Li, S.; Feng, Y. Initial Rotor Position Detection of Surface Mounted Permanent Magnet Synchronous Motor. *Proc. CSEE* **2011**, *31*, 48–54.
26. Yan-Jun, Y.U.; Gao, H.W.; Chai, F.; Cheng, S.K. Rotor magnetic polarity detection method for PMSM. *Elect. Mach. Control.* **2011**, *15*, 86–90.
27. Jeong, Y.S.; Lorenz, R.D.; Jahns, T.M.; Sul, S.K. Initial rotor position estimation of an interior permanent-magnet synchronous machine using carrier-frequency injection methods. *IEEE Trans. Ind. Appl.* **2005**, *41*, 38–45. [CrossRef]
28. Raute, R.; Caruana, C.; Staines, C.S.; Cilia, J.; Sumner, M.; Asher, G. Operation of a sensorless PMSM drive without additional test signal injection. In Proceedings of the 2008 4th IET Conference on Power Electronics, Machines and Drives, York, UK, 2–4 April 2008; pp. 616–620.

4

Commutation Torque Ripple Suppression Strategy of Brushless DC Motor Considering Back Electromotive Force Variation

Xinmin Li [1], Guokai Jiang [2], Wei Chen [3], Tingna Shi [4], Guozheng Zhang [1] and Qiang Geng [1,*]

1 School of Electrical Engineering and Automation, Tianjin Polytechnic University, Tianjin 300387, China; lixinmin@tju.edu.cn (X.L.); zhanggz@tju.edu.cn (G.Z.)
2 China Automotive Technology and Research Center Co. Ltd., Tianjin 300300, China; jiangguokai@tju.edu.cn
3 School of Artificial Intelligence, Tianjin Polytechnic University, Tianjin 300387, China; chen_wei@tju.edu.cn
4 College of Electrical Engineering, Zhejiang University, Hangzhou 310027, China; tnshi@tju.edu.cn
* Correspondence: gengqiang@tju.edu.cn.

Abstract: This paper presents a commutation torque ripple suppression strategy for brushless DC motor (BLDCM) in the high-speed region, which considers the back electromotive force (back-EMF) variation during the commutation process. In the paper, the influence of actual back-EMF variation on the torque and outgoing phase current during the commutation process is analyzed. A modified smooth torque mechanism is then reconstructed considering the back-EMF variation, based on which a novel torque ripple suppression strategy is further designed. Compared with the traditional strategy which controls the chopping duty cycle relatively smoothly in the commutation process, the proposed strategy dynamically regulates the chopping duty cycle, which makes it show a gradual decrease. This strategy can suppress the commutation torque ripple even in a long commutation process, and broaden the speed range of the commutation torque ripple reduction. Under the experimental conditions of this paper, the proposed strategy can effectively reduce the commutation torque ripple in the high-speed region, and avoid the outgoing phase current cannot be reduced to zero. The experimental results verify the correctness of the theoretical analysis and the feasibility of the proposed strategy.

Keywords: brushless DC motor; commutation torque ripple; back electromotive force

---

1. Introduction

The brushless DC motor (BLDCM) has advantages of simple structure, high power density, and high reliable operation [1–5]. However, the windings of the motor have inductances, and the transient process appears during the current exchanged between two phases, and this process will cause commutation torque ripple, which may reach about 50% of the average torque of the BLDCM [6–8]. If the commutation torque ripple is not suppressed by specific reduction strategy, the vibration and noise of motors will increase, and the promotion application of BLDCM will be restricted in a field that has strict requirements for torque ripple and noise [9–11]. When the BLDCM operates in its high-speed range, limited by the output voltage of the inverter, the incoming phase and the outgoing phase current are difficult to change rapidly, so the motor will generate greater commutation torque ripple in this region [12–14].

Pulse width modulation (PWM) techniques can be employed to suppress the commutation torque ripple by maintaining the non-commutation phase current constant [15–18], where only the outgoing phase modulation technique is applicable to the high-speed region in light of the relationship between the back electromotive force (back-EMF) and DC-link voltage. A voltage compensation method is proposed in [15] to control the incoming phase and the outgoing phase current slopes by the same degree. A three-segments modulation strategy is used in [16], where the action time of each segment is acquired based on the minimum commutation time. An integral sliding mode current controller is introduced to enhance the robustness of commutation torque ripple suppression in [17]. A three-phase PWM modulation technique is proposed to suppress commutation torque ripple in [18], where the torque observer and the calculation of commutation process are not necessary.

DC-link voltage boost techniques can also be adopted to reduce the commutation torque ripple by adding a DC–DC converter in front of the voltage source inverter (VSI) [19–23]. In [20], the single ended primary inductor converter (SEPIC) is introduced to adjust the required DC-link voltage during the commutation process. In [22], the Buck converter is added, by regulating the amplitude of the out-voltage and the current during the commutation interval, so the commutation torque can be reduced partly. In [23], a Z-source inverter is used to boost the DC-link voltage by the shoot-through vectors.

Normally, the torque during commutation process is considered to be proportional to the non-commutation phase current when the back-EMF is assumed to be constant. However, a constant non-commutation phase current will still generate a torque ripple under the actual back-EMF variation during the commutation process [24]. Furthermore, the torque ripple is related to the stator current and the commutation duration, which is much heavier in a long commutation process for the high-speed region.

A commutation torque ripple suppression strategy considering the back-EMF variation is designed in this paper for a high-speed region. Section 2 introduces the traditional strategy assuming the back-EMFs constant. Section 3 studies the impact of the back-EMF change on the torque control performance and the normal end of commutation in the traditional strategy, and then Section 4 designs a new strategy in light of the reconstructed smooth torque mechanism. In Section 5, the above theoretical analysis is experimentally verified. Conclusions come in Section 6.

2. Traditional Commutation Torque Ripple Suppression Strategy Ignoring Back-EMF Variation

The equivalent model of the BLDCM drive system is shown in Figure 1, where S_i and D_i, $i \in \{1, 2, \ldots, 6\}$, are metal-oxide-semiconductor field-effect transistor (MOSFET) and its anti-parallel diode respectively; N is the neutral point of three phases windings.

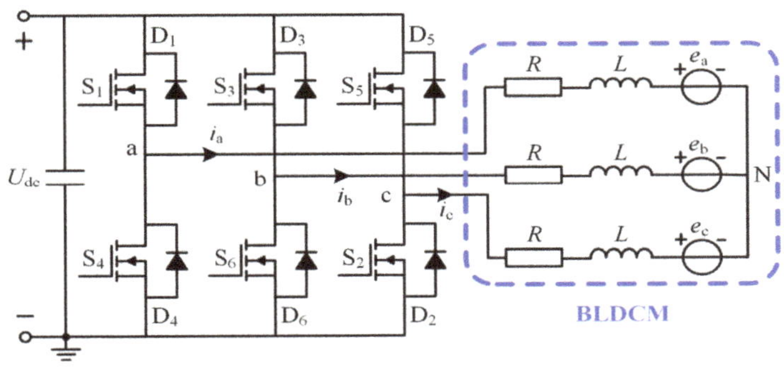

Figure 1. The equivalent model of the BLDCM drive system.

Assuming the three-phase stator windings are symmetrical, simultaneously neglecting the mutual inductance, and as shown in Figure 1 the three-phase windings are star-connected, the terminal voltage model can be expressed as

$$\begin{bmatrix} u_a \\ u_b \\ u_c \end{bmatrix} = \begin{bmatrix} R & 0 & 0 \\ 0 & R & 0 \\ 0 & 0 & R \end{bmatrix} \begin{bmatrix} i_a \\ i_b \\ i_c \end{bmatrix} + \begin{bmatrix} L & 0 & 0 \\ 0 & L & 0 \\ 0 & 0 & L \end{bmatrix} \frac{d}{dt} \begin{bmatrix} i_a \\ i_b \\ i_c \end{bmatrix} + \begin{bmatrix} e_a \\ e_b \\ e_c \end{bmatrix} + u_N \quad (1)$$

where u_a, u_b, and u_c are terminal voltages; i_a, i_b, and i_c are stator currents; e_a, e_b and e_c are phase back-EMFs, and u_N is the neutral point voltage of the motor.

The electromagnetic torque is given by

$$T_e = \frac{e_a i_a + e_b i_b + e_c i_c}{\omega_m} \tag{2}$$

where ω_m is the rotor mechanical angular velocity.

A BLDCM is normally driven by a six-step mode, where normal conduction periods and commutation periods exist. In a normal conduction period, only two windings are energized, and the other one is floating. In a commutation period, three-phase windings are all energized due to the stator inductance and the limited inverter voltage.

Figure 2 shows the modulation method for the high-speed region, where the shaded area and the non-shaded area represent commutation and normal conduction periods, respectively. The outgoing phase modulation starts to be used in the commutation period when Hall sectors change, and the non-commutation modulation is switched to be applied in the normal conduction period as the outgoing phase current is reduced to zero. The duty cycle of the chopping switch in the normal conduction period and the commutation period are denoted as d_{norm} and d_{cmt}, respectively; as shown in Figure 2, two duty cycles are not the same, and the value relation between these two duty cycles is $d_{cmt} < d_{norm}$. Taking the commutation process of $a^+c^- \rightarrow b^+c^-$ as an example, the commutation torque ripple in high-speed region is discussed as follows.

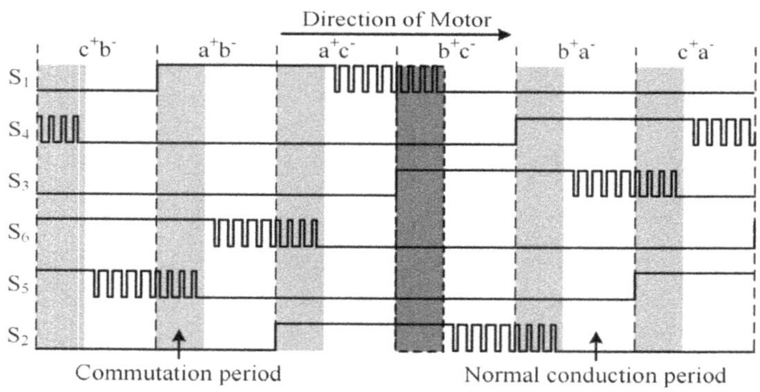

Figure 2. Modulation in high-speed region.

According to Figures 1 and 2, the terminal voltages are satisfied as $u_a = d_{cmt}U_{dc}$, $u_b = U_{dc}$ and $u_c = 0$ in the commutation process of $a^+c^- \rightarrow b^+c^-$, where d_{cmt} is the duty cycle of the commutation process. Substituting it into (1) and considering $i_a + i_b + i_c = 0$ in star-connected motors, the current equation can be obtained as

$$\begin{cases} 3L\frac{di_a}{dt} = (2d_{cmt} - 1)U_{dc} - (2e_a - e_b - e_c) - 3Ri_a \\ 3L\frac{di_b}{dt} = (2 - d_{cmt})U_{dc} - (2e_b - e_a - e_c) - 3Ri_b \\ 3L\frac{di_c}{dt} = -(1 + d_{cmt})U_{dc} - (2e_c - e_a - e_b) - 3Ri_c \end{cases} \tag{3}$$

If the back-EMF variation is neglected, the phase back-EMFs will be

$$\begin{cases} e_a = E \\ e_b = E \\ e_c = -E \end{cases} \tag{4}$$

where E is the magnitude of phase back-EMF.

By substituting (4) into (2), the torque can be simplified as

$$T_e = \frac{Ei_a + Ei_b - Ei_c}{\omega_m} = -\frac{2Ei_c}{\omega_m} = -\frac{60}{\pi}k_e i_c \tag{5}$$

According to (5), the torque is proportional to the non-commutation current during the commutation process on the assumption that back-EMF is unchanged, which means keeping the non-commutation current constant can effectively suppress the commutation torque ripple.

Substitute (4) into (3), the non-commutation current can be expressed as

$$3L\frac{di_c}{dt} = -(1 + d_{cmt})U_{dc} + 4E - 3Ri_c \tag{6}$$

To maintain the current unchanged, namely $di_c/dt = 0$, the required duty cycle should be

$$d_{cmt} = \frac{4E - 3Ri_c}{U_{dc}} - 1 = \frac{4E + 3RI}{U_{dc}} - 1 \tag{7}$$

3. Effect of Back-EMF Variation on Commutation Torque Ripple Suppression

The traditional high-speed commutation torque ripple suppression strategies usually neglect the change of back-EMF in the commutation process, and suppresses the commutation torque ripple by controlling the non-commutation phase current constant. The phase currents and back-EMFs in the commutation process of $a^+c^- \rightarrow b^+c^-$ are shown in Figure 3, where I is the current amplitude before the commutation, t_{cmt} is the commutation duration, namely the time from commutation initial time to the moment that the outgoing phase reduces to zero, t_{Hall} is the Hall sector's period.

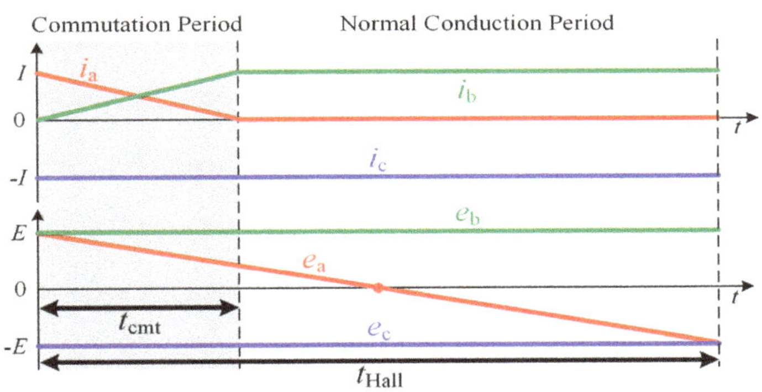

Figure 3. The variation of back-EMF during the commutation process.

It can be seen from Figure 3 that the actual back-EMFs during the commutation process are

$$\begin{cases} e_a = E - 2Et/t_{Hall} \\ e_b = E \\ e_c = -E \end{cases}, \quad (0 \le t \le t_{cmt}) \tag{8}$$

As the motor speed increases, the Hall sector's period gradually decreases, while the commutation duration is prolonged due to the limited DC-link voltage. Based on (8), when the commutation duration accounts for a large proportion of the Hall sector's period, there is a significant variation in the outgoing phase back-EMF during the commutation process. Therefore, the effect of back-EMF variation on the torque ripple suppression in the traditional strategy is analyzed as follows.

3.1. Effect of Back-EMF Variation on Torque

The actual back-EMF variation affects the suppression effect of commutation torque ripple. By substituting (8) into (2), the torque equation with back-EMF variation can be expressed as

$$T_e = -\frac{2Ei_c}{\omega_m} - \frac{2Ei_a}{\omega_m}\frac{t}{t_{Hall}} \tag{9}$$

According to (9), it can be seen that the torque during the commutation process is not only related to the non-commutation current, but also to the outgoing phase current. The smooth torque condition can be derived from (9) as

$$\frac{dT_e}{dt} = -\frac{2E}{\omega_m}\frac{di_c}{dt} - \frac{2Et}{\omega_m t_{Hall}}\frac{di_a}{dt} - \frac{2Ei_a}{\omega_m t_{Hall}} \tag{10}$$

If the traditional duty cycle (7) is employed to suppress the torque ripple, the smooth torque mechanism (10) can be simplified in light of (3) and (8) as

$$\frac{dT_e}{dt} = \frac{-2E}{3L\omega_m t_{Hall}}[3Lt\frac{di_a}{dt} + 3Li_a - 2Et] = \frac{-2E}{3L\omega_m t_{Hall}}[3L\frac{d(ti_a)}{dt} - 2Et] \tag{11}$$

Define T_{e0} as the torque at the initial moment of commutation. The torque during the commutation process can be obtained by integrating (11) as

$$T_e(t) = T_{e0} + \frac{2Et[Et - 3Li_a(t)]}{3L\omega_m t_{Hall}} \tag{12}$$

Based on (12), the torque waveform during the commutation process is shown in Figure 4. It can be seen that the torque change is zero only when $t = 3Li_a/E$, and a large torque ripple occurs at the end of commutation process, namely the moment i_a just reduces to zero.

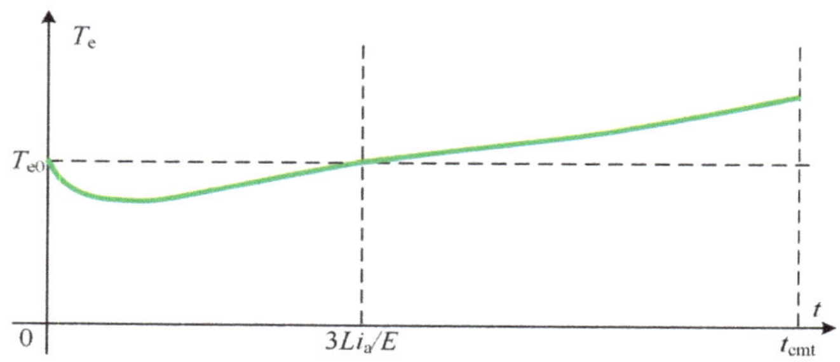

Figure 4. The effect of back-EMF variation on the torque during the commutation process.

3.2. Effect of Back-EMF Variation on Outgoing Phase Current

The actual back-EMF variation also affects the normal turn-off of outgoing phase current. If the traditional duty cycle (7) is used to suppress the torque ripple in high-speed region, the current differential equation can be acquired by substituting (8) into (3) as

$$L\frac{di_a}{dt} = -Ri_a - (U_{dc} - 2E - 2RI) + \frac{4E}{3t_{Hall}}t \tag{13}$$

To simplify the calculation, outgoing phase current term Ri_a can be neglected as the resistance is small and the outgoing phase is gradually reducing to zero. The outgoing phase current can be obtained by integrating (13) as

$$i_a = I - \frac{1}{L}mt + \frac{2E}{3Lt_{Hall}}t^2 = \frac{2E}{3Lt_{Hall}}(t - \frac{3t_{Hall}m}{4E})^2 + I - \frac{3t_{Hall}m^2}{8EL} \tag{14}$$

where $m = U_{dc} - 2E - 2RI$.

According to (14), the minimum value of outgoing phase current is $i_{a\_min} = I - 3t_{Hall}m^2/(8EL)$, which happens at $t = 3mt_{Hall}/(4E)$. If the minimum value is less than or equal to zero, the commutation process can be successfully finished as the outgoing phase current reduces to zero, as shown in the gray line in Figure 5. If the minimum value is larger than zero, the commutation process cannot be successfully finished as the outgoing phase current never reduces to zero, as shown in the black line in Figure 5.

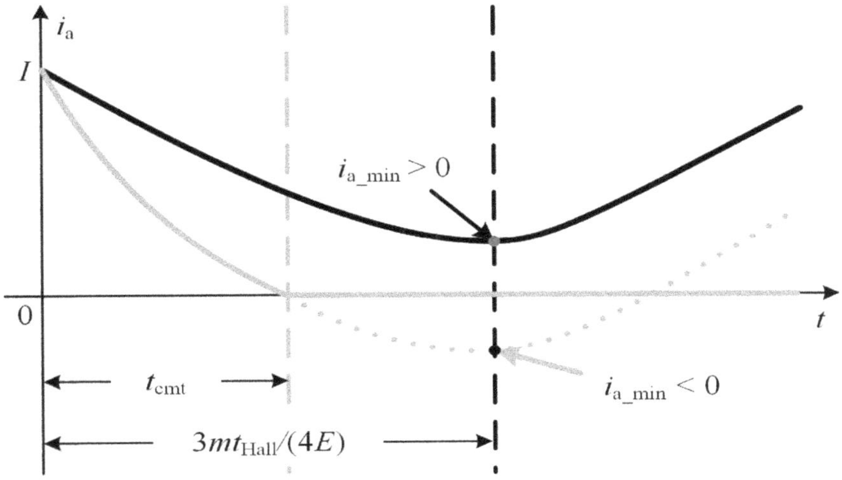

Figure 5. The effect of back-EMF variation on the outgoing phase current.

Therefore, the condition for the normal end of the commutation process is

$$i_{a\_min} = I - \frac{3t_{Hall}m^2}{8EL} \leq 0 \tag{15}$$

Since the variables in (15) are all larger than zero, it can be further simplified as

$$m \geq \sqrt{\frac{8ELI}{3t_{Hall}}} \tag{16}$$

Set p as the pole pairs and n as the motor speed, the Hall sector's period can be expressed as $t_{Hall} = 10/(np)$. Substitute it into (16), the speed range for the normal end of the commutation process in the traditional strategy is

$$n \leq \frac{U_{dc} - 2RI}{2(k_e + \sqrt{\frac{pk_eLI}{15}})} = \frac{f_1}{g_1} \tag{17}$$

where $k_e = E/n$ is phase back-EMF coefficient; f_1 and g_1 are numerator and denominator of the critical speed in the traditional strategy.

4. Proposed Commutation Torque Ripple Suppression Strategy Considering Back-EMF Variation

According to (10), the smooth torque mechanism with back-EMF variation is reconstructed as

$$t_{Hall}\frac{di_c}{dt} + t\frac{di_a}{dt} + i_a = 0 \tag{18}$$

Substitute (3) and (4) into (18), and it can be presented as

$$(2t - t_{Hall})d_{cmt}U_{dc} - (U_{dc} + 4E + 3Ri_a)t - (U_{dc} - 4E + 3Ri_c)t_{Hall} + \frac{4Et^2}{t_{Hall}} + 3Li_a = 0 \tag{19}$$

For commutation torque ripple suppression with back-EMF variation, the duty cycle is

$$d_{\mathrm{cmt}} = \frac{1}{(2t - t_{\mathrm{Hall}})U_{\mathrm{dc}}}[(U_{\mathrm{dc}} + 4E + 3Ri_{\mathrm{a}})t - \frac{4Et^2}{t_{\mathrm{Hall}}} + (U_{\mathrm{dc}} - 4E + 3Ri_{\mathrm{c}})t_{\mathrm{Hall}} - 3Li_{\mathrm{a}}] \tag{20}$$

If this duty cycle is employed, the torque gradient can always be zero over the whole commutation process, even with the changing back-EMF, and the torque ripple will be thus suppressed completely.

Moreover, the outgoing phase current in the proposed strategy is analyzed as follows. By substituting (20) into (3), it can be

$$L\frac{\mathrm{d}i_{\mathrm{a}}}{\mathrm{d}t} = -\frac{t_{\mathrm{Hall}}}{t_{\mathrm{Hall}} - 2t}(a + bi_a) \tag{21}$$

where $a = U_{\mathrm{dc}} - 2E + 2Ri_{\mathrm{c}} \geq 0$, $b = R - 2L/t_{\mathrm{Hall}}$.

If $b > 0$, the current in the interval during commutation process ($t \in [0,0.5t_{\mathrm{Hall}}]$) satisfies

$$L\frac{\mathrm{d}i_{\mathrm{a}}}{\mathrm{d}t} \leq -\frac{t_{\mathrm{Hall}}a}{t_{\mathrm{Hall}} - 2t} \tag{22}$$

By integrating (22), it can be presented as

$$i_{\mathrm{a}} \leq I + \frac{at_{\mathrm{Hall}}}{2L}\ln(1 - \frac{2t}{t_{\mathrm{Hall}}}) \tag{23}$$

As t gradually increases to $0.5t_{\mathrm{Hall}}$, $\ln(1 - 2t/t_{\mathrm{Hall}})$ will approach to negative infinity. Therefore, i_{a} can be reduced to zero in $t \in [0,0.5t_{\mathrm{Hall}}]$, and the commutation process can be successfully turned off.

If $b < 0$, the current in the interval during commutation process ($t \in [0,0.5t_{\mathrm{Hall}}]$) satisfies

$$L\frac{\mathrm{d}i_{\mathrm{a}}}{\mathrm{d}t} \leq -\frac{t_{\mathrm{Hall}}}{t_{\mathrm{Hall}} - 2t}(a + bI) \tag{24}$$

By integrating (24), it can be presented as

$$i_{\mathrm{a}} \leq I + \frac{t_{\mathrm{Hall}}(a + bI)}{2L}\ln(1 - \frac{2t}{t_{\mathrm{Hall}}}) \tag{25}$$

As t gradually increases to $0.5t_{\mathrm{Hall}}$, $\ln(1 - 2t/t_{\mathrm{Hall}})$ will also approach negative infinity. Therefore, if $a + bI \geq 0$, the outgoing phase current can be reduced to zero and the commutation process will end successfully. If $a + bI < 0$, i_{a} will be not decreased but increased as $\mathrm{d}i_{\mathrm{a}}/\mathrm{d}t > 0$ in light of (21), and the commutation cannot be shut off normally.

Based on the above analysis, the condition for $a + bI \geq 0$ is

$$U_{\mathrm{dc}} - 2E + 2Ri_{\mathrm{c}} + RI - \frac{2LI}{t_{\mathrm{Hall}}} \geq 0 \tag{26}$$

By ignoring the non-commutation phase current variation, the condition can be further simplified to be

$$U_{\mathrm{dc}} - 2E - RI - \frac{2LI}{t_{\mathrm{Hall}}} \geq 0 \tag{27}$$

According to (27), the speed range for the normal end of the commutation process in the proposed strategy is

$$n \leq \frac{U_{\mathrm{dc}} - RI}{2k_{\mathrm{e}} + \frac{pLI}{5}} = \frac{f_2}{g_2} \tag{28}$$

where f_2 and g_2 are numerator and denominator of the critical speed in the proposed strategy.

To compare the critical speeds of the traditional strategy and the proposed strategy, it can be obtained by subtracting the denominator of (28) from that of (17) as

$$g_1 - g_2 = \sqrt{\frac{4pLI}{15}}\left(\sqrt{k_e} - \sqrt{\frac{3pLI}{20}}\right) = \sqrt{\frac{4pLI}{15n}}\left(\sqrt{E} - \sqrt{\frac{9}{2\pi}\omega_e LI}\right) \quad (29)$$

where $w_e = \pi pn/30$ is the electrical angular velocity of the motor.

Since the winding inductance voltage is generally much smaller than the phase back-EMF, $g_1 - g_2 > 0$ is satisfied in light of (29). Moreover, $f_1 - f_2 < 0$ is also met based on (17) and (28). Consequently, $f_1/g_1 - f_2/g_2 < 0$ can be derived, which means the speed range of the proposed strategy is larger than that of the traditional strategy.

Above all, if $b > 0$ is satisfied under the rated speed, the outgoing phase current can be reduced to zero over the full speed range. if $b < 0$ is met under the rated speed, the proposed strategy still broadens the speed range for commutation torque ripple suppression compared with the traditional strategy.

5. Experimental Results

The control block of the proposed strategy is shown in Figure 6. In the normal conduction period, the proportional integral (PI) controller is employed to control the current, where i_p is the amplitude of the conduction phase current. In the commutation period, the duty cycle is obtained from (20) to guarantee $dT_e/dt = 0$ during the commutation process. The period switch is introduced to select the duty cycle and the modulation in normal conduction period or commutation period.

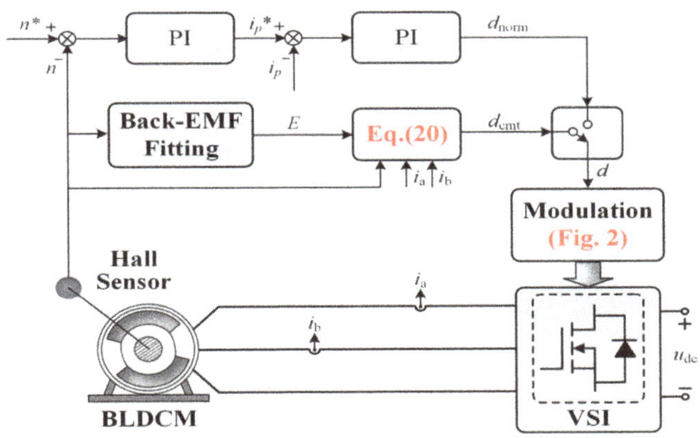

Figure 6. The control block of the proposed strategy.

To verify the correctness of the theoretical analysis and the effectiveness of the proposed strategy, the experimental system is established as shown in Figure 7.

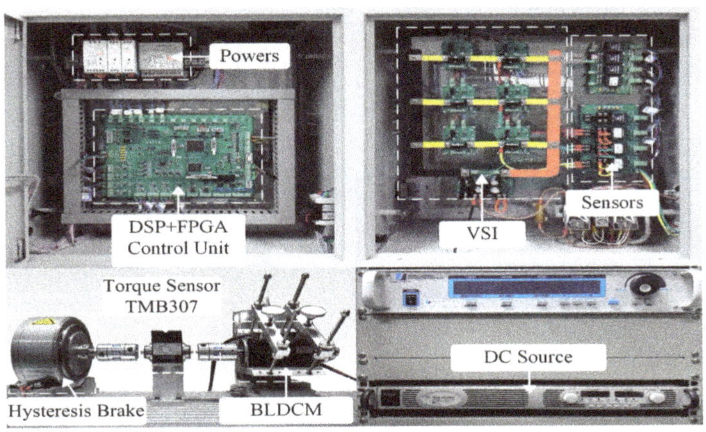

Figure 7. Experimental system.

In the experimental system, the torque sensor TMB307, whose measuring range is 10N·m and resolution is less than 0.01N·m, is used to measure the output mechanical torque of BLDCM. The control unit adopts the hybrid architecture of DSP (TMS320F28335) and FPGA (EP1C6Q240C8). MOSFET is manufactured by IR Corporation as IRFB4310-ZGPBF, whose switching frequency is 20kHz. The phase current is measured using the current sensor CSM025A. The measured torque, speed and duty cycle waveforms are output by the D/A converter. The rotor type of the experimental BLDCM is surface mounted, whose parameters are shown in Table 1.

Table 1. Parameters of BLDCM.

Parameter	Symbol	Value	Unit
Rated voltage	U_N	24	V
Rated current	I_N	14	A
Rated torque	T_N	3.2	N·m
Rated speed	n_N	600	r/min
Back-EMF coefficient	k_e	0.013	V/(r/min)
Phase resistance	R	0.2415	Ω
Phase inductance	L	0.387	mH
Pairs of pole	p	4	/

Based on IEC 60034-20-1, the torque ripple rate K_{rT} is defined as

$$K_{rT} = \frac{T_{high} - T_{low}}{T_{high} + T_{low}} \times 100\% \tag{30}$$

where T_{high} and T_{low} are the maximum and minimum values of the torque, respectively.

In the traditional strategy, the duty cycle is calculated by (7). Except for it, the modulations and the controller in the normal conduction period are all the same as that in the proposed strategy.

Figure 8 shows the waveforms of the motor under the rated load at 500 r/min, where the torque ripple rates of the traditional strategy and the proposed strategy are 7.644% and 4.376% respectively. It can be seen that the torque ripple is reduced to a certain degree in the traditional strategy. However, there is still a large torque fluctuation during the commutation process. Since the back-EMF variation is considered in the proposed strategy, the torque ripple rate is lower than 60% of the one in the traditional strategy.

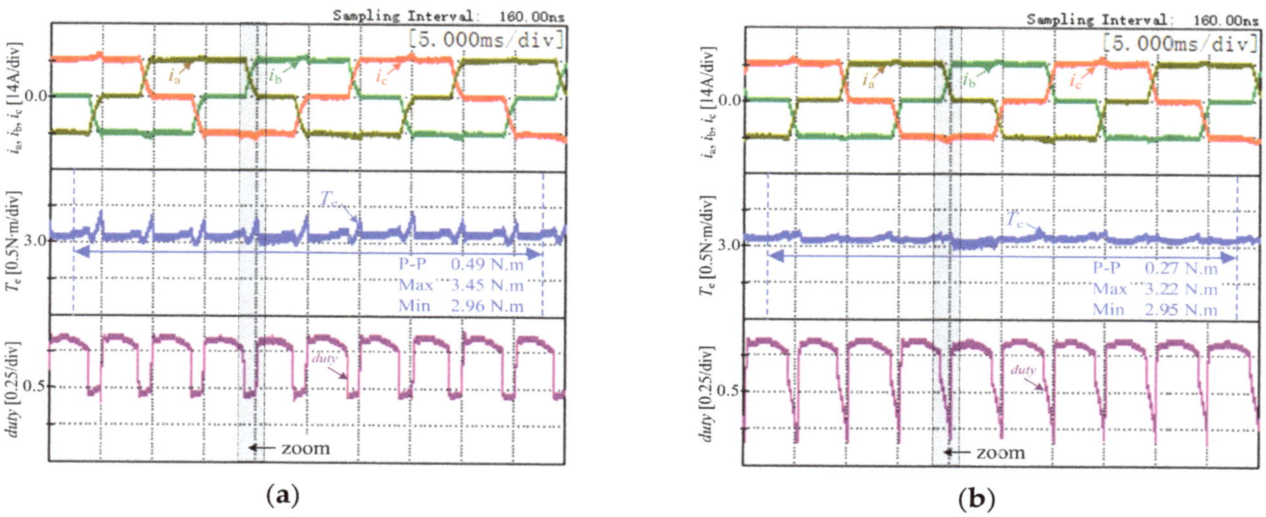

Figure 8. Experimental waveforms under the rated load at 500 r/min. (**a**) the traditional strategy. (**b**) the proposed strategy.

To account for the commutation process in detail, the enlarged waveforms of the dash area in Figure 8 is presented in Figure 9. In Figure 9a, the outgoing phase current is decreased to zero for about 1.2 ms and the motor torque is reduced first and then increased to a larger value during the commutation process with the constant duty cycle, which verifies the theoretical analysis in Figure 4. In Figure 9b, the outgoing phase current is decreased to zero for about 1 ms and the motor torque is almost constant with the reduced duty cycle.

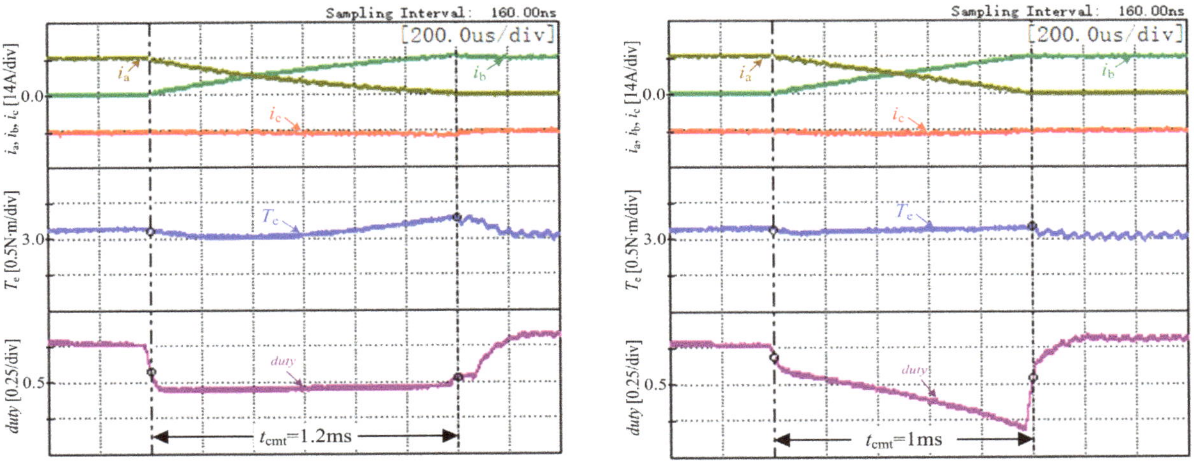

(a) (b)

Figure 9. Enlarged waveforms of the commutation process under the rated load at 500 r/min. (**a**) the traditional strategy. (**b**) the proposed strategy.

Figure 10 shows the waveforms of the motor under the rated load at 550 r/min, where the torque ripple rates of the traditional strategy and the proposed strategy are 14.928% and 4.685% respectively. The actual measured critical speed of the traditional strategy is 501 r/min, which is basically same as the theoretical value 497 r/min calculated by (17). When the motor speed is larger than the critical speed, in the traditional strategy the outgoing phase current will not be reduced to zero and the other currents are distorted during the commutation process, which generates a large torque fluctuation. As the experimental motor's parameters meet b > 0 in (21) in light of Table 1, it will never happen over the full speed range in the proposed strategy that the outgoing phase current cannot be reduced to zero.

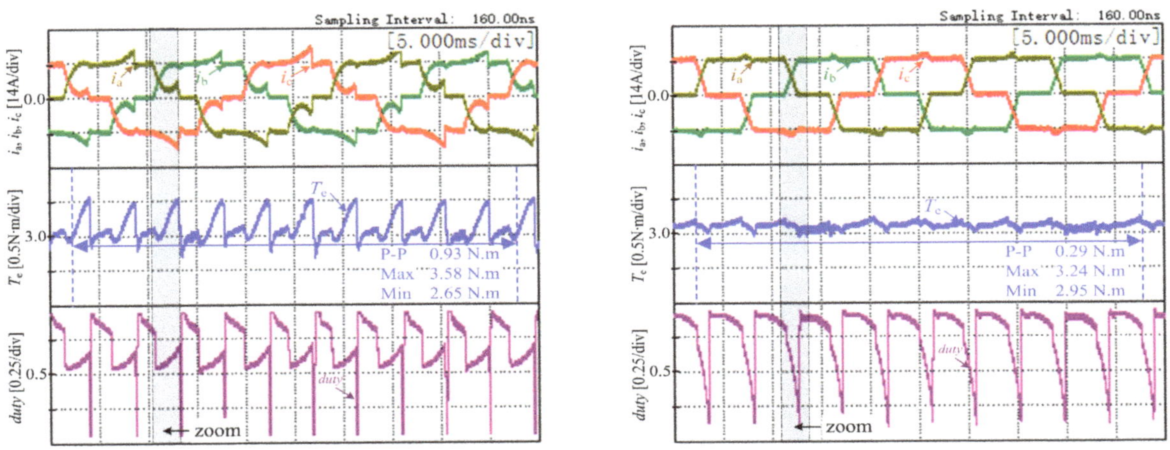

(a) (b)

Figure 10. Experimental waveforms under the rated load at 550 r/min. (**a**) the traditional strategy. (**b**) the proposed strategy.

To account for the commutation process in detail, the enlarged waveforms of the dash area in Figure 10 is presented in Figure 11. In Figure 11a, the reason for the current's distortion is that the waveform of outgoing phase current is approximately concave parabola and its minimum value is larger than zero, which causes the failure end of the commutation process. In this experiment, when the commutation time exceeds 2.5 ms, it is considered a failure and the commutation process is forcibly terminated. In addition, the torque is first decreased and then increased to a large value, which aggravates the torque ripple. These experimental results also verify the theoretical analysis in Figure 5. In Figure 11b, it can be seen that the proposed strategy can reduce the outgoing phase current to zero with the back-EMF variation considered, where the normal end of commutation can be acquired.

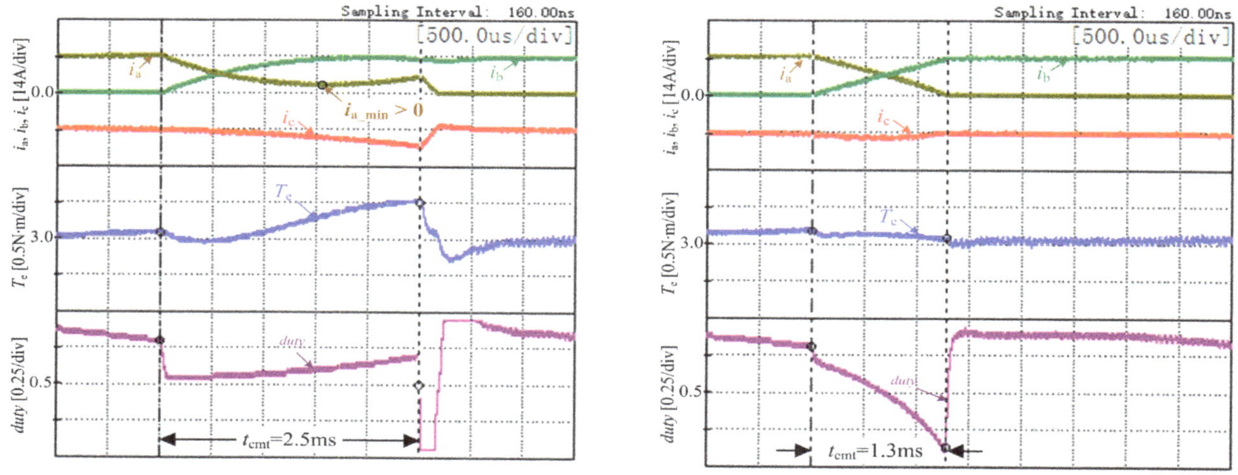

(a) (b)

Figure 11. Enlarged waveforms of the commutation process under the rated load at 550 r/min. (**a**) the traditional strategy. (**b**) the proposed strategy.

When the motor is operating under rated conditions (motor speed is 600 r/min, and load torque is 3.2 N·m), owing to the motor rated speed 600r/min is higher than the critical speed 497 r/min, the outgoing phase current will no longer be reduced to zero with the traditional strategy, and the traditional strategy is obviously invalid for rated conditions of the motor. Figure 12 shows the experimental waveforms with the proposed strategy under the rated load at 600 r/min. As shown in Figure 12, the torque ripple rate is 7.792%, which indicates the proposed strategy can suppress the torque ripple effectively, and the phenomenon that the outgoing phase current cannot be reduced to zero does not appear under rated conditions by the proposed strategy.

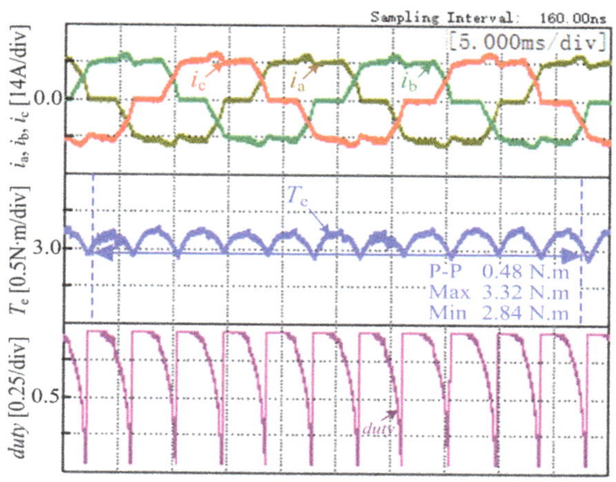

Figure 12. Experimental waveforms with the proposed strategy under the rated load at 600 r/min.

This paper considers the variation of the back-EMF in the commutation period, and the proposed strategy dynamically regulates the chopping duty cycle according to the variation trend of back-EMFs of three phases. Hence, the back-EMF coefficient is one of the important input parameters for the controller. To measure the back-EMF coefficient, driving by another motor at the speed 200 r/min, Figure 13 shows the measured back-EMFs of three phases in the open circuit windings. The back-EMF coefficient $k_e = 0.013$V/(r/min) listed in Table 1 can be calculated by this experimental result.

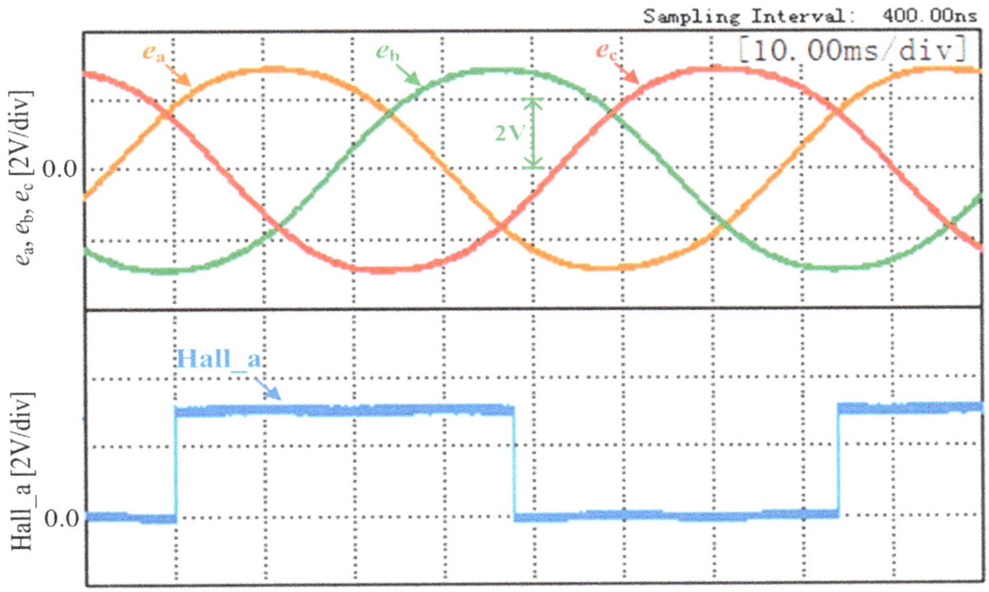

Figure 13. The measured three phases back-EMF and Hall signal of phase-a at 200r/min.

6. Conclusions

This paper aims to reduce the commutation torque ripple of BLDCM in its high-speed region, simultaneously broadening the valid speed range of the torque ripple reduction strategy without increasing the DC voltage. This paper firstly analyzes the problems of the traditional commutation torque ripple suppression for the high-speed region in a long commutation duration. A novel commutation torque ripple reduction strategy is then proposed based on a reconstructed smooth torque mechanism with back-EMF variation considered. Compared with the traditional strategy which controls the chopping duty cycle relatively smoothly in the commutation process, the proposed strategy dynamically regulates the chopping duty cycle, which make it show a gradual decrease. The proposed strategy has the following contributions:

1. The effect of back-EMF variation on torque and outgoing phase current during the commutation process is theoretically analyzed, and the speed range for the normal end of commutation is further deduced.
2. The smooth torque mechanism is reconstructed with the back-EMF variation considered, based on which a new commutation torque ripple suppression strategy is further designed. The proposed strategy can guarantee the torque gradient to be zero ($dT_e/dt = 0$) even under the changing back-EMF during the commutation process, which can more effectively reduce the commutation torque ripple.
3. The proposed strategy broadens the speed range for commutation torque ripple reduction, reducing the risk of phase current distortion and enhancing the reliability of the motor operation.

Author Contributions: Conceptualization, W.C., T.S., X.L and G.J.; methodology, X.L. and G.J.; software, G.J. and X.L.; validation, X.L., G.Z. and G.J.; formal analysis, G.Z. and Q.G.; writing—original draft preparation, G.J. and X.L.; writing—review and editing, G.Z., Q.G. and T.S.; funding acquisition, X.L., Q.G. and W.C.

References

1. Zhu, Z.Q.; Howe, D. Electrical Machines and Drives for Electric, Hybrid, and Fuel Cell Vehicles. *Proc. IEEE* **2007**, *95*, 746–765. [CrossRef]
2. Xia, C.; Jiang, G.; Chen, W.; Shi, T. Switching-Gain Adaptation Current Control for Brushless DC Motors. *IEEE Trans. Ind. Electron.* **2016**, *63*, 2044–2052. [CrossRef]
3. Ustun, O.; Kivanc, O.; Senol, S.; Fincan, B. On Field Weakening Performance of a Brushless Direct Current Motor with Higher Winding Inductance: Why Does Design Matter. *Energies* **2018**, *11*, 3119. [CrossRef]
4. Yoon, K.; Baek, S. Robust Design Optimization with Penalty Function for Electric Oil Pumps with BLDC Motors. *Energies* **2019**, *12*, 153. [CrossRef]
5. Priyadarshi, N.; Padmanaban, S.; Mihet-Popa, L.; Blaabjerg, F.; Azam, F. Maximum Power Point Tracking for Brushless DC Motor-Driven Photovoltaic Pumping Systems Using a Hybrid ANFIS-FLOWER Pollination Optimization Algorithm. *Energies* **2018**, *11*, 1067. [CrossRef]
6. Kamalapathi, K.; Priyadarshi, N.; Padmanaban, S.; Holm-Nielsen, J.; Azam, F.; Umayal, C.; Ramachandaramurthy, V. A Hybrid Moth-Flame Fuzzy Logic Controller Based Integrated Cuk Converter Fed Brushless DC Motor for Power Factor Correction. *Electronics* **2018**, *7*, 288. [CrossRef]
7. Carlson, R.; Mazenc, M.L.; Fagundes, J.C.D.S. Analysis of torque ripple due to phase commutation in brushless DC machines. *IEEE Trans. Ind. Appl.* **1992**, *28*, 632–638. [CrossRef]
8. Liu, Y.; Zhu, Z.Q.; Howe, D. Commutation-Torque-Ripple Minimization in Direct-Torque-Controlled PM Brushless DC Drives. *IEEE Trans. Ind. Appl.* **2007**, *43*, 1012–1021. [CrossRef]
9. Jung, S.; Kim, Y.; Jae, J.; Kim, J. Commutation Control for the Low-Commutation Torque Ripple in the Position Sensorless Drive of the Low-Voltage Brushless DC Motor. *IEEE Trans. Power Electron.* **2014**, *29*, 5983–5994. [CrossRef]
10. Li, H.; Zheng, S.; Ren, H. Self-Correction of Commutation Point for High-Speed Sensorless BLDC Motor with Low Inductance and Nonideal Back EMF. *IEEE Trans. Power Electron.* **2017**, *32*, 642–651. [CrossRef]
11. Cao, Y.; Shi, T.; Li, X.; Chen, W.; Xia, C. A Commutation Torque Ripple Suppression Strategy for Brushless DC Motor Based on Diode-Assisted Buck–Boost Inverter. *IEEE Trans. Power Electron.* **2019**, *34*, 5594–5605. [CrossRef]
12. Tan, B.; Hua, Z.; Zhang, L.; Fang, C. A New Approach of Minimizing Commutation Torque Ripple for BLDCM. *Energies* **2017**, *10*, 1735. [CrossRef]
13. Jiang, G.; Xia, C.; Chen, W.; Shi, T.; Li, X.; Cao, Y. Commutation Torque Ripple Suppression Strategy for Brushless DC Motors with a Novel Noninductive Boost Front End. *IEEE Trans. Power Electron.* **2018**, *33*, 4274–4284. [CrossRef]
14. Shi, T.; Cao, Y.; Jiang, G.; Li, X.; Xia, C. A Torque Control Strategy for Torque Ripple Reduction of Brushless DC Motor with Nonideal Back Electromotive Force. *IEEE Trans. Ind. Electron.* **2017**, *64*, 4423–4433. [CrossRef]
15. Song, J.H.; Choy, I. Commutation Torque Ripple Reduction in Brushless DC Motor Drives Using a Single DC Current Sensor. *IEEE Trans. Power Electron.* **2004**, *19*, 312–319. [CrossRef]
16. Shi, J.; Li, T. New Method to Eliminate Commutation Torque Ripple of Brushless DC Motor with Minimum Commutation Time. *IEEE Trans. Ind. Electron.* **2013**, *60*, 2139–2146. [CrossRef]
17. Xia, C.; Xiao, Y.; Chen, W.; Shi, T. Torque Ripple Reduction in Brushless DC Drives Based on Reference Current Optimization Using Integral Variable Structure Control. *IEEE Trans. Ind. Electron.* **2014**, *61*, 738–752. [CrossRef]
18. Lin, Y.; Lai, Y. Pulsewidth Modulation Technique for BLDCM Drives to Reduce Commutation Torque Ripple Without Calculation of Commutation Time. *IEEE Trans. Ind. Applicat.* **2011**, *47*, 1786–1793. [CrossRef]
19. Nam, K.Y.; Lee, W.T.; Lee, C.M.; Hong, J.P. Reducing torque ripple of brushless DC motor by varying input voltage. *IEEE Trans. Magn.* **2006**, *42*, 1307–1310. [CrossRef]
20. Shi, T.; Guo, Y.; Song, P.; Xia, C. A New Approach of Minimizing Commutation Torque Ripple for Brushless DC Motor Based on DC–DC Converter. *IEEE Trans. Ind. Electron.* **2010**, *57*, 3483–3490. [CrossRef]
21. Viswanathan, V.; Jeevananthan, S. Approach for torque ripple reduction for brushless DC motor based on three-level neutral-point-clamped inverter with DC-DC converter. *IET Power Electron.* **2015**, *8*, 47–55. [CrossRef]

22. Xiaofeng, Z.; Lu, Z. A New BLDC Motor Drives Method Based on BUCK Converter for Torque Ripple Reduction. In Proceedings of the 2006 CES/IEEE 5th International Power Electronics and Motion Control Conference, Shanghai, China, 14–16 August 2006; pp. 1–4.
23. Li, X.; Xia, C.; Cao, Y.; Chen, W.; Shi, T. Commutation Torque Ripple Reduction Strategy of Z-Source Inverter Fed Brushless DC Motor. *IEEE Trans. Power Electron.* **2016**, *31*, 7677–7690. [CrossRef]
24. Kang, B.H.; Kim, C.J.; Mok, H.S.; Choe, G.H. Analysis of torque ripple in BLDC motor with commutation time. In Proceedings of the IEEE International Symposium on Industrial Electronics (ISIE), Pusan, South Korea, 12–16 June 2001. [CrossRef]

Design and Analysis of Outer Rotor Permanent-Magnet Vernier Machines with Overhang Structure for In-Wheel Direct-Drive Application

Dong Yu, Xiaoyan Huang *, Lijian Wu and Youtong Fang

College of Electrical Engineering, Zhejiang University, Hangzhou 310027, China; 11410025@zju.edu.cn (D.Y.); lijianwu@zju.edu.cn (L.W.); youtong@zju.edu.cn (Y.F.)
* Correspondence: xiaoyanhuang@zju.edu.cn.

Abstract: This paper presents a novel outer rotor permanent-magnet vernier machine (PMVM) for in-wheel direct-drive application. The overhang structures of the rotor and flux modulation pole (FMP) are introduced. The soft magnetic composite (SMC) was adopted in the FMP overhang to allow more axial flux. The 3-D finite element analysis (FEA) was carried out to prove that the proposed machine can effectively utilize the end winding space to enhance the air-gap flux density. Hence the PMVM can offer 27.3% and 14.5% higher torque density than the conventional machine with no overhang structure and the machine with only rotor overhang structure, respectively. Nevertheless, the efficiency of the proposed machine is slightly lower than the conventional ones due to the extra losses from the overhang structures.

Keywords: permanent-magnet vernier machine; in-wheel direct-drive; outer rotor; overhang; soft magnetic composite

---

1. Introduction

The vernier machine was first proposed in the form of the vernier reluctance machine without any attention at that time [1]. Then, with the rapid development of the permanent magnet (PM) materials, the permanent-magnet vernier machine (PMVM) has been highlighted in recent years in various applications requiring high torque and power density, such as the in-wheel direct-drive system for the electric vehicle (EV) [2–6]. The PMVM can effectively utilize the magnetic flux harmonics to achieve higher output torque at a low speed due to the "magnetic gearing effect". Nevertheless, a relatively poor power factor (PF) is inevitable because of its inherent nature. Thus, an inverter with higher power rating is required, which will increase the size and cost of the whole system. The 3-D finite element analysis (FEA) and experiment results indicated that by replacing the surface-mounted PM with the spoke-type PM, the output torque capability and PF of the PMVMs can be significantly improved [7–10]. In addition, the PMVM with concentrated winding and the multitooth flux modulation poles (FMPs) can effectively shorten the length of the end winding, reduce the copper loss, and consequently improve the efficiency of the machine [11–14].

The PMVM with only the rotor overhang structure received more attention recently for applications with limited space in axial direction. By utilizing the end winding space, the rotor overhang structure can enhance the air-gap flux density and therefore the torque density. Both the analytical method and the finite element method (FEM) were carried out to analyze the effect of this structure [15–18]. Nevertheless, the conventional overhang structures only increase the axial stack length of the rotor. The effect could be compromised when the FMP remains the same as in the conventional design.

In this paper, a novel outer rotor PMVM with the overhang structure of the rotor and FMP will be proposed for the in-wheel direct-drive application. In Section 2, the operation principle and topology of the proposed machine are introduced. In Section 3, 3-D FEM simulations are carried out to verify the design of the proposed machine structure.

2. Operation Principle and Machine Configurations

The modulation principle is based on the "magnetic gearing effect", which means that a low-speed rotation of the PMs will cause dramatic variations in flux. Furthermore, the flux will interact with the high-speed rotating field generated by armature windings to produce the torque [19,20]. The FMPs are the key to cause these two rotating fields' modulation in a PMVM. To take advantage of the magnetic gearing effect, the number of FMPs (Z_s), the armature winding fundamental pole pair (p_s), and the PM pole pair (p_r) should meet the following equation:

$$p_r = Z_s \pm p_s \tag{1}$$

Figure 1 shows the configurations of the three PMVMs, M I, M II, and the proposed M III. All of them have the outer rotor structure and surface-mounted PMs. Each machine has the same Z_s, p_r, and p_s combination of 36/28/8 and their stator teeth are split into two FMPs, which makes the stator feasible to adopt concentrated winding to shorten the length of end winding. The double-layer concentrated windings with Y connection were applied in this design as shown in Figure 1g. In order to transmit torque at different speeds, the fundamental space harmonics velocity in the stator should be G_r times higher than that in the rotor, which is

$$G_r = \frac{\omega_s}{\omega_r} = \frac{p_s - Z_s}{p_s} \tag{2}$$

where ω_s and ω_r are the stator and rotor magnetic field rotating speeds, respectively.

The M II PMVM with the 10 mm bilateral rotor overhang structure is shown in Figure 1b. The length of overhang is determined according to the length of end winding. Apart from the difference in rotor, the stator of M II is also optimized for high torque density; the main differences are listed in Table 1.

In the proposed M III PMVM, both rotor overhang and FMP overhang are implemented to maximize the output torque. The FMPs have the same length of the rotor with overhang structure to provide the shortcut for the flux from the rotor overhang to the stator. Moreover, the FMPs of the proposed machine are partitioned from the stator to reduce the losses and cost of the SMC as shown in Figure 1c–f.

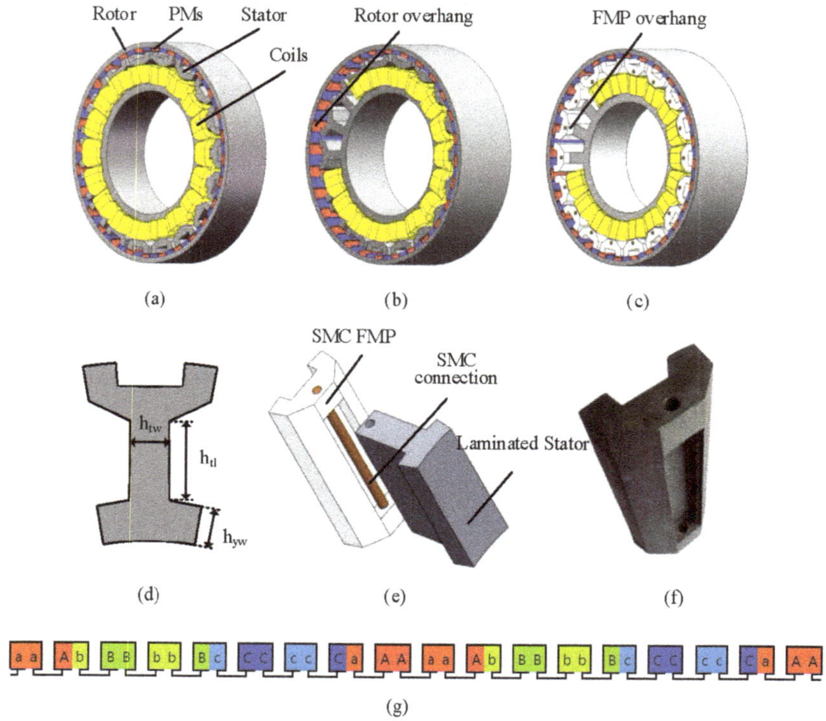

Figure 1. The configurations of the three machines. (**a**) M I: conventional PMVM. (**b**) M II: PMVM with rotor overhang. (**c**) M III: proposed PMVM with rotor and FMP overhang. (**d**) Cross-section of conventional stator tooth. (**e**) Side view of the proposed stator tooth. (**f**) FMP prototype. (**g**) Winding configuration.

Table 1. Main specifications of the three machines.

Parameters	M I	M II	M III
Number of stator slots		18	
Number of FMPs		36	
Number of rotor pole pairs		28	
Number of winding pole pairs		8	
Gear ratio		−28:8	
Rotor outer diameter (mm)		250	
Stator outer diameter (mm)		228	
Stator inner diameter (mm)	140	134	128
Air-gap length (mm)		1	
PM width (mm)		12	
PM thick (mm)		4	
Rotor axial length (mm)	60	80	80
Stator axial length (mm)	60	60	60
FMP axial length (mm)	60	60	80
Stator tooth width h_{tw} (mm)	10.0	11.5	13.0
Stator tooth length h_{tl} (mm)	20.3	22.5	24.4
Stator yoke width h_{yw} (mm)	8.0	9.0	10.0
Stator and rotor material		50ww350	
FMP material	50ww350	50ww350	Somaloy700
PM material	NdFeB38EH	NdFeB38EH	NdFeB38EH
Rated speed (rpm)		600	
Rated current (A)		20	
Current density (A/mm^2)		5	
Slot packing factor		0.6	
Turns per slot		40	
Winding resistance (Ω)		0.088	

The back electromotive force (EMF) and rated load torque increase versus overhang length of M II and M III are shown in Figure 2. It should be noted that when the length of the overhang is under 2 mm, the M II produces higher back EMF and torque than the M III. There are mainly two reasons: one is that when the overhang is short, the air-gap reluctance and the flux leakage of the rotor overhang is relatively small; and the other one is that the permeability of SMC is slightly smaller than the silicon steel as shown in Figure 3. As the length of the overhang increases, the flux linkage must pass through a longer air gap which mitigates the increased air-gap flux density. When the rotor overhang is long enough, the advantage of the proposed structure becomes obvious. The additional FMP overhang can effectively utilize the rotor overhang, which produces higher back EMF and torque in comparison with M II. Since the length of end winding of the machine is around 10.8 mm, a 10 mm overhang length is selected for M II and M III. In addition, the dimension of the stator is optimized for M II and M III to avoid saturation in the stator teeth while maintaining the same current density. Main parameters of the three PMVMs are summarized in Table 1.

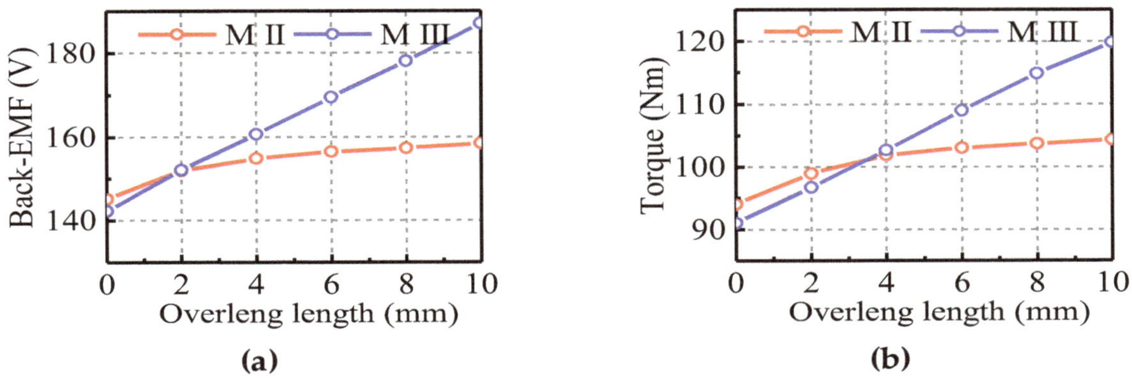

Figure 2. Back EMF (**a**) and torque (**b**) versus overhang length of M II and M III.

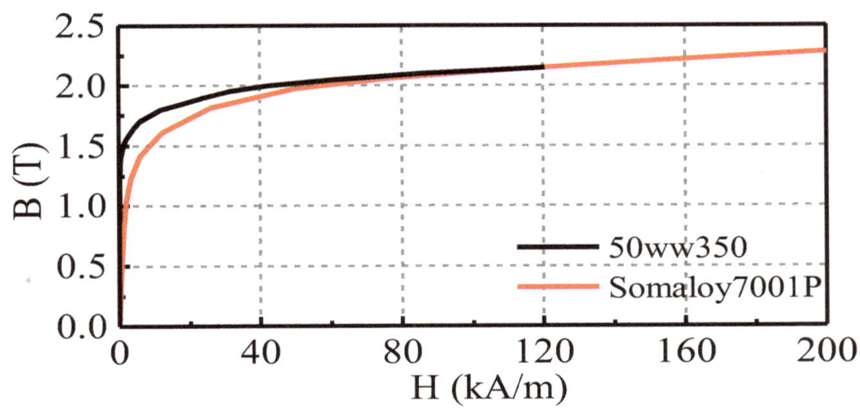

Figure 3. B–H curve of silicon steel and SMC.

3. FEA and Comparisons

As the overhang structure causes asymmetry in the axial direction, the 3-D FEM is essential for further analysis.

3.1. No-Load Characteristics

The no-load back EMF waveforms and their harmonic distributions of the three machines at the rotor speed of 600 rpm are demonstrated and compared in Figure 4. It can be seen that the back EMF amplitude of the proposed M III is the highest, 28.8%and 18.0% higher than the M I and the M II, respectively.

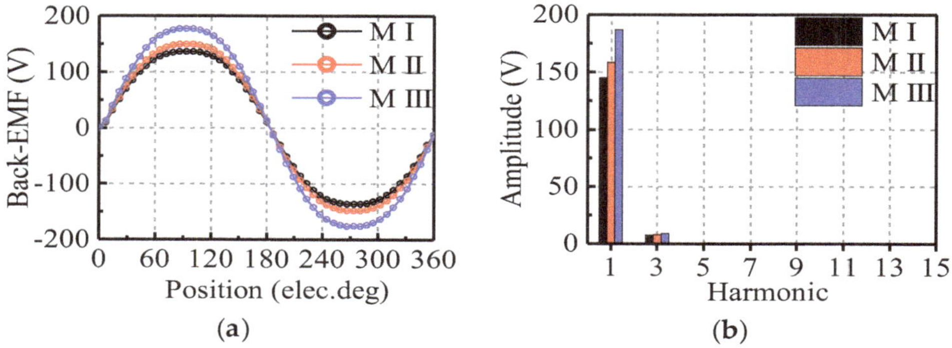

Figure 4. No-load back EMF of the three machines at 600 rpm. (**a**) Waveform. (**b**) Harmonic distributions.

The radial component of the air-gap flux density at the axial position of 0 mm and the corresponding space harmonic spectrum are shown in Figure 5a and c. In addition, the air-gap flux densities of M I and M II at the axial 0 mm position (the middle of the machines) are the same and the 28th harmonic orders are slightly higher than the M III's. This is mainly because of the low permeability of SMC applied to the FMPs, and this also could lead to the difference of back EMF between M II and M III when the overhang length is 0 mm as shown in Figure 2. The amplitude of the 28th harmonic order of air-gap flux density versus axial position are compared in Figure 5b. It should be noted that the air-gap flux density of M II drops apparently at the axial position near the rotor overhang, which is between 30 mm to 40 mm and −30 mm to −40 mm. However, the proposed M III can keep the high air-gap flux density until it reaches the edge of the overhang. As shown in Figure 6a, the PM flux linkage of the rotor overhang passes through the air gap and goes into the stator and contributes to the main flux linkage. However, this effect could be mitigated when the length of the overhang is increasing, and this problem is overcome by adding the FMP overhang. The flux linkage of the rotor overhang can successfully pass into the FMP overhang and then into the stator, in which way the reluctance is much smaller than the M II's, and less flux leakage occurs as shown in Figure 6b. Meanwhile, this also benefits from the SMC, which allows more axial flux. Therefore, the air-gap flux density of M III is improved, which contributes to the increase of back EMF and output torque.

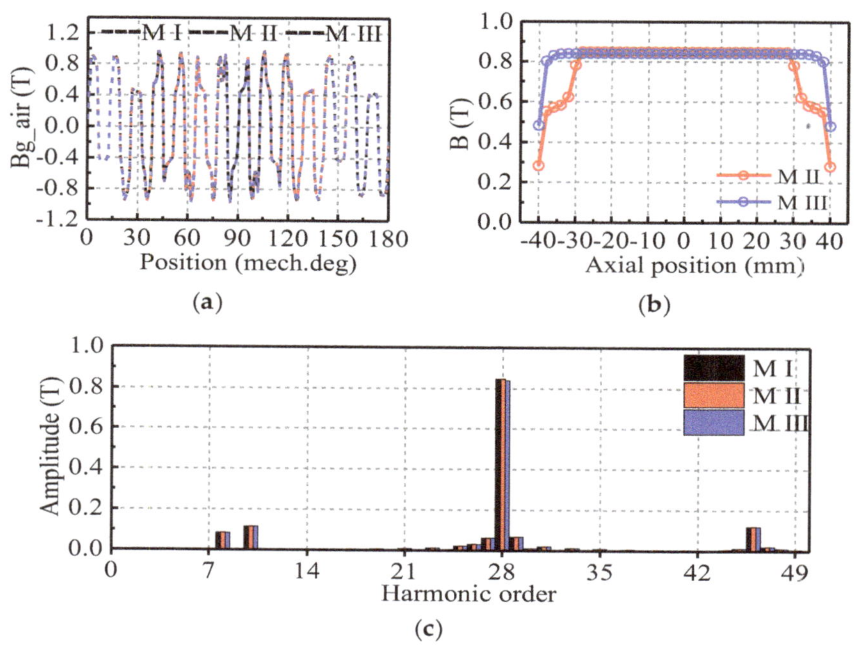

Figure 5. No-load air-gap magnetic flux density. (**a**) 0 mm waveform. (**b**) Air-gap flux density versus axial position. (**c**) 0 mm harmonic distributions.

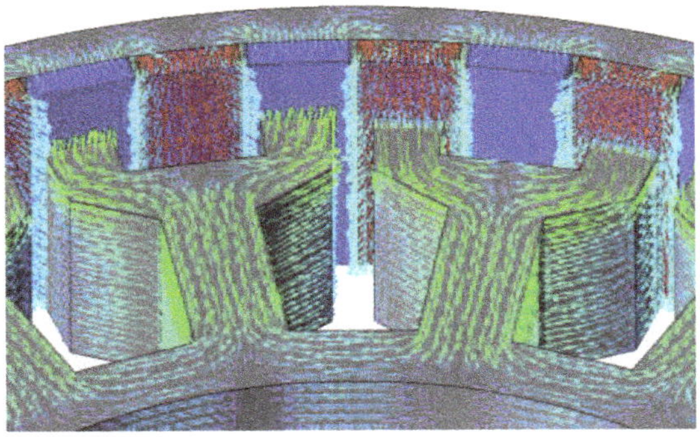

(a)

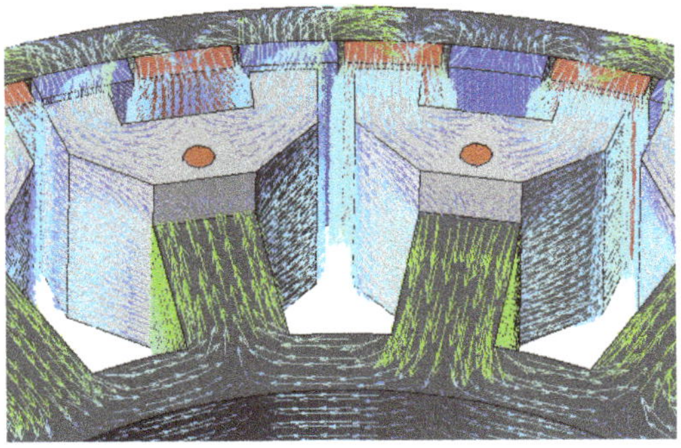

(b)

Figure 6. 3D magnetic flux density vector diagram at no-load condition. (**a**) M II. (**b**) M III.

3.2. *On-Load Characteristics*

The flux density distributions of the three PMVMs at the rated load (20 Arms current with $i_d = 0$ control) are compared in Figure 7. It can be seen that through the stator size adjustment, the maximum flux density in stator teeth of all the three machines are about 1.7 T, which is a typical value for the silicon steel 50ww350.

The characteristics of the three PMVMs at rated load condition are summarized in Table 2. The output waveforms of the three PMVMs at the rated load condition are shown in Figure 8. The proposed M III can offer higher output torque and power, which is 27.3% higher than M I and 14.5% higher than M II. The torque ripple of all the three machines is relatively low without any additional torque ripple reduction technique adopted, which is 1.52 Nm (1.6%), 1.70 Nm (1.8%), and 1.47 Nm (1.5%), respectively. The total weight (including stator, rotor, PMs, winding, and shell) and volume of the three machines are also calculated, and M III has the highest torque density in terms of Nm/kg and Nm/L.

Table 2. Main rated load performance of the three machines.

Items	M I	M II	M III
Average torque (Nm)	94.0	104.5	119.7
Torque ripple (Nm)	1.52 (1.6%)	1.70 (1.8%)	1.47 (1.5%)
Power (kW)	5.90	6.56	7.52
Loss (W)	412.4	442.2	572.5
Efficiency	93.5%	93.7%	92.9%
Weight (kg)	12.88	14.86	15.76
Torque density (Nm/kg)	7.30	7.03	7.60
Volume (L)	5.54	5.54	5.54
Torque density (Nm/L)	16.97	18.86	21.61

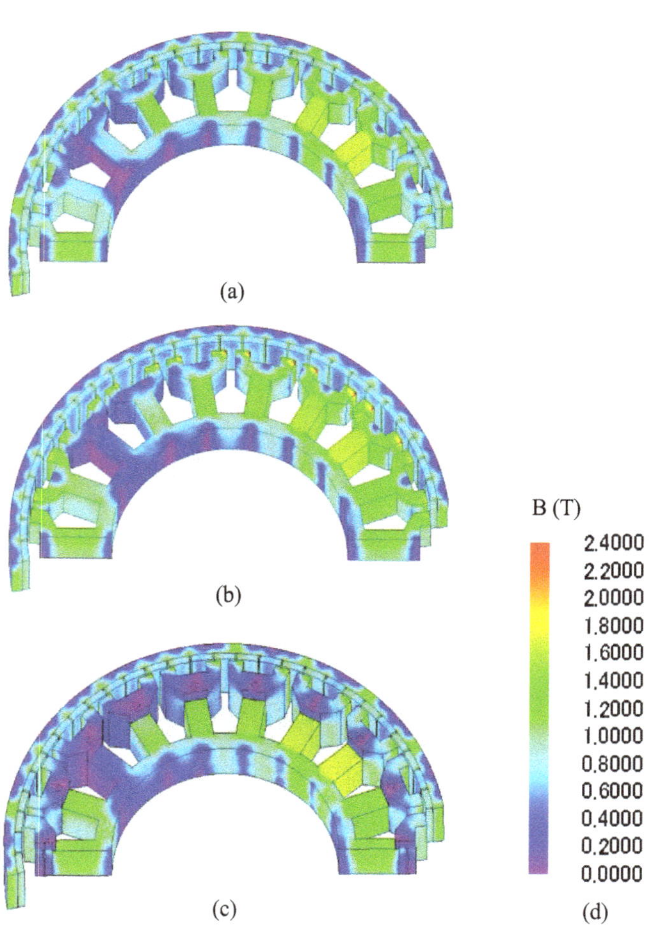

Figure 7. Flux density distributions of the three machines at rated load condition. (**a**) M I. (**b**) M II. (**c**) M III. (**d**) Scale bar.

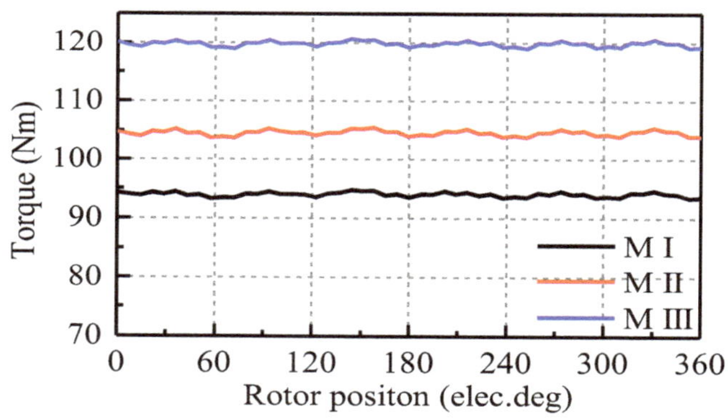

Figure 8. Output torque waveforms of the machines at rated load condition.

The main losses of the three PMVMs at rated load condition are listed in Figure 9. The copper loss of the three PMVMs is almost the same. The PM eddy current loss of these PMVMs cannot be neglected due to the high frequency. The losses in the rotor cores are very small while the stator core losses dominate. The efficiency of the three PMVMs can be approximated as

$$\eta = \frac{P_o}{P_o + P_{\text{loss}}} \tag{3}$$

where P_o is the output power and P_{loss} are the losses listed in Figure 9.

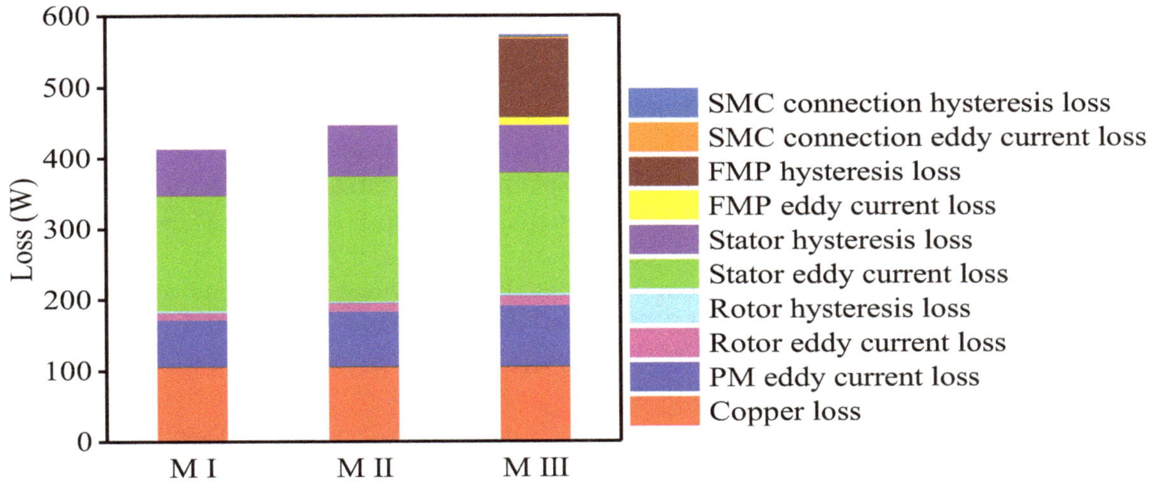

Figure 9. On-load loss comparison of the three PMVMs.

However, the hysteresis loss of the FMPs of M III is relatively high due to SMC inherent material characteristics, which lead to the increase of total loss. Besides the higher cost and relatively higher hysteresis loss of SMC material, only FMPs are made of SMC and the other parts of the stator cores are made of laminations in this paper.

4. Conclusions

In this paper, a novel outer rotor PMVM for in-wheel direct-drive application has been proposed, in which the overhang structure of the rotor and FMP are introduced. The FMP overhang structure is made of SMC to allow more axial flux. Detailed comparisons indicate that the proposed machine can effectively utilize the end winding space to enhance the air-gap flux density and offer higher torque density than the conventional machines with no overhang structure and only rotor overhang structure. Nevertheless, the efficiency of the proposed machine is slightly sacrificed due to the use of SMC.

Author Contributions: Conceptualization, D.Y. and X.H.; methodology, D.Y. and L.W.; software, D.Y. and X.H.; formal analysis, D.Y. and L.W.; data curation, D.Y. and Y.F.; writing, D.Y. and X.H.; supervision, L.W. and Y.F.; funding acquisition, X.H. and Y.F.

References

1. Lee, C.H. Vernier motor and its design. *IEEE Trans. Power Appl. Syst.* **1963**, *82*, 343–349. [CrossRef]
2. Toba, A.; Lipo, T.A. Generic torque-maximizing design methodology of surface permanent-magnet vernier machine. *IEEE Trans. Ind. Appl.* **2000**, *36*, 1539–1546.
3. Gu, C.; Zhao, W.; Zhang, B. Simplified Minimum Copper Loss Remedial Control of a Five-Phase Fault-Tolerant Permanent-Magnet Vernier Machine under Short-Circuit Fault. *Energies* **2016**, *9*, 860. [CrossRef]
4. Kim, B.; Lipo, T.A. Operation and Design Principles of a PM Vernier Motor. *IEEE Trans. Ind. Appl.* **2014**, *6*, 3656–3663. [CrossRef]
5. Kim, B.; Lipo, T.A. Analysis of a PM vernier motor with spoke structure. *IEEE Trans. Ind. Appl.* **2016**, *52*, 217–225. [CrossRef]
6. Jang, D.; Chang, J. A Novel Design Method for the Geometric Shapes of Flux Modulation Poles in the Surface-Mounted Permanent Magnet Vernier Machines. *Energies* **2017**, *10*, 1551. [CrossRef]
7. Zhang, Y.; Lin, H.; Fang, S.; Huang, Y.; Yang, H.; Wang, D. Air-gap flux density characteristics comparison and analysis of permanent magnet vernier machines with different rotor topologies. *IEEE Trans. Appl. Supercond.* **2016**, *26*, 1–5. [CrossRef]

8. Li, D.; Qu, R.; Lipo, T.A. High-power-factor vernier permanent-magnet machines. *IEEE Trans. Ind. Appl.* **2014**, *50*, 3664–3674. [CrossRef]
9. Li, X.; Chau, K.T.; Cheng, M. Comparative analysis and experimental verification of an effective permanent-magnet vernier machine. *IEEE Trans. Magn.* **2015**, *51*, 1–9.
10. Liu, G.; Chen, M.; Zhao, W.; Chen, Q.; Zhao, W. Design and analysis of five-phase fault-tolerant interior permanent-magnet Vernier machine. *IEEE Trans. Appl. Supercond.* **2016**, *26*, 1–5. [CrossRef]
11. Yang, H.; Lin, H.; Zhu, Z.-Q.; Fang, S.; Huang, Y. A Dual-Consequent-Pole Vernier Memory Machine. *Energies* **2016**, *9*, 134. [CrossRef]
12. Oner, Y.; Zhu, Z.Q.; Wu, L.J.; Ge, X.; Zhan, H.; Chen, J.T. Analytical on-load subdomain field model of permanent-magnet vernier machines. *IEEE Trans. Ind. Electron.* **2016**, *63*, 4105–4117. [CrossRef]
13. Liu, X.; Zhong, X.; Du, Y.; Chen, X.; Wang, D.; Ching, T.W. A New Magnetic Field Modulation Type of Brushless Double-Fed Machine. *IEEE Trans. Appl. Supercond.* **2018**, *28*, 1–5. [CrossRef]
14. Xu, L.; Liu, G.; Zhao, W.; Yang, X.; Cheng, R. Hybrid stator design of fault-tolerant permanent-magnet vernier machines for direct-drive applications. *IEEE Trans. Ind. Electron.* **2017**, *64*, 179–190. [CrossRef]
15. Seo, J.M.; Jung, I.S.; Jung, H.K.; Ro, J.S. Analysis of overhang effect for a surface-mounted permanent magnet machine using a lumped magnetic circuit model. *IEEE Trans. Magn.* **2014**, *50*, 1–7.
16. Yeo, H.K.; Park, H.J.; Seo, J.M.; Jung, S.Y.; Ro, J.S.; Jung, H.K. Electromagnetic and thermal analysis of a surface-mounted permanent-magnet motor with overhang structure. *IEEE Trans. Magn.* **2017**, *53*, 1–4. [CrossRef]
17. Min, S.G.; Bobba, D.; Sarlioglu, B. Analysis of overhang effects using conductor separation method in coreless-type PM linear machines. *IEEE Trans. Magn.* **2018**, *54*, 1–4. [CrossRef]
18. Ishikawa, T.; Sato, Y.; Kurita, N. Performance of novel permanent magnet synchronous machines made of soft magnetic composite core. *IEEE Trans. Magn.* **2014**, *50*, 1–4. [CrossRef]
19. Wn, F.; El-Refaie, A.M. Permanent magnet vernier machine: A review. *IET Electr. Power Appl.* **2019**, *13*, 127–137.
20. Okada, K.; Niguchi, N.; Hirata, K. Analysis of a Vernier Motor with Concentrated Windings. *IEEE Trans. Magn.* **2013**, *49*, 2241–2244. [CrossRef]

6

Study on the Electromagnetic Design and Analysis of Axial Flux Permanent Magnet Synchronous Motors for Electric Vehicles

Jianfei Zhao *, Qingjiang Han, Ying Dai and Minqi Hua

School of Mechatronic Engineering and Automation, Shanghai University, Baoshan District, Shanghai 200444, China
* Correspondence: jfzhao@shu.edu.cn.

Abstract: In order to provide a complete solution for designing and analyzing the axial flux permanent magnet synchronous motor (AFPMSM) for electric vehicles, this paper covers the electromagnetic design and multi-physics analysis technology of AFPMSM in depth. Firstly, an electromagnetic evaluation method based on an analytical algorithm for efficient evaluation of AFPMSM was studied. The simulation results were compared with the 3D electromagnetic field simulation results to verify the correctness of the analytical algorithm. Secondly, the stator core was used to open the auxiliary slot to optimize the torque ripple of the AFPMSM, which reduced the torque ripple peak-to-peak value by 2%. From the perspective of ensuring the reliability, safety, and driving comfort of the traction motor in-vehicle working conditions, multi-physics analysis software was used to analyze and check the vibration and noise characteristics and temperature rise of several key operating conditions of the automotive AFPMSM. The analysis results showed that the motor designed in this paper can operate reliably.

Keywords: AFPMSM; analytical algorithm; torque ripple; vibration noise; temperature field analysis

---

1. Introduction

Drive motors used in new energy vehicles require frequent start-ups and shutdowns, are subject to large accelerations or decelerations, and require high-speed, low-torque mode operation. Compared with radial flux permanent magnet synchronous motors (RFPMSM) [1–4], axial flux permanent magnet synchronous motors (AFPMSM) have the advantages of compact axial structure and high power density, which are suitable for new energy vehicles. As early as 2005, researchers have studied AFPMSMs. Methods of finite-element analysis and theoretical analysis were combined into a multi-dimensional optimization program to optimize the design of high-power coreless stator AFPM generators in 2005 [5]. In 2006, the author of Reference [6] studied the torque ripple component of permanent magnet motors and proposed a method to minimize the torque ripple of surface permanent magnet motors. In order to ensure the reliability, safety, and ride comfort of the AFPMSM for electric vehicles, it is also possible to find an effective and reliable method from the control strategy. In Reference [7], a sliding mode vector control system based on cooperative control is established to drive AFPMSM. In Reference [8], the current control method is used to drive the AFPMSM, and internal model control was introduced. This paper focuses on the design and optimization of the AFPMSM for electric vehicles.

The design of AFPMSM usually requires 3D finite element method (FEM) analysis. Considering that multiple design variables need to be analyzed simultaneously, the time to design AFPMSM is significantly increased. The analytical algorithms to simplify analysis time are used to solve this problem [9–11]. A surrogate assisted multi-objective optimization algorithm was applied to significantly reduce function calls, effectively reducing motor design time [9]. Under the same magnetic energy of the two models' permanent magnets, an AFPM motor is equivalent to a linear synchronous permanent magnet motor [10]. In order to obtain the key parameters of accurate surface mount permanent magnet synchronous motors, an analysis model based on Maxwell's equation and magnetic equivalent circuit was introduced [11]. The analysis algorithm can effectively reduce the development time when designing the motor. The AFPMSM for electric vehicles has a complicated structure, and the use of FEM will greatly increase the analysis time. For AFPMSM, its structure is complex, and the analytical algorithm can effectively reduce the development time in motor design. Therefore, the electromagnetic design method based on analytical formula studied in this paper has certain advantages.

Similar to RFPMSMs, AFPMSMs also have a torque ripple problem during operation. The existence of torque ripple will cause torsional vibration of the transmission system, which will adversely affect the motor control and torque output quality. Therefore, it is necessary to analyze and optimize its torque characteristics. The authors of Reference [12,13] optimized the torque ripple from the control method. The authors of Reference [14,15] optimized the torque ripple from the motor itself. In this paper, the stator core was used to open the auxiliary slot to optimize the torque ripple.

In recent years, AFPMSM's multi-physics analysis has also made great progress. The authors of Reference [16–18] used the steady-state heat conduction equation method, the T-type thermal network model method, and the AFPMSM full prediction heat transfer coefficient model with magnet geometry parameters to perform thermal analysis on AFPMSM. The authors of Reference [19–22] conducted a detailed study of the AFPMSM's degaussing fault identification. The method was proposed for detecting and locating asymmetric demagnetization defects in AFPMSM, which can be used for real-time condition monitoring of demagnetization defects or as a virtual temperature sensor for magnets. [19]. A time harmonic analysis model was proposed to study the demagnetization and rotor eccentricity of a single stator dual rotor AFPMSM [20]. A forward model was established for demagnetization detection of an AFPMSM. [21]. In addition, the eddy current losses of AFPMSMs have been analyzed in detail in Reference [22–24]. Because electric vehicles require high system efficiency, strong system environment adaptability and low noise, based on these characteristics, the current study analyzed and checked the vibration noise and temperature rise of several key operating conditions of automotive AFPMSMs based on multi-physics analysis software to ensure that the motor can always run smoothly.

In summary, this paper covers the study and design of a dual-stator-single-rotor structure of the new energy vehicle traction AFPMSM (the motor parameters are shown in Table 1; the motor structure diagram is shown in Figure 1). The AFPMSM for electric vehicles was based on a fast electromagnetic design method based on analytical formulas and a refined optimization design simulation technology based on machine-electric-thermal-structure multi-physics analysis, providing a complete solution for designing and optimizing AFPMSM for electric vehicles.

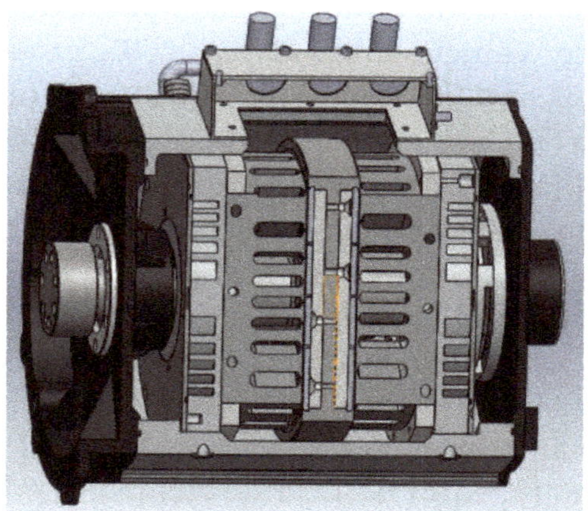

Figure 1. Double stator-single rotor type axial flux permanent magnet synchronous motor (AFPMSM).

Table 1. Motor parameters.

Motor Parameters	Symbol	Value
Pole-pairs	p	6
Slot number	Q	27×2
Air gap length	δ	1 mm
Rated torque	T_N	300 Nm
Rated power	P_N	50 kW
Stator outer diameter	D_{so}	250 mm
Phase number	m	3
Rated speed	n	1600 rpm
Permanent magnet Remanence	B_r	1.089 T
Coercivity	HC	2 A/m

2. Magnetic Path Analysis Algorithm for AFPMSM

2.1. Modeling and Analysis of Magnetic Circuit Analysis Algorithm

The AFPMSM has a unique magnetic circuit structure, and its electromagnetic design is quite different from that of the RFPMSM. In this paper, the model of AFPMSM was simplified by the 2D multi-loop equivalent method. Based on this, the magnetic circuit analysis algorithm was used to design the motor and obtain the electromagnetic parameters of the motor. The specific method is as follows. Take the motor stator N different diameters and expand it into a linear motor, and then superimpose the simulation results of N linear motors to get the final result, as shown in Figure 2. The radius D_{iave} taken by the i-th linear motor is as follows:

$$D_{iave} = D_{out} - \frac{2i-1}{N}\frac{D_{out} - D_{in}}{2}, i = 1, 2, \cdots, N \tag{1}$$

where D_{out} and D_{in} are the outer diameter and inner diameter of the stator.

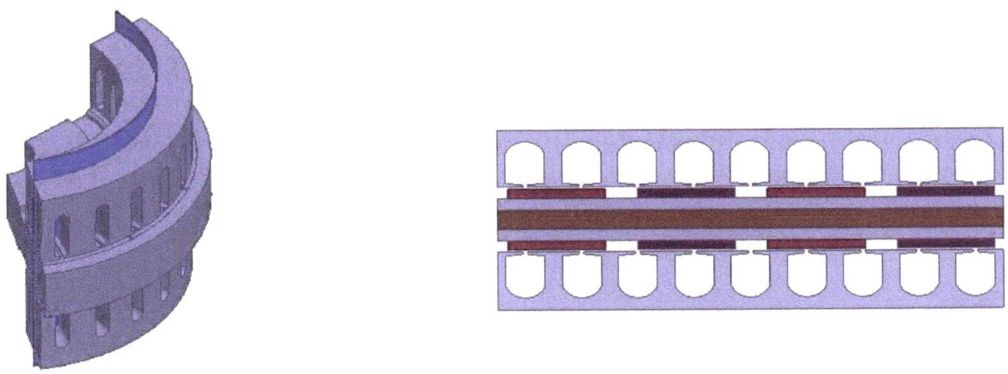

(**a**) (**b**)

Figure 2. Double stator-single rotor type AFPMSM. (**a**) AFPMSM 3D model. (**b**) Equivalent linear motor model.

When calculating the value of the electromagnetic field under ideal conditions, take $N = 1$. When calculating the loss and other information, take $N > 1$.

Then, use the equivalent magnetic network circuit as shown in Figure 3 to calculate the key electromagnetic parameters of the motor. In Figure 3, R_y is the equivalent reluctance of the stator yoke, R_t is the magnetic reluctance of the tooth, F_l is the magneto motive force generated by the stator winding, R_s is the slot reluctance, R_g is the air gap reluctance, and F_{PM} is the magneto motive force generated by the permanent magnet.

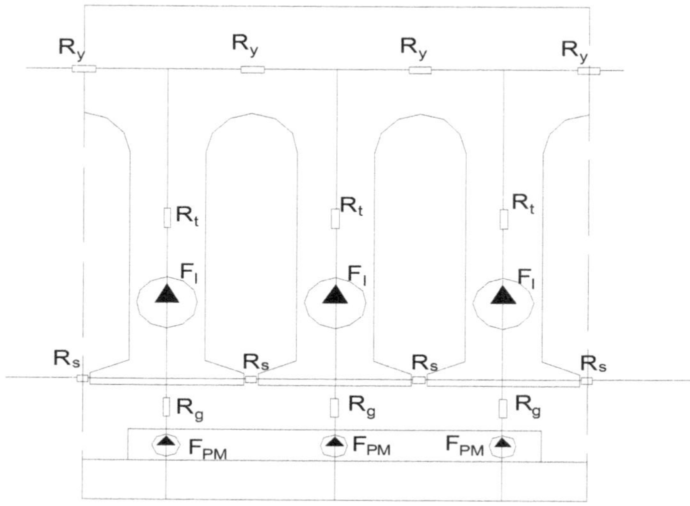

Figure 3. Equivalent magnetic network model of the AFPMSM.

The AFPMSM is equivalent to a linear motor in this paper, and the magnetic circuit analysis method (MCAM) model was used to analyze the air gap magnetic density distribution of AFPMSM [25–28]:

$$B_{mg}(r,x) = \sum_{n=1,3,5,\cdots}^{\infty} \frac{-\frac{8B_r}{n\pi}\sin\left(\frac{\alpha_p(r)n\pi}{2}\right)e^{\frac{-n\pi g'}{\tau_p(r)}}\cos\left(\frac{n\pi x}{\tau_p(r)}\right)}{\left(e^{\frac{-n\pi g'}{\tau_p(r)}}+1\right)+\frac{\mu_{PM}\left(e^{\frac{-n\pi g'}{\tau_p(r)}}+1\right)\left(e^{\frac{2n\pi l_{PM}}{\tau_p(r)}}+1\right)}{\mu_0\left(e^{\frac{2n\pi l_{PM}}{\tau_p(r)}}-1\right)}} \quad (2)$$

where r is the radius and x is the perimeter at different radii and is equal to $2\pi r$. $\tau_p(r)$ is the pole pitch of the permanent magnet when the radius is r, and $\tau_p(r) = \pi r/p$. $\alpha_p(r)$ is the polar arc coefficient when the radius is r, and $\alpha_p(r) = w_{PM}(r)/\tau_p(r)$, where $w_{PM}(r)$ is the width of the permanent magnet when the radius is r. l_{PM} is the thickness of the permanent magnet, μ_0 is the vacuum permeability, μ_{PM} is the permanent magnet permeability, and g' is the air gap length.

Since the open slot affects the air gap flux density, the air gap permeability function $\lambda(x)$ is introduced to consider the effect of the open slot on the air gap flux distribution. The air gap magnetic density distribution $B_{mag}(r,x)$ including the slotting effect is

$$B_{mag}(r,x) = \lambda(x)B_{mg}(r,x) \quad (3)$$

when the motor is used as a drive motor; the no-load phase voltage is generated only by the flux generated by the permanent magnet. According to Equations (2) and (3), it can be calculated as:

$$e_{i,PM} = -N_{ph}k_{db}\frac{\Delta\Phi_{i,PM}}{\Delta t} \quad (4)$$

where N_{ph} is the number of turns in series for each phase, K_{db} is the fundamental winding factor, and $\Phi_{i,PM}$ is the air gap flux obtained by numerical integration of the air gap flux density distribution,

$$\Phi_{i,PM} = \int_0^{\tau_{p,i}}\int_0^{\frac{l_s}{N}} B_{agap,i}(x)dxdl \quad (5)$$

where $\tau_{p,i}$ is the permanent magnet pole pitch of the i-th equivalent plane, l_s is the length of the stator punch and $l_s = (D_{out}-D_{in})/2$, and $B_{agap,i}(x)$ is the air gap magnetic density function in the i-th equivalent plane.

Therefore, the no-load back electromotive force (EMF) of the AFPMSM is

$$e_{PM}(t) = \sum_{i=1}^{N} e_{i,PM}(t) \tag{6}$$

This paper presents a method to reduce the cogging torque. The cogging torque is calculated from the no-load air gap flux density distribution. Assume that the cogging torque of the axial flux motor unit is

$$T_{cog,i}(\theta) = \frac{\partial W}{\partial \theta} = \frac{\partial\left(\iiint_V B^2_{agap,i} dV\right)}{2\mu_0 \partial \theta} \tag{7}$$

where V is the air gap volume, θ is the rotor position angle, and W is magnetic energy.

During calculation of the cogging torque, it is assumed that the magnetic flux density distribution of each axial flux motor unit does not change in the radial direction. Analogizing the back EMF calculation, the cogging torque generated by the axial flux motor is

$$T_{cog}(\theta) = \sum_{i=1}^{N} T_{cog,i}(\theta) \tag{8}$$

2.2. *Comparative Analysis of MCAM and 3D FEM*

As described in Section 2.1, based on Matrix Laboratory programming, the magnetic circuit analysis method compared with the finite element method as follows.

The air gap magnetic density of the motor has a crucial influence on the saturation degree of the motor, the output power requirement, and the loss level. Therefore, it is important to analyze the air gap magnetic density of the AFPMSM. Figure 4 shows that the air-gap magnetic density distribution waveforms calculated by the FEM and the MCAM are basically the same, and the FEM can more accurately consider the cogging and magnetic flux leakage of the motor, so the calculation accuracy is higher than the analytical method. However, the MCAM takes a short time and can be used for screening the initial design of the AFPMSM.

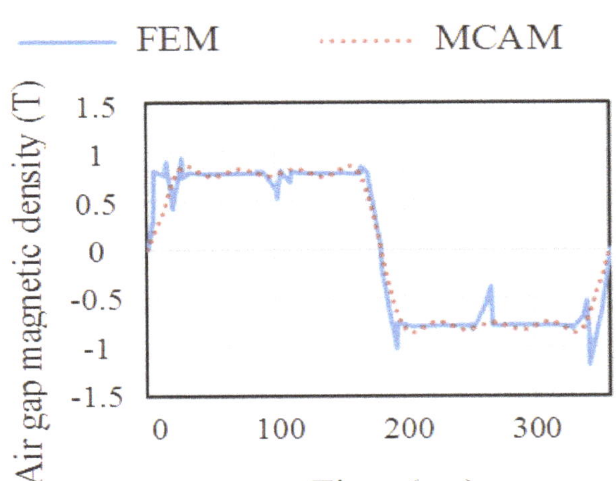

Figure 4. No-load air gap magnetic density of AFPMSM.

Figure 5 is a comparison of the analytical results of the FEM and MCAM of the AFPMSM back EMF. It can be seen from Figure 6a,b that the MCAM has omissions in the analysis of the third harmonic magnetic density, and the analysis results of the fundamental wave and other harmonics are in good agreement with the finite element method.

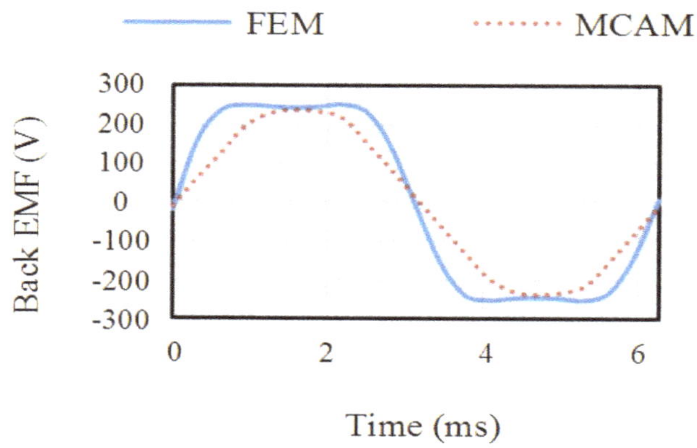

Figure 5. No-load back electromotive force (EMF) of the AFPMSM.

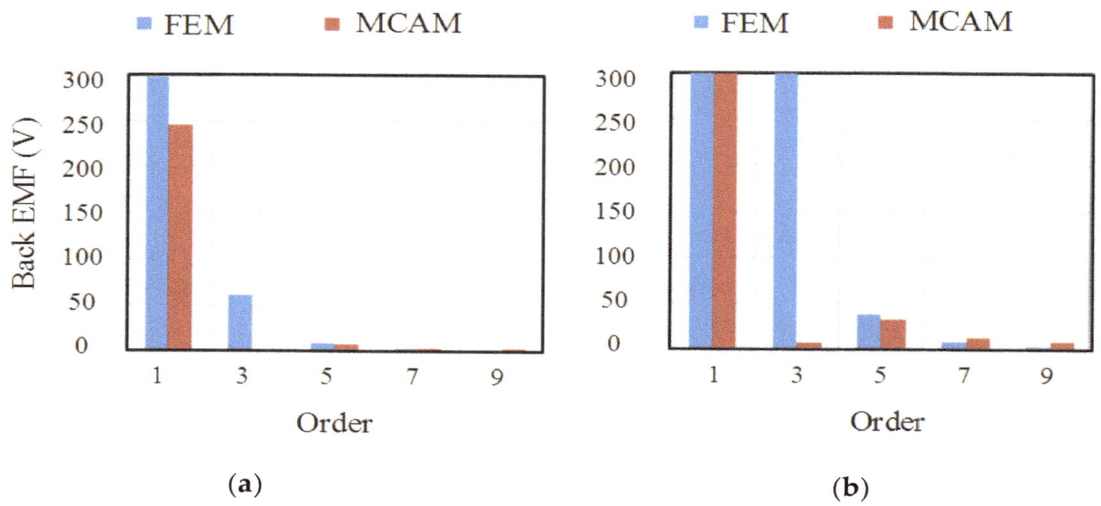

Figure 6. Comparison of back-EMF harmonic analysis results of the AFPMSM. (**a**) Overall picture. (**b**) Partial enlargement picture.

Figure 7 shows that that the MCAM is basically consistent with the analysis of the positioning torque and the finite element analysis results, and the maximum amplitude difference is 0.2 Nm, which satisfies the initial design accuracy requirements of the motor.

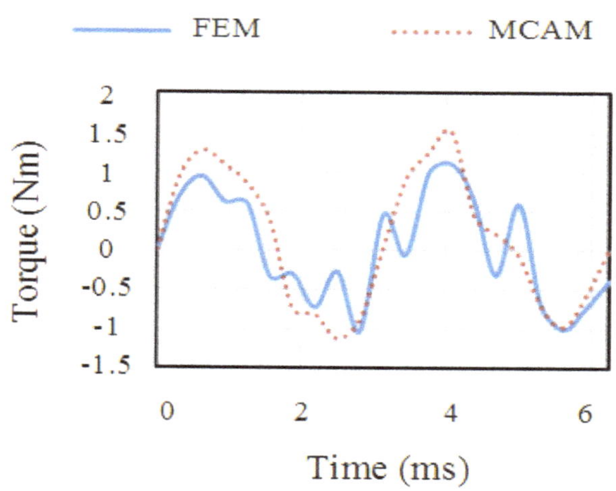

Figure 7. Analysis of positioning torque of AFPMSM.

In summary, the air gap magnetic density, the no-load back EMF, and the positioning torque calculated by the FEM and the MCAM are basically the same. Since the FEM can fully consider the motor cogging effect and the magnetic flux leakage coefficient, the accuracy is higher than the MCAM. However, when the distribution trend is basically the same, the time used by the MCAM is much less than that of the FEM. Therefore, the MCAM studied in this paper can be used to screen the initial design of the AFPMSM.

3. Torque Ripple Optimization of AFPMSM

Electric vehicles generally operate in torque mode. The output of the motor torque is controlled according to the depth at which the driver steps on the throttle. The presence of torque ripple adversely affects the torque output. Therefore, the torque ripple needs to be optimized. The effect of opening the auxiliary slot on the motor stator on the torque ripple is equivalent to increasing the number of armature slots, i.e., changing the pole slot fit. If the position and size of the open auxiliary slot are properly selected, the torque ripple of the motor can be weakened. First, theoretical analysis is used to determine the number of auxiliary slots, slot width, and position distribution on each tooth, and then the slot depth is continuously adjusted in the three-dimensional simulation. The slot depth should suppress the torque ripple as much as possible without affecting the performance of the motor. The number of auxiliary slots is derived below. The motor in this paper has 12 poles and 27 slots. According to the literature [23]:

$$\{i(k+1)|i \in N^*\} \tag{9}$$

$$\left\{i \cdot \frac{2p}{GCD(z,2p)}\middle|i \in N^*\right\} \tag{10}$$

where p is the pole pairs, z is the number of stator slots, and $GCD(z, 2p)$ is the greatest common divisor (GCD) of z and $2p$.

To attenuate torque ripple, the number k of stator toothed auxiliary slots should be such that the intersection of set (9) and set (10) is reduced. The AFPMSM studied in this paper has $p = 6$, $z = 27$, and $k = 2$, so the two sets are {3, 6, 9 . . . } and {4, 8, 12 . . . }. Obviously, there is no intersection between the two sets, so the torque ripple is significantly reduced compared to the absence of the auxiliary slot. If $k = 1$, the intersection of the above two sets is {4, 8, 12 . . . }. In summary, $k = 2$ and the width of the auxiliary slot is equal to the slot width.

In summary, the motor of the present invention opens two auxiliary slots, the auxiliary slots are evenly distributed on the stator teeth, and the width of the auxiliary slots is the same as the width of the stator slots. The finite element analysis model is shown in Figure 8a, and a partial enlarged view of the auxiliary slot is shown in Figure 8b.

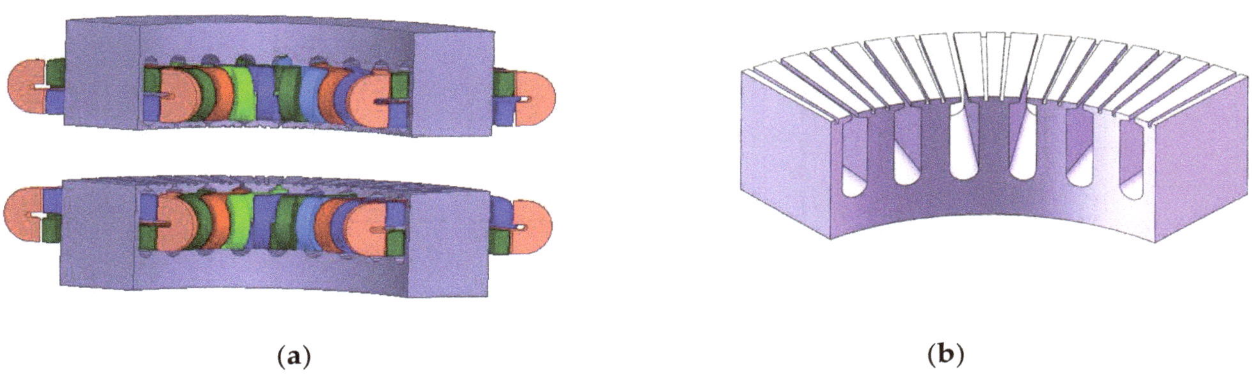

Figure 8. AFPMSM auxiliary slot schematic. (**a**) Finite element analysis model. (**b**) Partial enlarged view of the auxiliary slot.

In order to verify the feasibility of opening the auxiliary slot to suppress the torque ripple, this section analyzes the load back EMF and load torque of constant torque maximum speed operating point and constant power zone maximum speed operating point under rated load and peak load before and after optimization. The maximum speed of the constant torque is 1600 rpm, and the maximum speed of the constant power is 3000 rpm.

3.1. Comparative Analysis of Rated Load Characteristics at 1600 rpm

Under rated load conditions, it can be seen from Figure 9 that the third, fifth, seventh, and ninth harmonics of the load back electromotive force of the auxiliary slot method are weakened when the rotational speed is 1600 rpm; the weakening of the fifth and seventh harmonics is more significant.

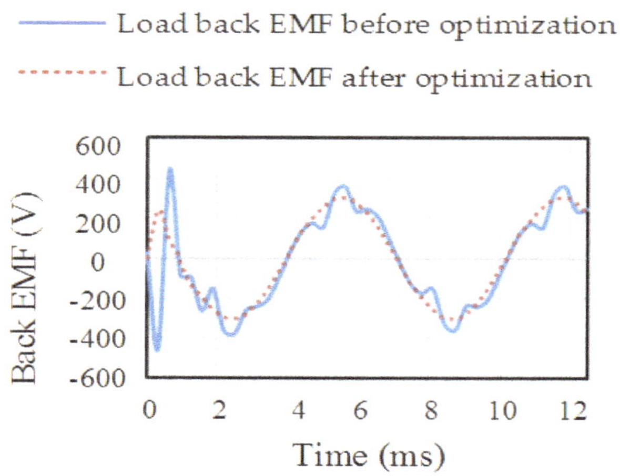

Figure 9. Back EMF of the AFPMSM at the rated load.

Figure 10 shows that when the rotation speed is 1600 rpm, the peak-to-peak load torque of the auxiliary slot method is 10 Nm, and the peak-to-peak value of the load torque without the auxiliary slot method is 16 Nm. The peak-to-peak value of the torque ripple after opening the auxiliary slot is reduced by about 4%, so the optimization method of opening the auxiliary slot at rated load can effectively attenuate the torque ripple under constant torque conditions.

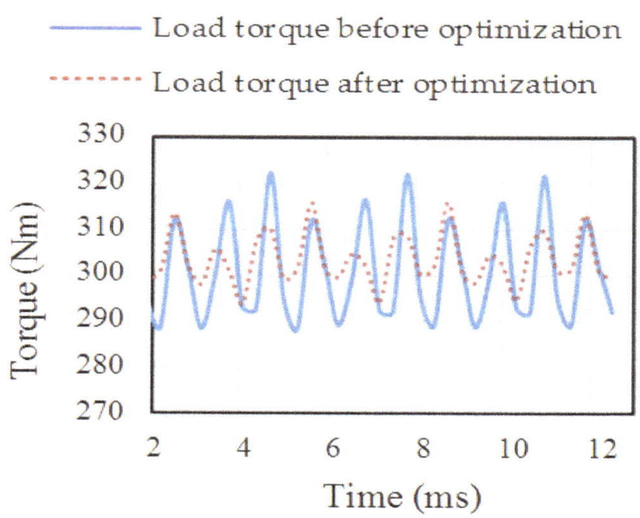

Figure 10. Torque of the AFPMSM at the rated load.

3.2. Comparative Analysis of Peak Load Characteristics at 1600 Rpm

Under the peak load condition, it can be seen from Figure 11 that the seventh and ninth harmonics of the load back electromotive force of the auxiliary slot method are weakened when the rotation speed is 1600 rpm; the third and fifth harmonics are higher than the no auxiliary slot method.

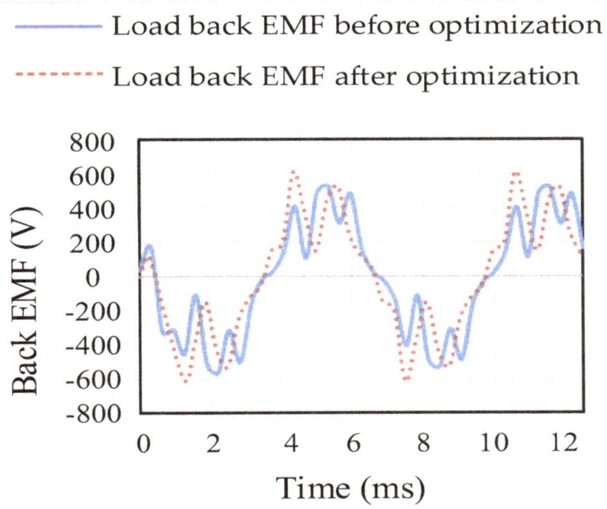

Figure 11. Back EMF of the AFPMSM at the peak load.

Figure 12 shows that the load torque peak-to-peak value of the open auxiliary slot method is 26 Nm when the rotation speed is 1600 rpm, and the load torque peak-to-peak value of the no auxiliary slot method is 32 Nm. The torque ripple of the open auxiliary slot method is reduced by about 2%, so the optimization method of opening the auxiliary slot at the peak load can also effectively attenuate the torque ripple under constant torque conditions.

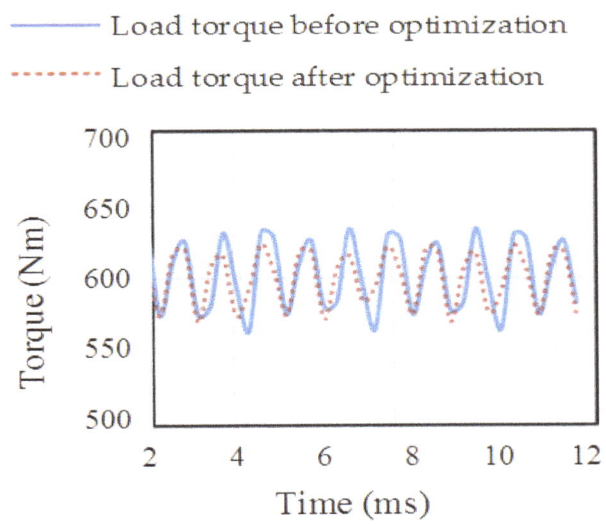

Figure 12. Torque of the AFPMSM at the peak load.

3.3. Comparative Analysis of Rated Load Power Characteristics at 3000 rpm

Under the rated load condition, as can be seen from Figure 13, when the rotational speed is 3000 rpm, the fifth, seventh, and ninth harmonics of the load back EMF of the auxiliary slot method are all weakened; the weakening of the fifth, seventh, and ninth harmonics is more significant.

Figure 14 shows that the load torque peak-to-peak value of the auxiliary slot optimization method is 3.5 Nm when the rotation speed is 3000 rpm, and the peak-to-peak value of the load torque before optimization is 10 Nm. The torque ripple of the auxiliary slot method is reduced by about 4%, so the torque ripple optimization method of the auxiliary slot can effectively attenuate the load torque ripple under constant power conditions.

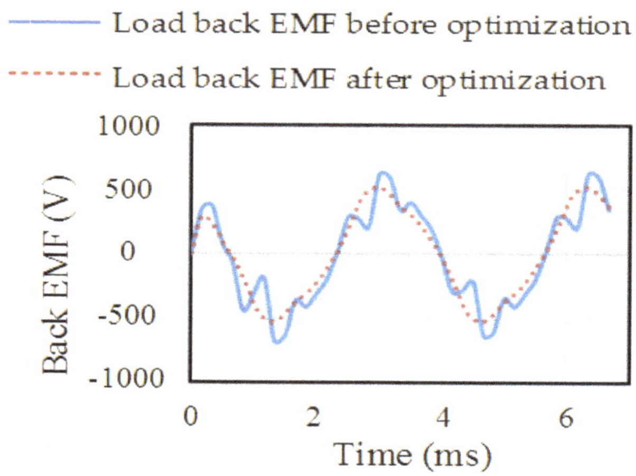

Figure 13. Back EMF of the AFPMSM at the rated load power.

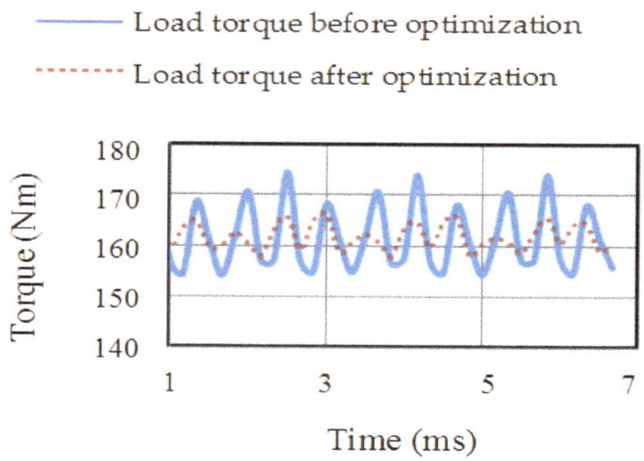

Figure 14. Torque of the AFPMSM at the rated load power.

3.4. Comparative Analysis of Peak Load Power Characteristics at 3000 rpm

Figure 15 shows that the load back EMF waveform is significantly optimized under the peak load power speed of 3000 rpm. Figure 16 shows that the load torque peak-to-peak value of the auxiliary slot optimization method is 10 Nm when the rotation speed is 3000 rpm, and the peak-to-peak value of the load torque before optimization is 16 Nm. The torque ripple of the auxiliary slot method is reduced by about 2%. Therefore, the torque ripple optimization method of the auxiliary slot can effectively attenuate the load torque ripple under constant power conditions and peak load conditions.

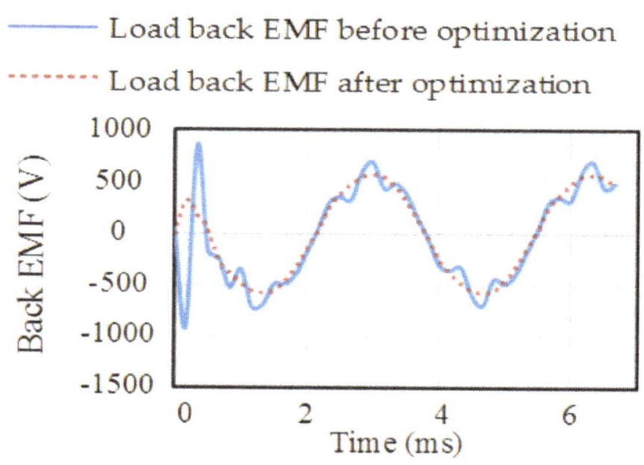

Figure 15. Back EMF of the AFPMSM at the peak load power.

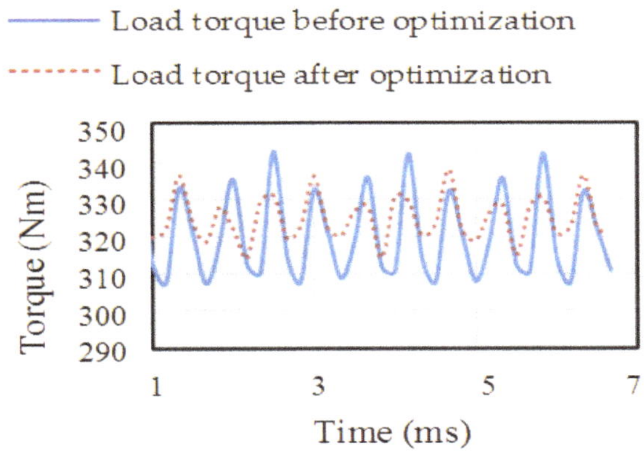

Figure 16. Torque of the AFPMSM at the peak load power.

In summary, at the constant torque maximum speed operating point, the rated load and peak load torque ripple ratio are reduced by 4% and 2%, respectively, before optimization. At the constant power maximum speed operating point, the rated load and peak load torque ripple are reduced by 4% and 2%, respectively, before optimization. Therefore, the AFPMSM opening auxiliary slot can effectively weaken the torque ripple of the motor, which is a reference for scholars studying the axial flux motor optimization design.

4. Multi-Physics Analysis

The traction motor of electric vehicle pursues high density and light weight in its design, and the working environment is complex and changeable. This is easy for causing electromagnetic noise and excessive temperature rise, which affects the reliability, safety and ride comfort of the whole vehicle. Therefore, multi-physics analysis of the motor is required. Controlling the electromagnetic noise within a certain range can improve the ride comfort; in addition, the heating problem of the motor directly affects the safety and reliability of the drive system, so thermal analysis of the motor is necessary. Figure 17 shows the finite element simulation model obtained and optimized above.

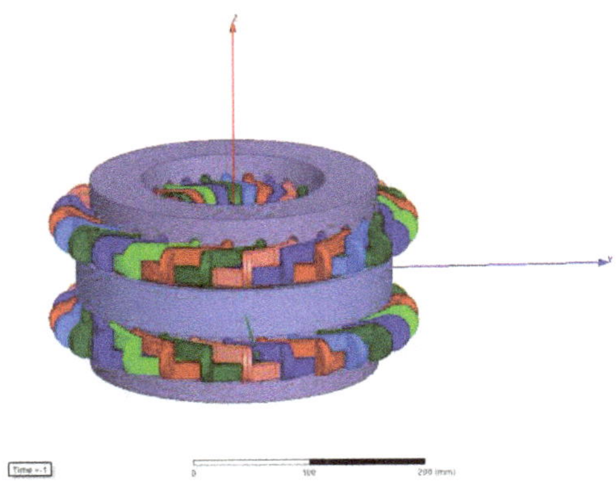

Figure 17. AFPMSM finite element simulation model.

The stator and rotor structure of AFPMSM is different from the traditional RFPMSM. In order to ensure the comprehensiveness of the analysis, the electromagnetic interference generated by the electromagnetic force of the stator and the electromagnetic force of the rotor is analyzed.

4.1. *Electromagnetic Noise Generated by the AFPMSM Stator Structure*

Firstly, a 3D finite element modal simulation model of the AFPMSM stator structure was established. The stator core structure has an elastic modulus of 205 GPa, a Poisson's ratio of 0.27 and a density of 7305 kg/m^3. Figure 18 shows the stator structure finite element simulation model and mesh segmentation diagram.

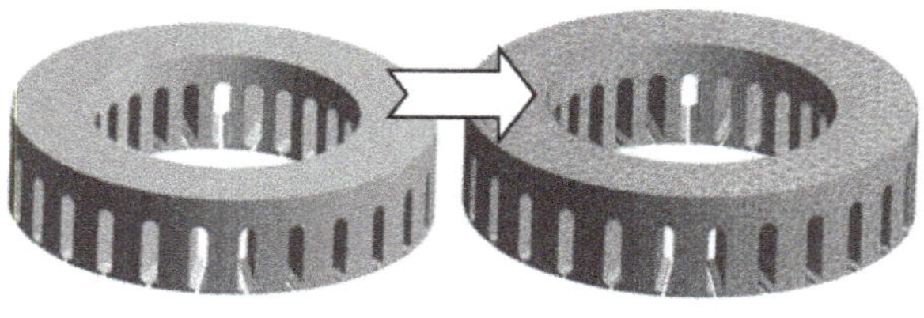

Figure 18. Finite element simulation model of the stator core.

In the finite element modal calculation, the AFPMSM stator model uses the additional mass method to make the winding end equivalent to a ring, thus calculating the natural frequency of the motor stator. The AFPMSM stator structure mode shape obtained by simulation is shown in Figure 19, and the modal frequency is shown in Table 2. It can be seen from Table 2 that using the winding as an additional mass analysis stator mode can increase the stiffness of the stator, increase the frequency range, and reduce the possibility of motor resonance.

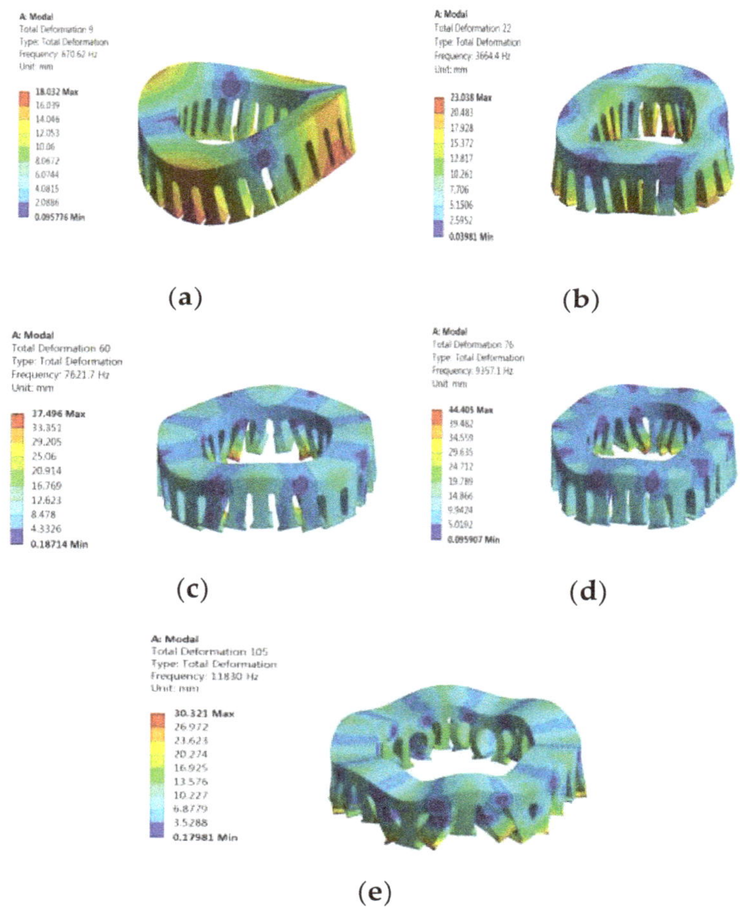

Figure 19. Finite element simulation model of the stator core. (**a**) Second order model. (**b**) Third order model. (**c**) Fourth order model. (**d**) Fifth order model. (**e**) Sixth order model.

Table 2. Modal frequency of stator core (Hz).

Modal Order	Order 2	Order 3	Order 4	Order 5	Order 6
Frequency	671	3664	7621	9357	11,830

The air domain model of motor electromagnetic noise radiation is established by A-weighting. The radiation characteristics of motor noise are analyzed based on the cylindrical air domain model. The results of sound pressure level analysis at each frequency are shown in Figures 20 and 21.

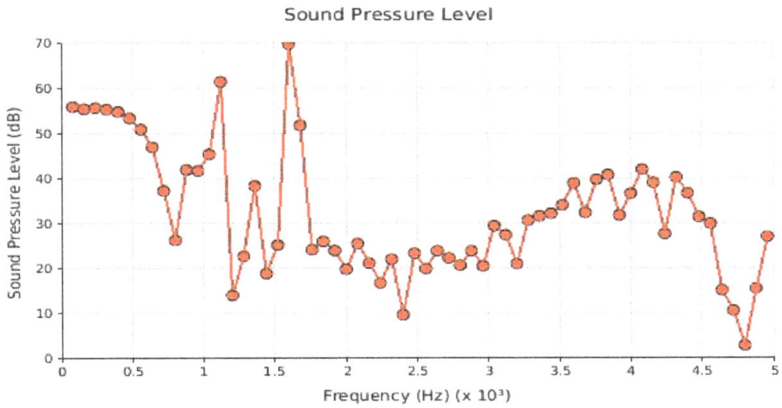

(a)

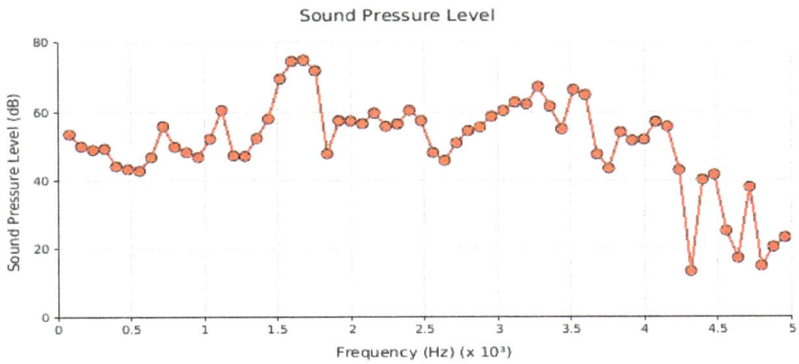

(b)

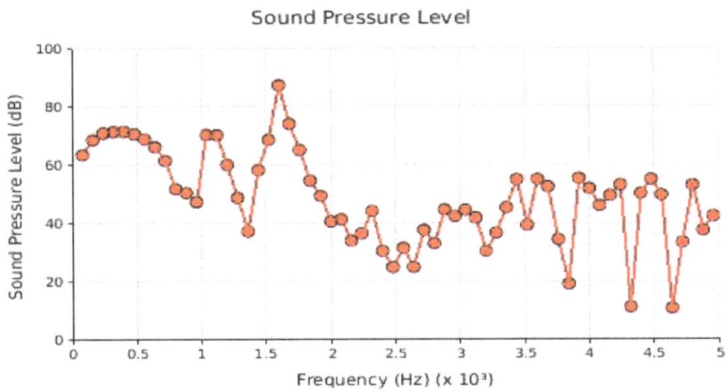

(c)

Figure 20. Stator system electromagnetic noise at 1600 rpm. (**a**) Electromagnetic noise at no load. (**b**) Electromagnetic noise at rated load. (**c**) Electromagnetic noise at peak load.

The sound pressure level of the electromagnetic noise at each working point of the motor can be calculated by the sound pressure level summation formula, as shown below:

$$L = 10\lg(\sum_{i=1}^{n} 10^{0.1L_i}) \tag{11}$$

where L_i is the sound pressure level value at the frequency and n is the number of frequency points.

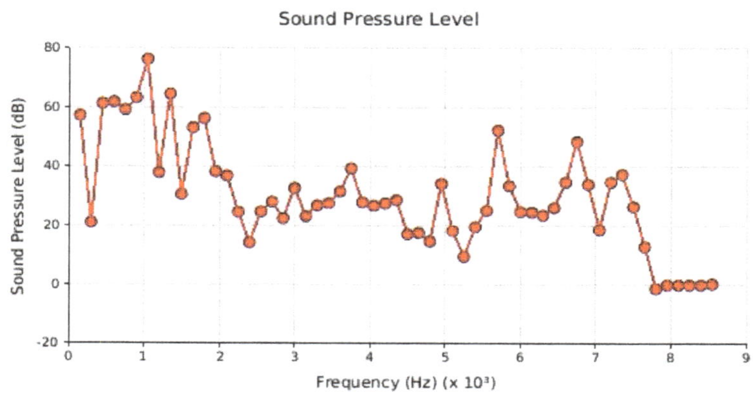

(**a**)

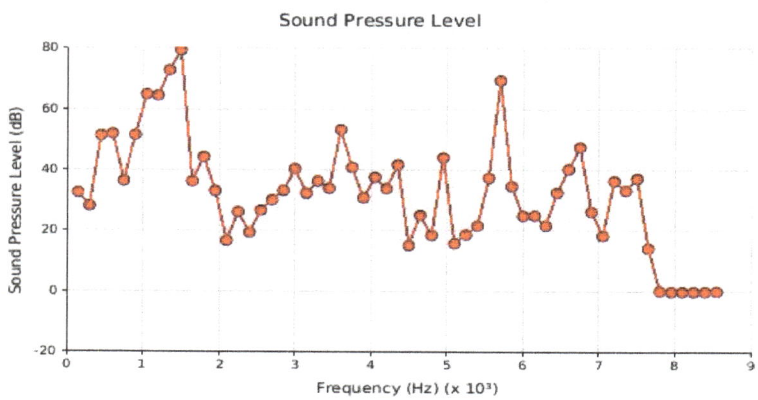

(**b**)

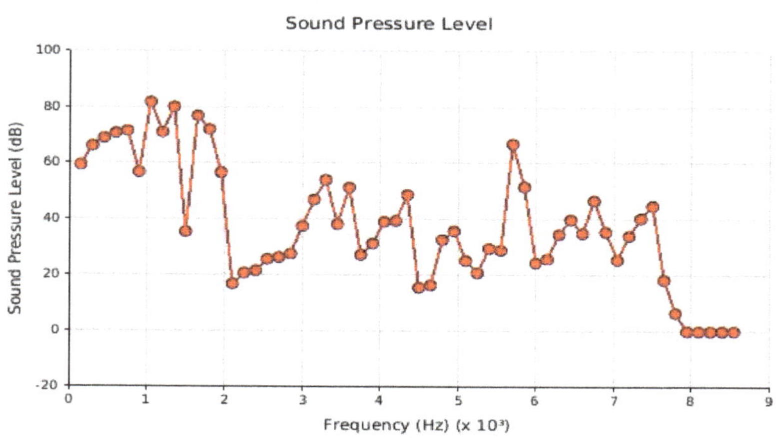

(**c**)

Figure 21. Stator system electromagnetic noise at 3000 rpm. (**a**) Electromagnetic noise at no load. (**b**) Electromagnetic noise at rated load. (**c**) Electromagnetic noise at peak load.

According to the sensitive range of the human ear to frequency, the sound pressure components in the range of 0 to 6000 Hz were selected for summation. The electromagnetic noise of the no-load, rated load, and peak load of the motor 1600 r/min were calculated to be 71 dB, 80 dB, and 88 dB, respectively; the electromagnetic noise of no-load, rated load, and peak load of 3000 r/min are 77 dB, 81 dB, and 86 dB respectively. The peak load noise of the motor 3000 r/min is large, but the torque required by the general vehicle during high-speed cruising is not large, therefore, the electromagnetic noise of the working point has little effect on the performance of the whole vehicle, and certain sound-absorbing or sound-insulating materials can be used for noise reduction processing.

4.2. Electromagnetic Noise Generated by the AFPMSM Rotor Structure

The finite element modal simulation model of the automotive AFPMSM rotor structure was established, and the natural frequency and vibration mode of the rotor structure were obtained, as shown in Figure 22.

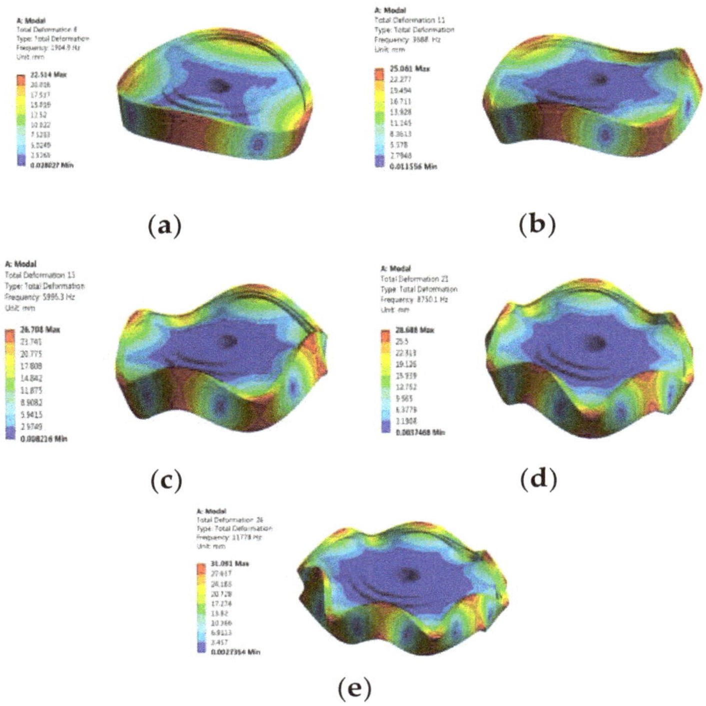

Figure 22. Rotor core axial flux mode shape. (**a**) Second order model. (**b**) Third order model. (**c**) Fourth order model. (**d**) Fifth order model. (**e**) Sixth order model.

The natural frequencies of the second-order mode and the third-order mode of the rotor of the AFPMSM are 1905 Hz and 3688 Hz, respectively, and the generated electromagnetic vibration amplitude is small, so electromagnetic vibration does not cause large vibration noise.

The electromagnetic noise of the rotor structure of the motor was simulated, and the noise sound pressure level results of the respective frequency components were obtained as shown in Figures 23 and 24. Since the peak load electromagnetic noise is greater than the rated load electromagnetic noise, the rotor section only shows the electromagnetic noise under peak load conditions.

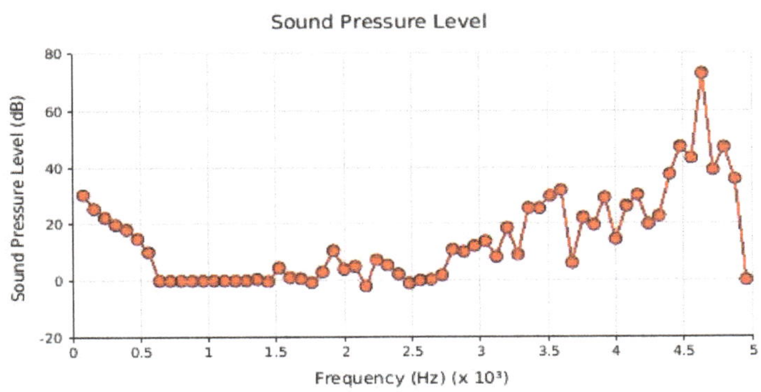

Figure 23. Peak load electromagnetic noise at 1600 rpm.

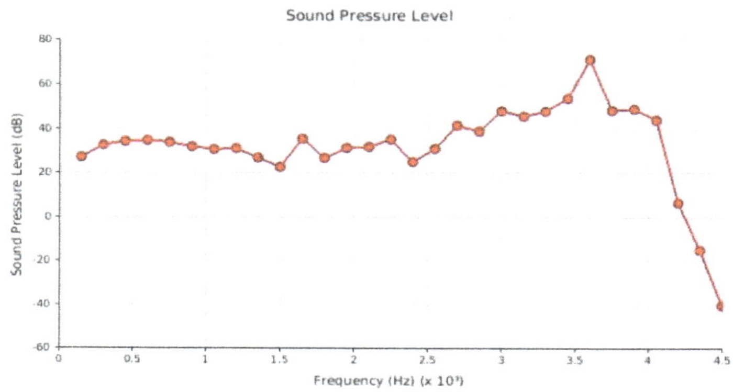

Figure 24. Value-loaded electromagnetic noise at a speed of 3000 rpm.

Calculated by Formula (11), the peak load electromagnetic noise of 1600 r/min is 72 dB; the peak load electromagnetic noise of the motor speed of 3000 r/min is 71 dB. The electromagnetic noise generated by the stator structure is the main source of AFPMSMs.

When the electromagnetic performance of the vehicle motor was evaluated, the magnetic field saturation of the constant torque working point was the largest, and the magnetic field distortion of the working point of the constant power maximum speed was the largest, which is the typical working condition of the traction motor for the vehicle. Therefore, the maximum speed operating point of the constant torque zone and the maximum speed operating point of the constant power zone are often selected as the key working points of the electromagnetic performance. In summary, the rated load and peak load electromagnetic noise of the stator structure at the constant torque operating point are 80 dB and 88 dB, respectively, and the rated load and peak load electromagnetic noise at the constant power maximum speed operating point are 81 dB and 86 dB, respectively. The peak load electromagnetic noise of the rotor structure at the constant torque operating point is 72 dB, and the peak load electromagnetic noise at the constant power maximum speed operating point is 71 dB. According to biology, the sound range that does not affect human health ranges from 0 to 90 dB. The maximum noise of the motor designed in this paper is 88 dB, which is within a reasonable range.

5. Thermal Analysis of AFPMSM

The problem of heating and cooling of the motor for vehicles directly affects the safety and reliability of the drive system. The accurate prediction of the temperature rise of the motor and the design of the efficient cooling system are of great significance to the safety and reliability of the new energy vehicles.

5.1. AFPMSM Temperature Field Equivalent Simplified Calculation Model

To simplify the analysis, an equivalent model of the stator winding was established, introducing an equivalent conductor that replaced the winding copper wire and an equivalent insulating layer of all the insulating materials in the slot, as shown in Figure 25.

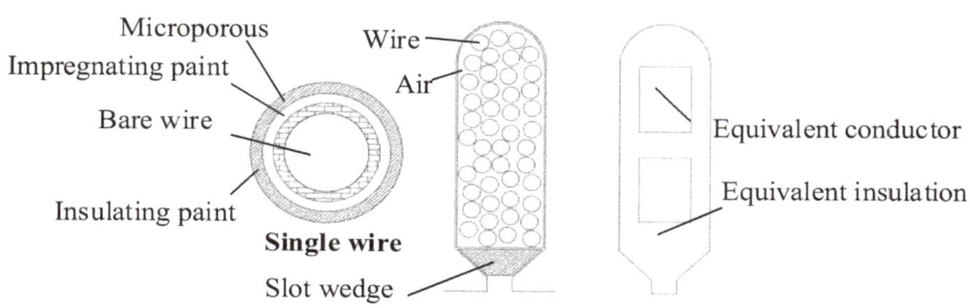

Figure 25. Equivalent winding model.

The equivalent thermal conductivity of the slot insulation is calculated as

$$\lambda_{eq} = \sum_{i=1}^{n} d_i / \left(\sum_{i=1}^{n} \frac{d_i}{\lambda_i} \right) \tag{12}$$

where λ_{eq} is the equivalent thermal conductivity of the insulating material, d_i is the equivalent thickness of each insulating material in the groove, λ_i is the thermal conductivity of the corresponding insulating material.

The AFPMSM in this paper uses a fully enclosed end cap cooling structure. To simplify the analysis, the following assumptions were made:

1. The insulating paint of the winding is evenly distributed, and the winding is completely dip-coated;
2. Since the inter-strand insulation of the armature windings cannot be completely considered and wound, the inner region of the stator slots is equivalently treated. The cross-sectional areas of the upper and lower windings are equal, and the thickness of the insulating layer, the windings and the left and right sides, and the windings of the windings and the upper and lower sides are equal;
3. Ignoring radiation heat dissipation and contact thermal resistance between the rotor and the rotating shaft;
4. The loss of each part of the motor does not change with temperature;
5. The heat generated by the heat source is mainly carried away by the end cover cooling water, and the heat exchange between the outer surface of the casing and the surrounding air is negligible.

The solution domain model of the motor based on the above assumptions is shown in Figure 26.

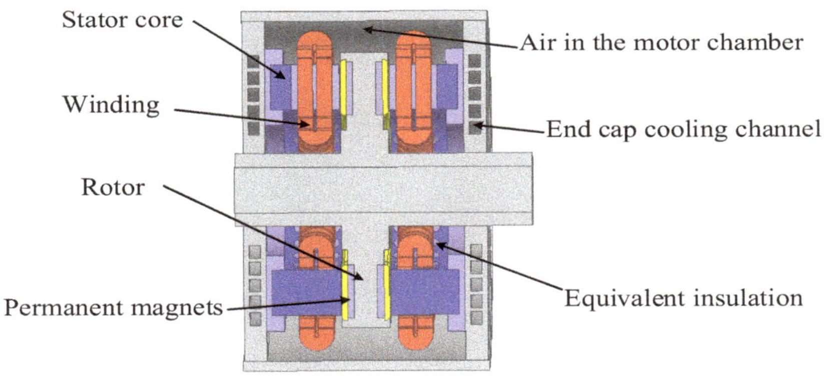

Figure 26. Motor temperature field solution domain model.

The axial flux motor is meshed, and the split model of each part of the motor is shown in Figure 27. The whole motor is divided into 17,346,364 individual cell grids and 3,412,225 nodes.

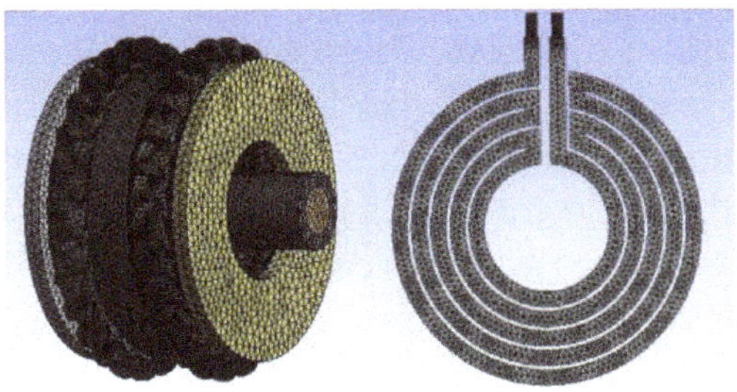

Figure 27. Schematic diagram of the prototype fluid-solid coupling simulation model.

5.2. Temperature Field Analysis under AFPMSM Rated Conditions

Thermal simulation of the motor was conducted based on ANSYS/Workbench. The inlet water flow rate was 0.8 m/s, the ambient temperature was 30 °C, and the inlet coolant temperature was 25 °C. The rated torque operating point of 50 kW and 1600 r/min was selected, and the motor temperature distribution as shown as Figure 28 is when the motor reached the thermal steady state for a long time.

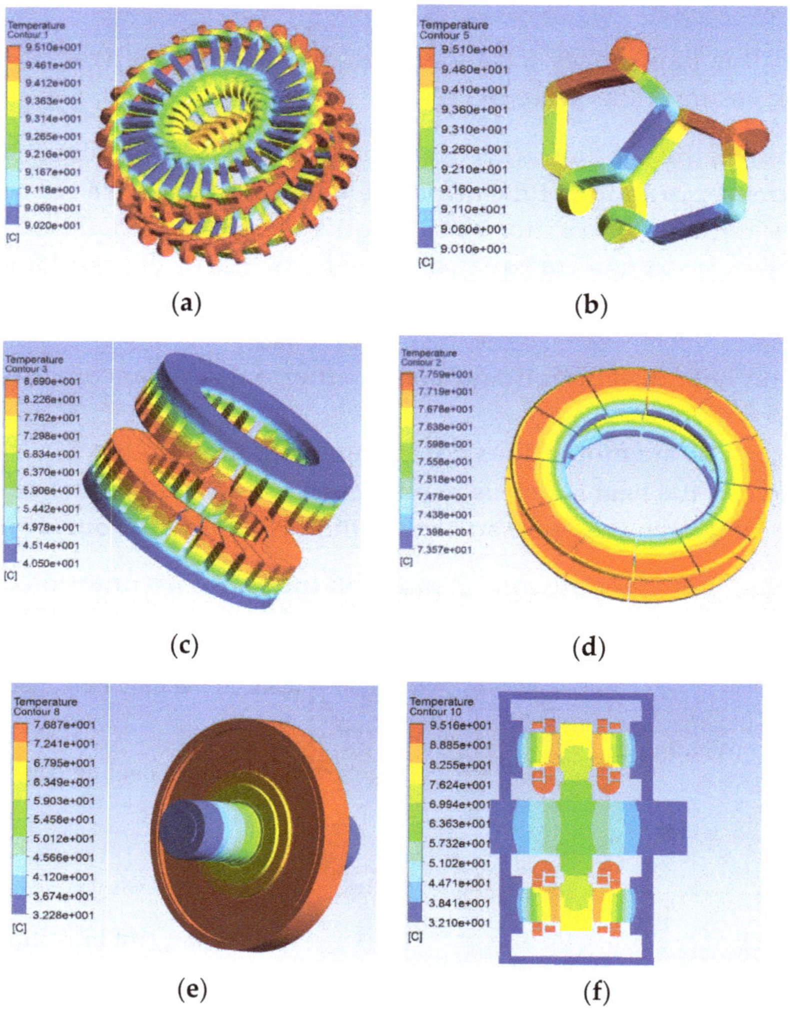

Figure 28. Temperature map of each part of the AFPMSM. (**a**) Winding temperature cloud map. (**b**) Upper and lower winding temperature cloud map. (**c**) Stator core temperature cloud map. (**d**) Permanent magnet temperature map. (**e**) Rotor and shaft temperature cloud map. (**f**)Whole machine section temperature cloud map.

As seen from the thermal simulation results, the temperature rise at the end of the stator winding is high. There is a direct temperature difference between the upper and lower windings in the same slot, and the temperature near the stator yoke winding is lower. The temperature of the stator teeth is significantly higher than the stator yoke. The temperature at the outer diameter of the permanent magnet is significantly higher than the inner diameter. The maximum temperature of the motor appears at the outer end of the winding (near the casing) at 95 °C. The maximum temperature of the stator core is the stator tooth at 87 °C. The maximum temperature of the permanent magnet is at the outer diameter of the permanent magnet at 78 °C. The internal temperature of the whole machine is, from the highest to the lowest, the winding end, the stator tooth, the permanent magnet, the rotor disk, the stator disk, and the casing. The analysis shows that the temperature rise of the motor's long-term rated operation is within a reasonable range, and the motor can operate safely and reliably.

5.3. Temperature Field Analysis of AFPMSM under Peak Operating Conditions

Considering the operating conditions of the actual electric vehicle, the temperature rise of the motor is further checked by operating the peak torque for 60 s after assuming that the rated operation of the motor reaches a thermal steady state. The temperature rise of the motor components over time during the 60 s operation of the motor with peak torque is shown in Figure 29.

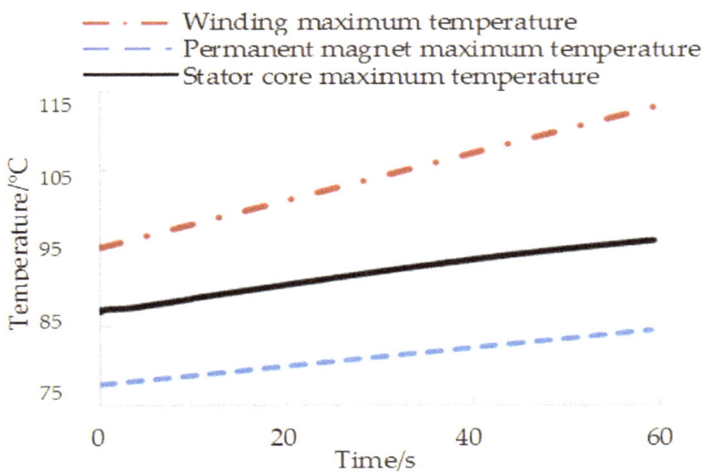

Figure 29. Temperature of main components of the motor changes with time.

Figure 29 shows that the winding temperature rises the fastest when the motor peak torque is running, and the maximum temperature of the winding reaches 113 °C after running for 60 s, which is 18 °C higher than the rated working temperature; the maximum temperature of the stator core is 95.9 °C, which is 9 °C higher than the rated working condition. The maximum temperature of the permanent magnet is 84.68 °C, which is 7 °C higher than the rated working temperature. The winding temperature rises the fastest. Because the copper consumption of the peak torque condition is relatively large, the thermal conductivity of the insulating material is poor, and the generated heat is difficult to dissipate in a short time. In this paper, the motor uses H-class insulation. The results of motor thermal analysis show that the temperature rise of the motor meets the technical requirements.

6. Conclusions

This paper first proposes a 2D multi-loop equivalent method to simplify the AFPMSM model and, based on this, establishes the equivalent magnetic network model and air gap magnetic density distribution function of AFPMSMs. Secondly, by comparing the torque ripple values under multi-operating conditions before and after the open stator auxiliary slot, the feasibility of reducing the AFPMSM torque ripple is verified by the open stator auxiliary slot. Finally, the experimental results of AFPMSM in multi-physics such as vibration, noise, and temperature rise are obtained by finite element software and simulation experiments. The following conclusions can be drawn:

1. The electromagnetic design method based on the analytical method was used to analyze the key electromagnetic parameters such as air gap magnetic density harmonics, positioning torque and no-load back EMF of AFPMSM. The correctness of the analytical calculation was verified by 3D electromagnetic field simulation.
2. The stator core was used to open the auxiliary slot to suppress the torque ripple, and the maximum torque ripple peak-to-peak value was reduced by 2%.
3. Based on the multi-physics simulation software, the vibration and noise characteristics of the motor were analyzed from the electromagnetic force and the modal state. The electromagnetic noise characteristics of the no-load, rated load, and peak load operating points at different speeds were simulated. The noise, vibration and harshness (NVH) of the motor was verified.

4. The thermal simulation based on fluid-structure coupling checked the temperature rise of the AFPMSM after the rated operating point reached the thermal steady state and transient operation for 1 min, which ensured the safety and reliability of the motor running under vehicle working conditions.

In summary, the AFPMSM design and research method proposed in this paper verifies the superior performance of AFPMSM as a motor for vehicles. On the other hand, this paper provides a complete solution for the design and optimization of AFPMSMs for electric vehicles, which can provide relevant researchers with AFPMSM performance reference indicators.

Author Contributions: Conceptualization, J.Z., Y.D., and Q.H.; methodology, J.Z.; software, Q.H. and M.H.; validation, J.Z., Y.D., and Q.H.; formal analysis, J.Z. and Y.D.; investigation, Y.D. and M.H.; resources, J.Z.; data curation, Q.H. and M.H.; writing—original draft preparation, J.Z., Q.H., and Y.D.; writing—review and editing, J.Z. and Q.H.; visualization, Q.H. supervision, J.Z.; project administration, J.Z.; funding acquisition, J.Z.

References

1. Chen, Y.; Zhang, B. Minimization of the Electromagnetic Torque Ripple Caused by the Coils Inter-Turn Short Circuit Fault in Dual-Redundancy Permanent Magnet Synchronous Motors. *Energies* **2017**, *10*, 1798. [CrossRef]
2. Wang, W.; Wang, W. Compensation for Inverter Nonlinearity in Permanent Magnet Synchronous Motor Drive and Effect on Torsional Vibration of Electric Vehicle Driveline. *Energies* **2018**, *11*, 2542. [CrossRef]
3. Pietrusewicz, K.; Waszczuk, P.; Kubicki, M. MFC/IMC Control Algorithm for Reduction of Load Torque Disturbance in PMSM Servo Drive Systems. *Appl. Sci.* **2019**, *9*, 86. [CrossRef]
4. Jia, Y.F.; Chu, L.; Xu, N.; Li, Y.K.; Zhao, D.; Tang, X. Power Sharing and Voltage Vector Distribution Model of a Dual Inverter Open-End Winding Motor Drive System for Electric Vehicles. *Appl. Sci.* **2018**, *8*, 254. [CrossRef]
5. Wang, R.J. Optimal design of a coreless stator axial flux permanent-magnet generator. *IEEE Trans. Magn.* **2005**, *41*, 55–64. [CrossRef]
6. Aydin, M.; Huang, S.; Lipo, T.A. Torque quality and comparison of internal and external rotor axial flux surface-magnet disc machines. *IEEE Trans. Ind. Electron.* **2006**, *53*, 822–830. [CrossRef]
7. Zhao, J.F.; Hua, M.Q.; Liu, T.Z. Research on a Sliding Mode Vector Control System Based on Collaborative Optimization of an Axial Flux Permanent Magnet Synchronous Motor for an Electric Vehicle. *Energies* **2018**, *11*, 3116. [CrossRef]
8. Darba, A.; Esmalifalak, M.; Barazandeh, E.S. Implementing SVPWM technique to axial flux permanent magnet synchronous motor drive with internal model current controller. In Proceedings of the 2010 4th International Power Engineering and Optimization Conference, Shah Alam, Malaysia, 23–24 June 2010; pp. 126–131.
9. Lim, D.K.; Cho, Y.S.; Ro, J.S.; Jung, S.Y.; Jung, H.K. Optimal Design of an Axial Flux Permanent Magnet Synchronous Motor for the Electric Bicycle. *IEEE Trans. Magn.* **2016**, *52*, 1–4. [CrossRef]
10. Kim, J.S.; Lee, J.H.; Kim, Y.J.; Jung, S.Y. Characteristics analysis method of axial flux permanent magnet motor based on 2-D finite element analysis. *IEEE Trans. Magn.* **2017**, *53*, 1–4. [CrossRef]
11. Hemeida, A.; Sergeant, P. Analytical Modeling of Surface PMSM Using a Combined Solution of Maxwell–s Equations and Magnetic Equivalent Circuit. *IEEE Trans. Magn.* **2014**, *50*, 1–13. [CrossRef]
12. Xia, C.; Zhao, J.; Yan, Y.; Shi, T. A novel direct torque control of matrix converter-fed PMSM drives using duty cycle control for torque ripple reduction. *IEEE Trans. Ind. Electron.* **2013**, *61*, 2700–2713. [CrossRef]
13. Vafaie, M.H.; Dehkordi, B.M.; Moallem, P.; Kiyoumarsi, A. Minimizing torque and flux ripples and improving dynamic response of PMSM using a voltage vector with optimal parameters. *IEEE Trans. Ind. Electron.* **2015**, *63*, 3876–3888. [CrossRef]
14. Wang, D.; Lin, H.; Yang, H.; Wang, K. Cogging torque optimization of flux memory pole-changing permanent magnet machine. *IEEE Trans. Appl. Supercond.* **2016**, *26*, 1–5. [CrossRef]

15. Lai, C.; Feng, G.; Mukherjee, K.; Kar, N.C. Investigations of the influence of PMSM parameter variations in optimal stator current design for torque ripple minimization. *IEEE Trans. Energy Convers.* **2017**, *32*, 1052–1062. [CrossRef]
16. Wahsh, S.; Shazly, J.; Yassin, A. Steady state heat conduction problems of AFPMSM using 3D Finite Element. In Proceedings of the 18th International Middle-East Power Systems Conference, Helwan Univ, ON, Egypt, 27–29 December 2016; pp. 949–953.
17. Chen, Q.; Liang, D.; Gao, L.; Wang, Q.; Liu, Y. Hierarchical thermal network analysis of axial-flux permanent-magnet synchronous machine for electric motorcycle. *IET Electr. Power Appl.* **2018**, *12*, 859–866. [CrossRef]
18. Rasekh, A.; Sergeant, P.; Vierendeels, J. Fully predictive heat transfer coefficient modeling of an axial flux permanent magnet synchronous machine with geometrical parameters of the magnets. *Appl. Thermal Eng.* **2017**, *110*, 1343–1357. [CrossRef]
19. De Bisschop, J.; Sergeant, P.; Dupré, L. Demagnetization fault detection in axial flux PM machines by using sensing coils and an analytical model. *IEEE Trans. Magn.* **2017**, *53*, 1–4. [CrossRef]
20. De Bisschop, J.; Sergeant, P.; Hemeida, A.; Vansompel, H.; Dupre, L. Analytical Model for Combined Study of Magnet Demagnetization and Eccentricity Defects in Axial Flux Permanent Magnet Synchronous Machines. *IEEE Trans. Magn.* **2017**, *53*, 1–12. [CrossRef]
21. De Bisschop, J.; Abdallh, A.; Sergeant, P.; Dupré, L. Identification of Demagnetization Faults in Axial Flux Permanent Magnet Synchronous Machines Using an Inverse Problem Coupled With an Analytical Model. *IEEE Trans. Magn.* **2014**, *50*, 1–4. [CrossRef]
22. Hemeida, A.; Sergeant, P.; Vansompel, H. Comparison of Methods for Permanent Magnet Eddy-Current Loss Computations With and Without Reaction Field Considerations in Axial Flux PMSM. *IEEE Trans. Magn.* **2015**, *51*, 1–11. [CrossRef]
23. Scheerlinck, B.; De Gersem, H.; Sergeant, P. 3D Eddy-Current and Fringing-Flux Distribution in an Axial-Flux Permanent-Magnet Synchronous Machine with Stator in Laminated Iron or SMC. *IEEE Trans. Magn.* **2015**, *51*, 1–4. [CrossRef]
24. Chen, Q.; Liang, D.; Jia, S.; Ze, Q.; Liu, Y. Loss Analysis and Experiment of Fractional-Slot Concentrated-Winding Axial Flux PMSM for EV Applications. In Proceedings of the 10th IEEE Annual Energy Conversion Congress and Exposition, Portland, ON, USA, 23–27 September 2018; pp. 4329–4335.
25. Fei, W.; Luk, P.C.K. Torque ripple reduction of axial flux permanent magnet synchronous machine with segmented and laminated stator. In Proceedings of the 2009 IEEE Energy Conversion Congress and Exposition, San Jose, CA, USA, September 2009; pp. 132–138.
26. Woolmer, T.; McCulloch, M. Analysis of the yokeless and segmented armature machine. In Proceedings of the 2007 IEEE International Electric Machines & Drives Conference, Antalya, Turkey, 3–5 May 2007; pp. 704–708.
27. Aydin, M.; Huang, S.; Lipo, T. Axial flux permanent magnet disc machine: A review. In Proceedings of the Symposium on Power Electronics, Electrical Drives, Automation, and Motion (SPEEDAM), Capri, Italy, 16–18 June 2004; pp. 61–71.
28. Hosseini, S.M.; Agha-Mirsalim, M.; Mirzaei, M. Design, prototyping, and analysis of a low cost axial-flux coreless permanent-magnet generator. *IEEE Trans. Magn.* **2008**, *44*, 75–80. [CrossRef]

7

Direct Torque Control of PMSM with Modified Finite Set Model Predictive Control

GuangQing Bao [1,2,3], WuGang Qi [1,2,3,*] and Ting He [4]

1 College of Electrical Engineering and Information Engineering, Lanzhou University of Technology, Lanzhou 730050, China; gqbao@lut.cn
2 Key Laboratory of Gansu Advanced Control for Industrial Processes, Lanzhou University of Technology, Lanzhou 730050, China
3 National Demonstration Center for Experimental Electrical and Control Engineering Education, Lanzhou University of Technology, Lanzhou 730050, China
4 Gansu Natural Energy Research Institute, Lanzhou 730046, China; 13679261064@163.com
* Correspondence: qwg_0218@163.com

Abstract: A direct torque control (DTC) with a modified finite set model predictive strategy is proposed in this paper. The eight voltage space vectors of two-level inverters are taken as the finite control set and applied to the model predictive direct torque control of a permanent magnet synchronous motor (PMSM). The duty cycle of each voltage vector in the finite set can be estimated by a cost function, which is designed based on factors including the torque error, maximum torque per ampere (MTPA), and stator current constraints. Lyapunov control theory is introduced in the determination of the weight coefficients of the cost function to guarantee stability, and thus the optimal voltage vector reference value of the inverter is obtained. Compared with the conventional finite control set model predictive control (FCS-MPC) method, the torque ripple is reduced and the robustness of the system is clearly improved. Finally, the simulation and experimental results verify the effectiveness of the proposed control scheme.

Keywords: direct torque control; finite control set mode predictive control; duty cycle; maximum torque per ampere; permanent magnet synchronous motor

---

1. Introduction

Permanent magnet synchronous motor (PMSM) direct torque control (DTC) has been widely used in industry due to its simple control structure, fast dynamic response, and high efficiency [1]. However, the traditional DTC control results in large torque ripple due to the insufficient switching frequency of the inverters and two nonlinear hysteresis comparators [2,3].

To solve the problem of torque ripple, many scholars have put forward many improved methods. An improved method is calculating the effective voltage vector action time in real time to ensure the minimum torque ripple by the current torque error [4–6]. This method reduces the torque ripple to a certain extent, but the calculation process is complicated. At present, voltage space vector modulation (SVM) is also introduced into DTC [7,8]. This method can effectively reduce the torque ripple, but it cannot eliminate the steady-state error of the torque. At the same time, the calculation process is highly dependent on the parameters of the motor and has poor robustness.

As a real-time online optimization control method, model predictive control (MPC) has received extensive attention in the field of electric drives and power converters due to its high dynamics and resistance to parameter disturbance [9,10]. The method of finite control set MPC (FCS-MPC) directly utilizes the discreteness of inverter output voltage and the finiteness of switching state; at the same time, it does not need modulation and has a small amount of computation, which makes it a hot research topic in the field of power electronics [11,12]. In [13], on the premise that the flux linkage and torque of the motor are limited within the target range, a longer prediction step is selected to optimize the switching frequency of the inverter. A strategy of separate control of transient and steady state are adopted to reduce the switching frequency of inverter [14]. Mayne [15] proposed an FCS-MPC algorithm with model error compensation which enhances the robustness of the system. In [16], FCS-MPC method is applied in a grid-connected converter, and the stability of the control system is proved by Lyapunov stability theory. In addition, the virtual voltage vector is also integrated into FCS-MPC architecture, and the optimal voltage vector is preselected by using the results of Lyapunov stability theory. This method reduces the number of possible voltage vectors from 38 to 10, thus reducing the computational burden of the controller.

The cost function of FCS-MPC is the key to realizing the optimal selection of the voltage vector, which can be established according to the control targets, the system model based predictive variables, and the reference variables [17]. Depending on the control situation, it is also necessary to include system constraints. The cost function can contain multiple control objectives, variables, and constraints. Correct design of reasonable weight coefficients is very important for selecting the voltage vectors and governing the switching sequence of the inverter based on predictive current control of the PMSM with the FCS model [18,19].

The main contribution of this paper is to combine Lyapunov function theory with FCS-MPC to reduce torque ripple. Among them, the operating conditions of MTPA and torque tracking are guaranteed by using cost function. The Lyapunov function theory is introduced in the calculation of the finite set voltage vector duty cycle to obtain the optimal voltage vector. The simulation and experimental results show that the method achieves fast torque response and torque ripple minimization.

2. Discrete Mathematical Model of PMSM and Drive

The main circuit of the electric drive system under consideration is illustrated in Figure 1, which also corresponds to the experimental system layout. The PMSM has the characteristics of being multivariable and nonlinear and has strong coupling [20]. In order to simplify the mathematical model of PMSM, the following assumptions are made: (1) Y-shaped connection of stator windings, symmetrical distribution of three-phase windings, and space difference of each winding is 120°. (2) Eddy current loss, hysteresis loss, and the change of motor parameters are neglected. (3) It is assumed that permanent magnets on the rotor generate a main magnetic field in the stator–rotor airgap (the magnetic field is distributed sinusoidally along the circumference of the airgap) 22. So the continuous time model for PMSM in the d–q coordinate system can be described as:

$$L_d \frac{di_d}{dt} = (u_d - Ri_d + \omega L_q i_q), \tag{1}$$

$$L_q \frac{di_q}{dt} = (u_q - Ri_q - \omega L_d i_d - \omega \psi_f), \tag{2}$$

$$T_e = 1.5p(\psi_f + (L_d - L_q)i_d)i_q, \tag{3}$$

where u_d, u_q, i_d and i_q are the stator voltages and currents, R is the stator winding resistance, L_d and L_q are the d- and q-axes inductances, ψ_f is the flux linkage of the rotor, p is the number of pairs of poles, ω is the mechanical angular speed, and T_e is the electromagnetic torque. i_q is proportional to the electromagnetic torque and i_d is proportional to the reactive power. A predictive current control scheme is thus formed in which the reference current is generated by an external speed control loop.

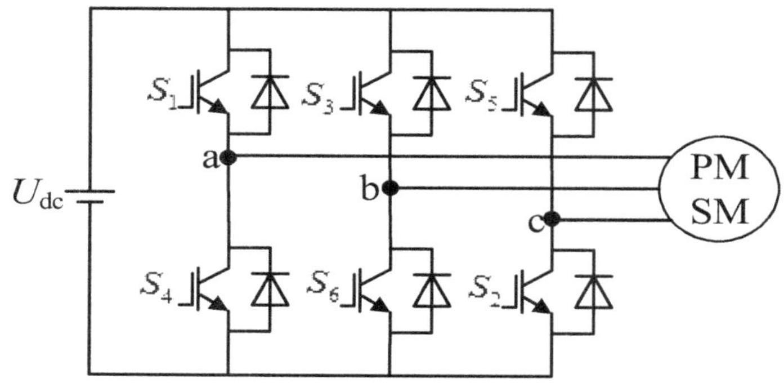

Figure 1. The main circuit of the permanent magnet synchronous motor (PMSM) drive.

Discretize the mathematical model of PMSM to obtain the necessary conditions for model prediction. Since the sampling time T_s is sufficiently small, the Euler approximation is used for the stator current derivative di/dt of the sampling time T_s.

$$\frac{di}{dt} = \frac{i(k+1) - i(k)}{T_s}, \tag{4}$$

where T_s is the sampling period. Then, the discrete current model of the PMSM is obtained:

$$L_d i_d^{k+1} = (L_d - T_s R) i_d^k + T_s \omega L_q i_q^k + T_s u_d^k, \tag{5}$$

$$L_q i_q^{k+1} = (L_q - T_s R) i_q^k - T_s \omega L_d i_d^k - T_s \omega \psi_f + T_s u_q^k, \tag{6}$$

$$T_e^{k+1} = 1.5 p i_q^{k+1} [\psi_f + (L_d - L_q) i_d^{k+1}]. \tag{7}$$

The commutation process of two-level voltage source inverters is accomplished by use of DC bus. The switching state can be represented by the switching signals S_a and S_b and switching on and off on different bridge arms. The upper and lower switches of each bridge arm of the inverter cannot be turned on at the same time to prevent short circuit S_c, as shown in Table 1:

Table 1. Inverter switching states.

	S_a	S_b	S_c
0	S_1 on S_4 off	S_3 on S_6 off	S_2 on S_5 off
1	S_1 off S_4 on	S_3 off S_6 on	S_2 off S_5 on

From the above table, the vector form function of the switching state of the three-phase bridge arm (S) is as follows:

$$S = \frac{2}{3}(S_a + aS_b + a^2 S_c), \tag{8}$$

where $a = e^{j2\pi/3}$. The relationship between output voltage vector with the switch state can be defined as:

$$U_{out} = SU_{dc} = \frac{2}{3} U_{dc}(S_a + aS_b + a^2 S_c), \tag{9}$$

where U_{dc} is the DC source voltage and U_{out} is the output voltage of the inverter.

Considering the different switching states of the inverters, eight different voltage vectors are obtained, as shown in Table 2. It is noted that $V0 = V7$, resulting in a finite set of seven different voltage vectors in the plane.

Table 2. Switching state and voltage vector.

S_a	S_b	S_c	**Voltage Vector U_{out}**
0	0	0	$V0 = 0$
0	0	1	$V1 = \frac{2}{3}U_{dc}e^{j\frac{4\pi}{3}}$
0	1	0	$V2 = \frac{2}{3}U_{dc}e^{j\frac{2\pi}{3}}$
0	1	1	$V3 = \frac{2}{3}U_{dc}e^{j\pi}$
1	0	0	$V4 = \frac{2}{3}U_{dc}$
1	0	1	$V5 = \frac{2}{3}U_{dc}e^{j\frac{5\pi}{3}}$
1	1	0	$V6 = \frac{2}{3}U_{dc}e^{j\frac{\pi}{3}}$
1	1	1	$V7 = 0$

3. Predictive Control Based on Duty Cycle

Within one sampling interval, the cost function values that correspond to the seven voltage vectors are calculated by the FCS. The switching state that minimizes the cost function is chosen as the switching state of the inverter. The process of model prediction optimization is shown in Figure 2, where x, x_p, and x^* are the torque of the real response, the predicted value, and the reference value, respectively. If the corresponding $x(k)$ is the best value at time k, the cost function value $x(k + 1)$ corresponding to the seven voltage vectors is calculated at time $(k + 1)$, and the seven calculated cost functions values are compared with reference values, where the closest is to be the optimal solution at the moment $(k + 1)$. V_4 is the control signal of the voltage vector corresponding to the optimal value of the time $(k + 1)$. In the same way, V_3 should be selected as the optimal control signal at time $(k + 2)$ [12].

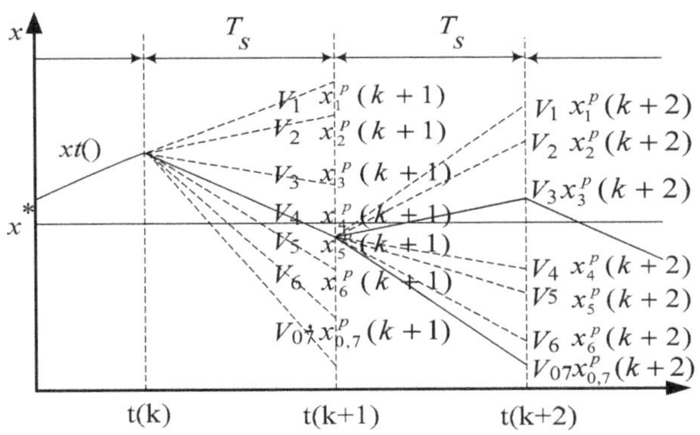

Figure 2. Optimization process diagram of model predictive control (MPC).

3.1. Design of Cost Function

The maximum torque/current control (MTPA) consumes the minimum stator current at the same electromagnetic torque output, which is usually used in interior PMSM systems. The space vector analysis of PMSM in the d–q coordinate system is shown in Figure 3. Assuming that the stator current vector i_s leads q axis β, the current component in the d–q coordinate system is as follows:

$$\begin{cases} i_d = -|i_s|\sin\beta \\ i_q = |i_s|\cos\beta \\ i_s = \sqrt{i_d^2 + i_q^2} \end{cases}, \tag{10}$$

$$T_e = 1.5p[\psi_f i_s \cos\beta - (Ld - Lq)i_s^2 \cos\beta \sin\beta]. \tag{11}$$

If the amplitude of the stator current of the PMSM is a constant value, it can be concluded that the electromagnetic torque of the motor is only related to β. In order to find the maximum value of the torque, that is, to satisfy the MTPA condition, and calculate Equation (11) with β as a variable, the following relationship is obtained:

$$\frac{dT_e}{d\beta} = -\psi_f is \sin\beta - (L_d - L_q)i_s[\cos\beta^2 - \sin\beta^2] = 0. \quad (12)$$

Because i_s and β satisfy the Pythagorean theorem, the relationship of i_d and i_q under the MTPA condition can thus be derived as follows:

$$i_d + \frac{L_d - L_q}{\psi_f}(i_d^2 - i_q^2) = 0. \quad (13)$$

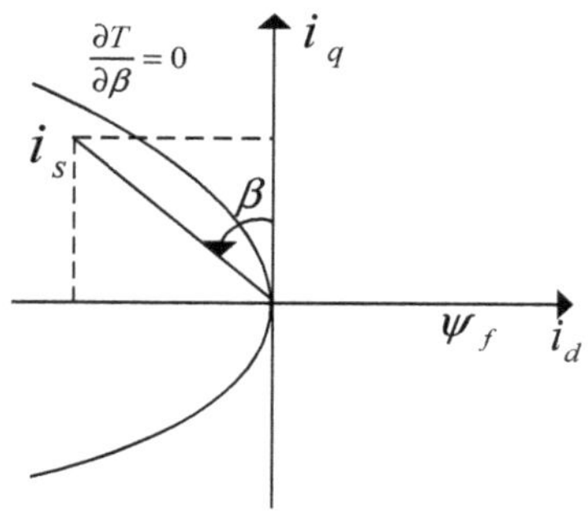

Figure 3. Space vector analysis of the PMSM.

The most important part of MPC is the cost function design. The cost function is expressed in orthogonal coordinates and considering the torque tracking, MTPA condition, and current constraint:

$$J(k) = K_T J_T(k) + K_M J_M(k) + K_C(J_{C1}(k) + J_{C2}(k)), \quad (14)$$

where K_T, K_M, and K_C are the weighting factors of the cost function and are positive real numbers. The main aim of the MPC is to minimize the torque error. Thus, the cost for the error $J_T(k)$ is:

$$J_T(k) = \left(T_e^* - T_e^{k+1}\right)^2. \quad (15)$$

Since high currents lead to large losses, $J_M(k)$ is to have a low absolute current:

$$J_M(k) = \left(i_d^k + \frac{L_d - L_q}{\psi_F}\left(\left(i_d^k\right)^2 - \left(i_q^k\right)^2\right)\right)^2. \quad (16)$$

Tertiary control goals can be added in order to ensure the operational safety of the system. $J_{C1}(k)$ denotes the current limit, and the current amplitude of the system must be smaller than the maximum allowable value $I_{\max}$.

$$J_{C1}(k) = \begin{cases} 0 & \left(I_{\max} \geq \sqrt{\left(i_d^k\right)^2 + \left(i_q^k\right)^2}\right) \\ \left(I_{\max} - \sqrt{\left(i_d^k\right)^2 + \left(i_q^k\right)^2}\right)^2 & \left(I_{\max} < \sqrt{\left(i_d^k\right)^2 + \left(i_q^k\right)^2}\right) \end{cases} \quad (17)$$

To make the point of the cost function converge to the parabola on the left side of the i_q axis, as shown in Figure 3. Adding current constraints $J_{C2}(k)$ to the cost function:

$$J_{C2}(k) = \begin{cases} 0, \left(i_d^k \leq 0\right) \\ \left(i_d^k\right)^2, \left(i_d^k > 0\right) \end{cases}. \quad (18)$$

To ensure current constraints and avoid control variables converging to the wrong MTPA curve, $K_C \gg K_T, K_M$ should be satisfied. For the choice of K_T and K_M, $K_T + K_M = 1$ can be made. A larger value of K_T leads to faster convergence of the torque, while a larger value of K_M indicates that the MTPA state converges quickly. $K_T > K_M$ can be selected [21] to achieve a fast torque response.

3.2. Duty Cycle Calculation

The duty cycle of each voltage vector in a finite set is calculated by Lyapunov function to eliminate torque and current ripple. Without considering the current limitation, that is, ignoring $J_{C1}(k)$ and $J_{C2}(k)$, the Lyapunov function is expressed as follows:

$$V(\mathrm{k}+1) = k_T J_T(k+1) + k_M J_M(k+1). \tag{19}$$

In order to optimize the operation of the PMSM, the condition $V(k+1) = 0$ must be fulfilled to ensure $T_e^{k+1} = T_e^*$ and its current operating point is on the MTPA curve. In addition, the condition $dV(k+1)/dt = 0$ indicates that the PMSM is operating at the optimum state.

Lemma 1. *Defining the function of the current:*

$$f(i_d, i_q) = i_d + (L_d - L_q)(i_d^2 {-}_q^2)/\psi_f \tag{20}$$

If the torque of the PMSM is kept at a nonzero constant value, the function $f(i_d, i_q)$ is a strict monotonic function along the constant torque curve. The correlative poof is introduced in the Appendix A.

With the help of the function $f(i_d, i_q)$, the stator current trajectory is shown in Figure 4. Point A is the initial point; the trajectory first reaches the point B along the constant torque curve and then reaches the point C along the MTPA curve with the condition $V(k) = 0$. When the current trajectory is along $\overrightarrow{AB}$ to $\overrightarrow{BC}$, the Lyapunov function is strictly decreasing.

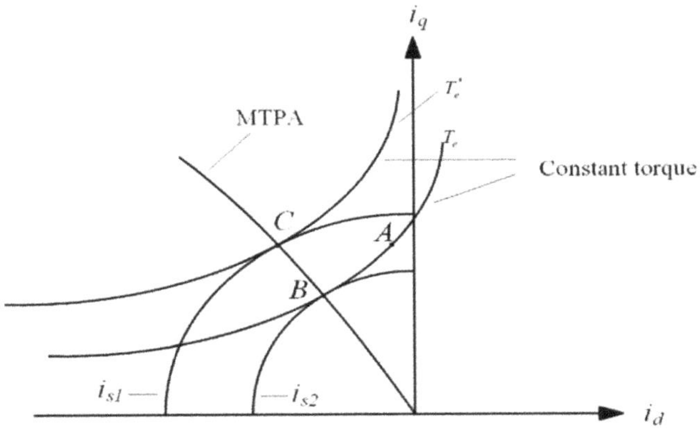

Figure 4. Stator current vector trajectory.

Lemma 2. *Under the initial condition of the PMSM, if the back EMF of the PMSM is within the voltage limit range of the inverter, there exists a feasible voltage vector u_{dq}^* that fulfills:*

$$dV(k+1)/dt \leq 0. \tag{21}$$

Lemma 3. *For a given current dynamics:*

$$\frac{di_{dq}^{k+1}}{dt} = Ai_{dq}^{k+1} + Bu_{dq}^{k+1} + E, \tag{22}$$

where $A = \begin{bmatrix} -\frac{R}{L_d} & \frac{\omega L_q}{L_d} \\ -\frac{\omega L_d}{L_q} & -\frac{R}{L_q} \end{bmatrix}, B = \begin{bmatrix} \frac{1}{L_d} & 0 \\ 0 & \frac{1}{L_q} \end{bmatrix}, E = \begin{bmatrix} 0 \\ -\frac{-\omega\psi_f}{L_q} \end{bmatrix}.$

If the back EMF of the PMSM is within the voltage limit range of the inverter, then at least one voltage vector is satisfied:

$$\frac{dV(k+1)}{dt} = \frac{\partial V(k+1)}{\partial i_{dq}^{k+1}}(Ai_{dq}^{k+1} + Bu_{dq}^{k+1} + E) \leq 0. \tag{23}$$

The proofs of Lemma 2 and Lemma 3 are introduced in the Appendix A.

In order to minimize the ripple of the PMSM current and torque, the Lyapunov function is expected to keep $V = 0$ and $dV(k+1)/dt = 0$ in a steady state. The duty cycle of each voltage vector satisfying the desired requirement is calculated as follows:

$$T_{duty,k+1}^{j} = \begin{cases} T\sigma\,, & \frac{dV(K+1)}{dt} > 0 \\ 0\,, & \frac{dV(K+1)}{dt} = 0 \\ \frac{-V(K+1)}{dV(k+1)/dt}\,, & \frac{dV(K+1)}{dt} < 0 \end{cases}, \tag{24}$$

Here, $T_\sigma \ll T_s$ is a time constant. The voltage vector with $dV(k+1)/dt > 0$ has to be taken into account in the FCS-MPC, and then the current reaches the limit, T_σ is introduced, and the current constraint plays a major role in the cost function. When there is disturbance in the PMSM system, such as machine parameter drift and voltage error in the inverter, we also need to select a smaller T_σ to compromise the disturbance at this point. When the PMSM model and the prediction are accurate, $T_\sigma = 0$ is the ideal value for the ripple reduction. All voltage vectors $u_{dq,k+1}^{j}$ with $dV(k+1)/dt < 0$ are candidates to minimize the cost function.

3.3. Finite Control Set Model Prediction

Finally, the optimum output voltage which can be realized by the SVPWM is obtained by Equation (23). Only when voltage $u_{dq,k+1}^{j}$ acts on the PMSM will it affect the torque, that is, to ensure the implementation of MTPA and $dV(k+1)/dt = 0$. It is noteworthy that setting the optimal duty cycle T_{duty}^{k+1} equal to T_s in the transient state of the proposed control strategy can minimize the cost function. Therefore, under the same sampling frequency, the torque response of the new strategy is as fast as that of the traditional FCS-MPC. In steady state, $dV(k+1)/dt = 0$ can be maintained. Therefore, current ripple and torque ripple can be minimized. Figure 5 is a block diagram of the predictive direct torque control algorithm of the FCS-MPC proposed in this paper.

$$\begin{bmatrix} u_d^{k+1} \\ u_q^{k+1} \end{bmatrix} = (1 - T_{duty}^{k+1})\begin{vmatrix} Ri_d^{k+1} - \omega L_q i_q^{k+1} \\ Ri_q^{k+1} - \omega L_d i_d^{k+1} + \omega\psi_F \end{vmatrix} + T_{duty}^{k+1}\begin{vmatrix} u_d^{j,k+1} \\ u_q^{j,k+1} \end{vmatrix} \tag{25}$$

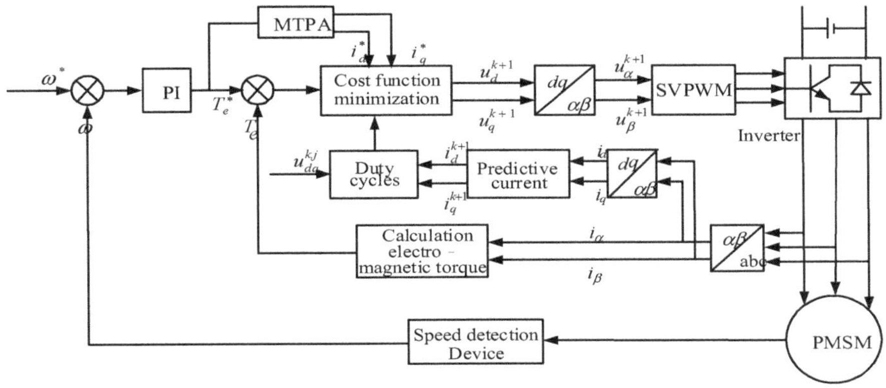

Figure 5. Block diagram of the proposed MPC.

4. Simulation Results

In order to investigate the effectiveness and feasibility of the proposed method, the FCS-MPC of the surface-mounted PMSM based on MATLAB/Simulink software was built. Then, it was compared with that of the classical FCS-MPC method. The parameter settings are shown in Table 3.

Table 3. Surface-mounted permanent magnet synchronous motor (SPMSM) parameters for simulation.

Variable	Parameter	Value
P_N	Rated power	6.5 kW
n_N	Rated speed	1500 rpm
U_N	Rated voltage	306 V
I_N	Rated current	12.3 A
R	Stator resistance	1.01 Ω
L	L_d, Lq	15 mH
p	Number of pairs of poles	4
J	Rotary inertia	0.01535 kg·m^2
ψ_f	Flux linkage of the rotor	0.175 Wb

4.1. Steady-State Operation

Figure 6 shows the simulation results of traditional FCS-MPC and improved FCS-MPC under steady-state conditions. The reference value of motor speed is set to 500 rpm, and the load torque is 20 N·m. As can be seen from the figure, compared with conventional FCS-MPC, the torque ripple of the improved method is significantly reduced. In Figure 6, the average values of the torque ripples of proposed (0.9 N·m) are smaller compared to those of conventional FCS-MPC (3 N·m) under the same conditions. It is noted that, in this paper, torque ripples are calculated by the following equation [22]:

$$T_{rip} = \sqrt{\frac{1}{N}\sum_{i=1}^{N}(T_e - T_{ave})^2}, \tag{26}$$

where N is the number of samples and T_{ave} is the average value of the torque.

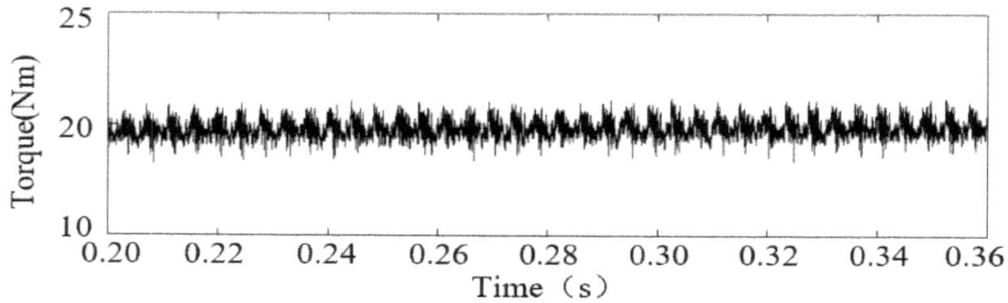

(**a**) Conventional finite control set MPC (FCS-MPC)

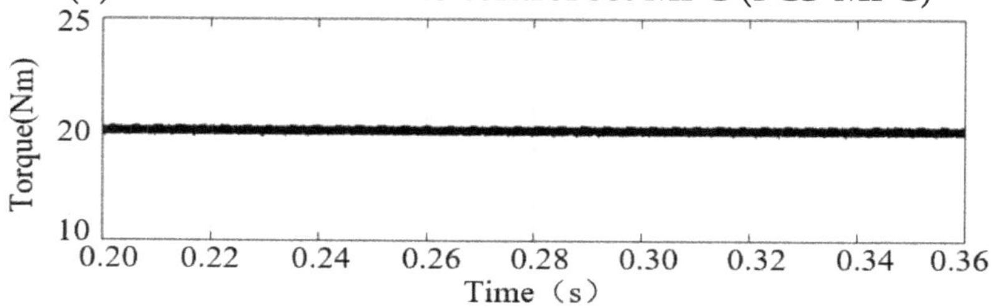

(**b**) Proposed FCS-MPC

Figure 6. Torque steady-state response simulation.

4.2. Dynamic Response

In order to compare the torque dynamic, a magnified view of the torque response is shown in Figure 7. From Figure 7, we can see that the torque suddenly increases from 10 to 20 N·m in only 0.003 s, and in the process of torque mutation, the conventional FCS-MPC method has obvious overshoot. Compared with the conventional FCS-MPC, the improved method has faster dynamic response because the motor adopts optimized duty cycle modulation in the whole control cycle.

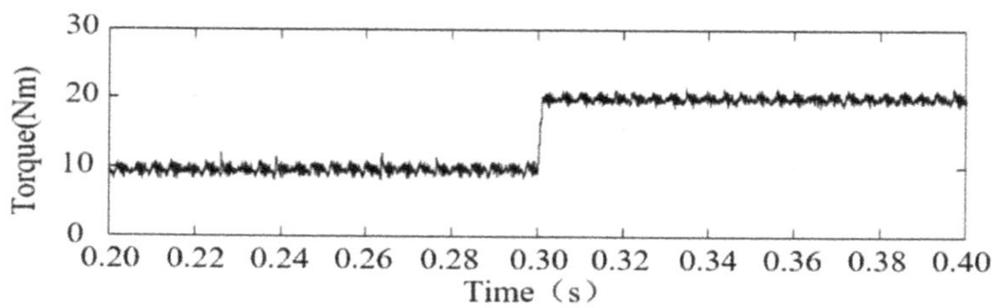

(**a**) Conventional FCS-MPC

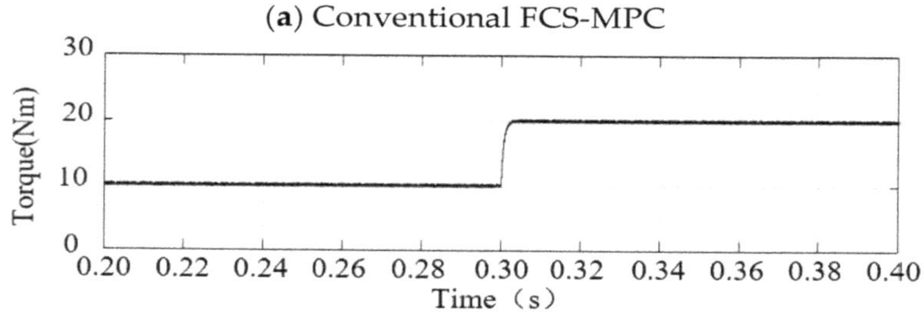

(**b**) Proposed FCS-MPC

Figure 7. Torque dynamic simulation at $t = 0.3$ s.

4.3. Motor Parameter Robustness

As we all know, a temperature rise of the motor will cause the stator internal resistance to increase. The influence of internal resistance on the torque is discussed in this paper. The torque waveforms of the two control methods in the condition of rated stator resistance (the blue solid line) and 1.5-times the rated value (the red dashed line) are given in Figure 8a,b, respectively. It is noticeable that slight torque ripples occur under the conventional FCS-MPC method when the resistance increases, while the proposed FCS-MPC control has better robustness in which the torque is insensitive to the resistance change. However, when entering a steady state, the two control systems are not affected by internal resistance changes.

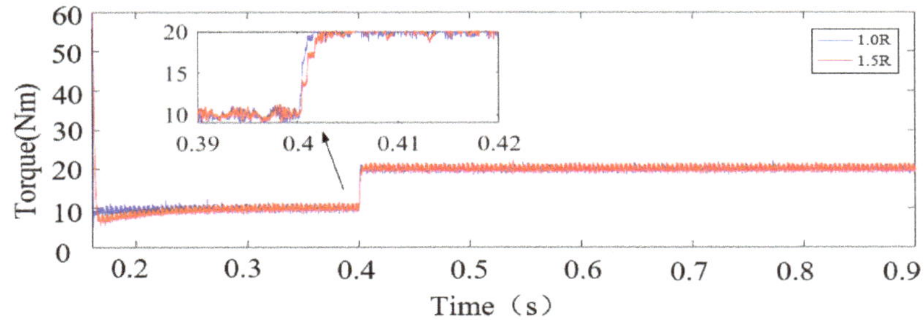

(**a**) Traditional FCS-MPC

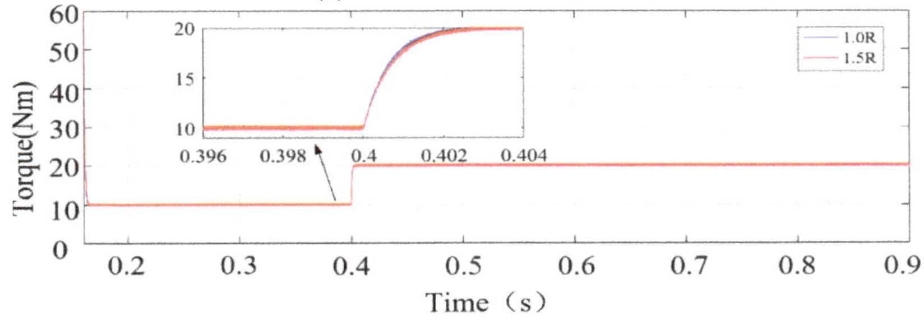

(**b**) Proposed FCS-MPC

Figure 8. Torque ripple simulation when the resistance is changed.

Almost all permanent magnet devices, including permanent magnet motors, will exhibit the demagnetization phenomenon in varying degrees due to factors such as the working environment and usage time. The effect of changes in the motor flux on torque is also discussed. Figure 9a,b shows the torque waveforms of the two control methods when the rotor flux is 0.9-times the rated value, equal to the rated value, and 1.1-times the rated value, to simulate permanent magnet motors that have been used for a long time, used normally, and newly manufactured. As shown in Figure 9, the blue point line is 0.9-times the rated flux, the red solid line is the rated flux, and the green dotted line is 1.5-times the rated flux. Evidently, the conventional FCS-MPC for the change of flux is the change of the maximum torque. However, the improved predictive control is insensitive to the flux change.

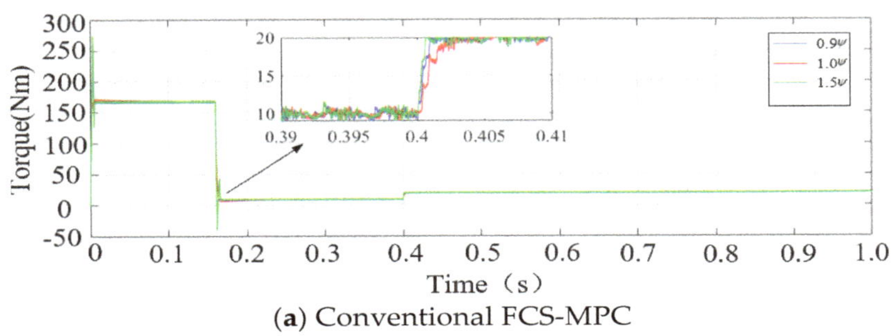

(**a**) Conventional FCS-MPC

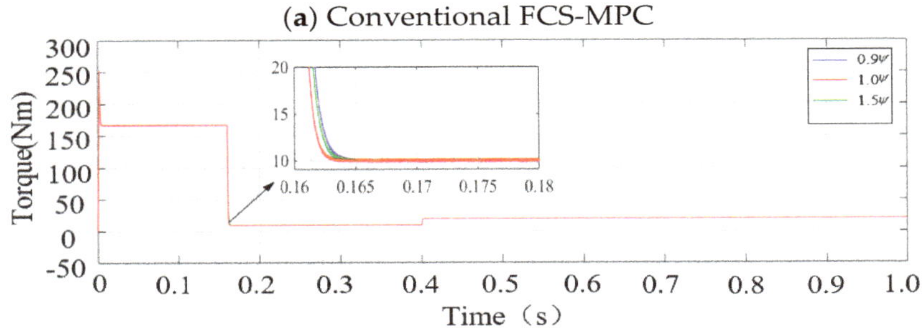

(**b**) Proposed FCS-MPC

Figure 9. Torque ripple simulation when the flux is changed.

For a mechanical load, the oscillation of torque is very bad for the shaft and load of the motor. Therefore, it can be seen from the simulation waveforms of the above two sections that the robustness of predictive control to parameter changes is better than that of conventional FCS-MPC. In theory, the switch meter of direct torque control is fixed in advance and cannot be adjusted according to the running state of the motor. Therefore, the robustness of direct predictive control to parameter changes is better than that of direct torque control, which is also consistent with the previous prediction results.

5. Experimental Test

The experiment aimed to verify the torque performance of this method with surface-mounted permanent magnet synchronous motor (SPMSM) and interior permanent magnet synchronous motor (IPMSM). The experimental platform as shown in Figure 10 was built by using dSPACE1102. The load of the PMSM is provided by the load machine.

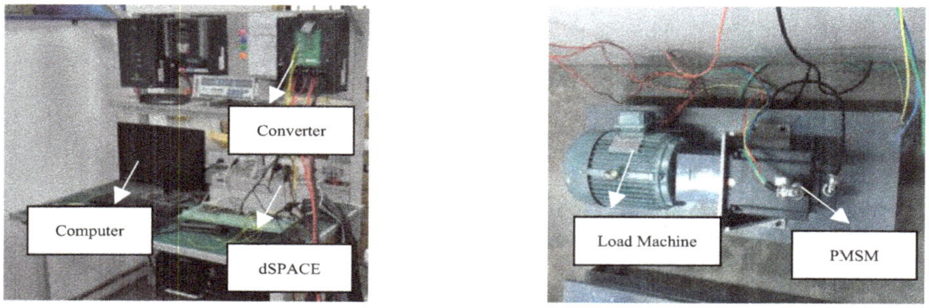

Figure 10. The controlled 6.5 kW PMSM drive test rig.

5.1. Experiment on SPMSM

The parameters of the SPMSM are shown in Table 3. The steady-state performance test results of conventional FCS-MPC and proposed FCS-MPC are shown in Figure 11. In the experiment, the reference load torque was 20 N·m, and the results show that the proposed FCS-MPC significantly reduced the torque ripple under the same average switching frequency, while the conventional FCS-MPC had a lot of torque ripples (6 N·m) compared to proposed FCS-MPC (2.8 N·m). This is consistent with the simulation results.

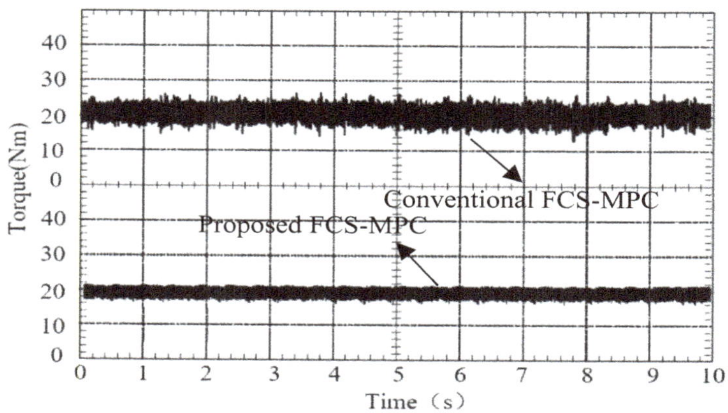

Figure 11. Torque steady-state responses (SPMSM).

In order to test the dynamic performance of the proposed FCS-MPC method, a load torque of 25 N·m was suddenly added when the steady speed of the motor was 500 rpm. Figure 12 shows that the response times of proposed FCS-MPC and conventional FCS-MPC are approximately the same; however, the proposed FCS-MPC strategy torque ripple is smaller than that of the conventional FCS-MPC. Meanwhile, the proposed FCS-MPC method enables the motor to quickly recover to the reference speed value compared to conventional FCS-MPC.

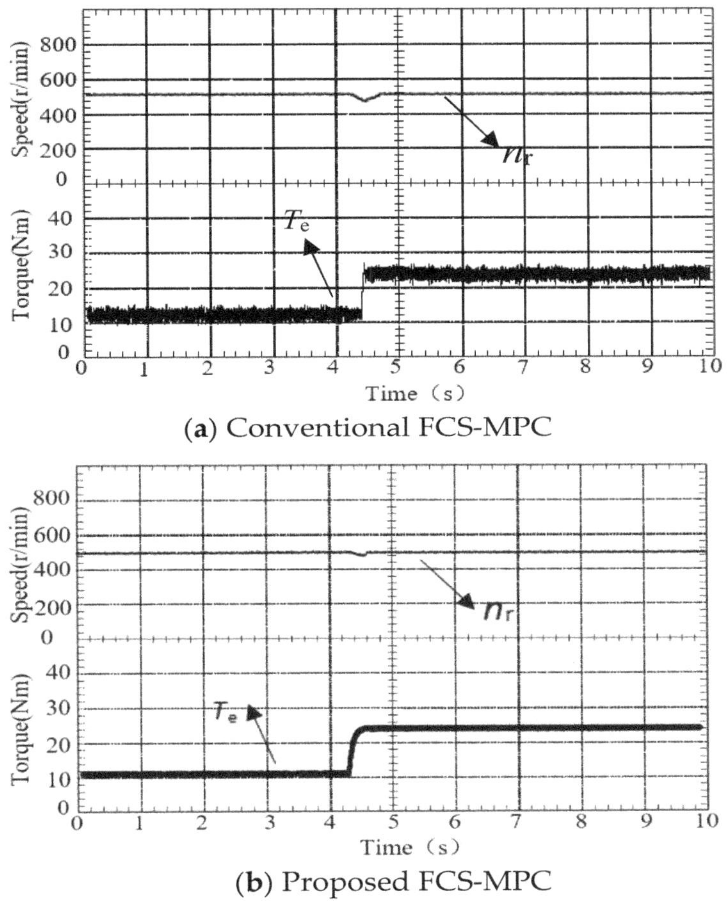

Figure 12. Speed dynamic and torque dynamic responses (SPMSM).

5.2. Experiment on IPMSM

Some parameters of the experiment motor are shown in Table 4. The following experiments were carried out in three aspects: torque steady-state responses, speed dynamics, and torque dynamics responses.

Table 4. Interior permanent magnet synchronous motor (IPMSM) parameters for simulation.

Variable	Parameter	Value
P_N	Rated power	5.2 kW
n_N	Rated speed	1500 rpm
U_N	Rated voltage	380 V
R	Stator resistance	0.39 Ω
L	Ld	0.88 mH
L	Lq	1.62 mH
p	Number of pairs of poles	4
ψ_f	Flux linkage of the rotor	0.163 Wb

Figure 13 shows electromagnetic torque waveforms of two methods, with the load torque reference value set to 30 N·m. As can be seen from Figure 13, the method of proposed FCS-MPC can significantly reduce torque ripple by comparing torque waveforms; the average torque ripple of proposed method is only 3 N·m while that of conventional FCS-MPC is 7 N·m.

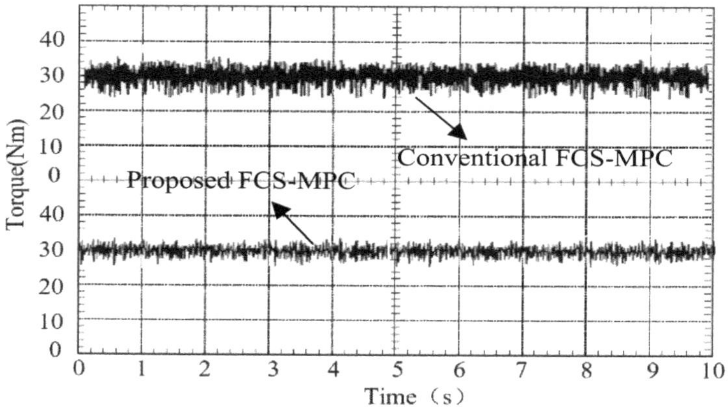

Figure 13. Torque steady-state responses (IPMSM).

Figure 14 shows the waveforms of speed and electromagnetic torque of two methods when the starting moment of the motor and step change of load torque with load torque increasing from 10 to 20 N·m. It can be seen from Figure 14 that at the moment of starting the motor, the speed and electromagnetic torque of the conventional and proposed FCS-MPC methods increase sharply, reaching the given reference value quickly; however, compared with the conventional FCS-MPC, the strategy of proposed FCS-MPC has smaller torque ripple. When the load torque changes abruptly, the proposed FCS-MPC can also track the change of torque quickly, and the corresponding speed is faster than conventional FCS-MPC.

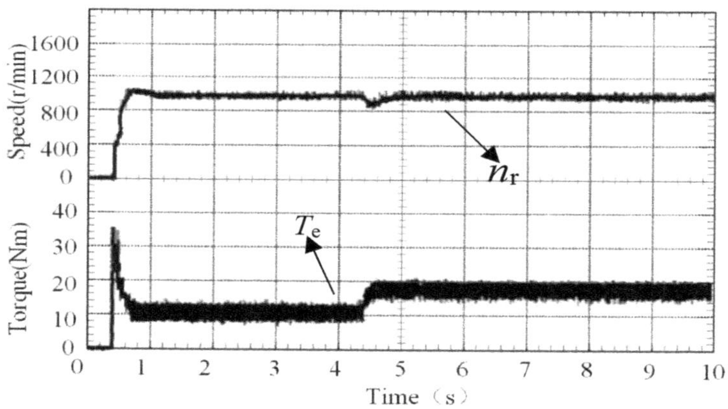

(**a**) Conventional FCS-MPC

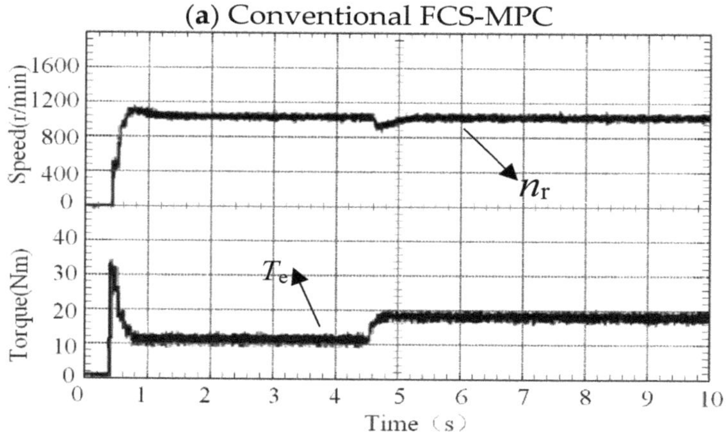

(**b**) Proposed FCS-MPC

Figure 14. Speed dynamic and torque dynamic responses (IPMSM).

Through the analysis of the above experiments, compared with the conventional FCS-MPC control system, the proposed FCS-MPC in this paper effectively reduces the torque ripple and improves the following performance of the motor torque.

6. Summary

In this paper, a new scheme of direct torque control for PMSM based on a finite control set (FCS) model is proposed. The eight voltage vectors of the two-level converter are utilized as an FCS for the torque prediction of the PMSM. The cost function is used to estimate the duty cycle of each voltage vector. Thus, the optimal voltage vector can be obtained from eight voltage vectors and their duty cycles. Compared with the classical FCS-MPC method, the proposed method has smaller torque ripple and excellent dynamic performance.

Author Contributions: Conceptualization and methodology, G.B. and W.Q.; software, T.H.; validation, G.B. and W.Q.; writing—original draft preparation, T.H.; writing—review and editing, G.B. and W.Q.; funding acquisition, G.B. All authors have read and agreed to the published version of the manuscript.

Appendix A

Proof of Lemma 1. Finding partial derivatives of $f(i_d, i_q)$ and $T_e = 1.5pi_q[\psi_f + (L_d - L_q)i_d]$ based on current vectors i_d and i_q, respectively:

$$\frac{\partial f(i_d, i_q)}{\partial i_{dq}} = [1 + \frac{2(L_d - L_q)}{\psi_f} i_d, -\frac{2(L_d - L_q)}{\psi_f} i_q] \tag{A1}$$

$$\frac{\partial T_e}{\partial i_{dq}} = [1.5p(L_d - L_q)i_q, 1.5p(\psi_f + (L_d - L_q)i_q)] \tag{A2}$$

Considering the torque T_e as constant, the trajectory of the operating point moves along the direction of the α which fulfills $(T_e/i_{dq}) \cdot \alpha = 0$. The vector α can be described as:

$$\alpha = \varepsilon[-1.5p(\psi_f + (L_d - L_q)i_q), 1.5p(L_d - L_q)i_q)]^T, \tag{A3}$$

where ε is a positive real number.

$$\frac{\partial f(i_d, i_q)}{\partial i_{dq}} \cdot \alpha = 1.5p\varepsilon\left[\psi_F + 3(L_d - L_q)i_d\right] - \frac{3p\varepsilon(L_d - L_q)^2}{\psi_f}(i_d^2 + i_q^2) < 0. \tag{A4}$$

Therefore, the function $f(i_d, i_q)$ is a strict monotonic function along the constant torque curve. □

Proof of Lemma 2. The Lyapunov function is derived as:

$$\frac{dV(k+1)}{dt} = \frac{\partial V(k+1)}{\partial i_{dq}^{k+1}} \cdot \frac{di_{dq}^{k+1}}{dt}. \tag{A5}$$

Therefore, the trajectory of the current makes the value of Lyapunov function decrease and there exists a $\triangle i_{dq}^{k+1}$ which fulfills:

$$\frac{\partial V(k+1)}{di_{dq}^{k+1}} \Delta i_{dq}^{k+1} \leq 0. \tag{A6}$$

Therefore, there exists a current derivative:

$$\frac{di_{dq}^{k+1}}{dt} = \mu \cdot \Delta i_{dq}^{k+1}, \tag{A7}$$

so that:

$$\frac{dV(k+1)}{dt} = \frac{\partial V(k+1)}{\partial i_{dq}^{k+1}} \cdot \mu \cdot \Delta i_{dq}^{k+1} \leq 0. \tag{A8}$$

Here, μ can be a very small positive constant. □

Proof of Lemma 3. Any reference vector $u_{dq}^{*}(x)$ within the region of feasibility [having a magnitude of less than $(2/3)U_{dc}$] is contained within one of the six nonzero switching sectors of width $(\pi/3)$ with vertices (v_d^0, v_q^0), (v_d^i, v_q^i), and (v_d^j, v_q^j), where $i, j \in \{1, \ldots, 6\}$ are the nonzero switching states to the left and right of the reference vector and (v_d^0, v_q^0) is one of the two zero vectors. Containment within a switching sector ensures the existence of coefficients γ and η satisfying $\gamma, \eta \geq 0$ and $\gamma + \eta \leq 1$ such that the reference vector is expressible as a convex combination of the realizable inputs, given by:

$$u_{dq}^{*}(x) = \gamma v_{dq}^{i} + \eta v_{dq}^{j} + (1 - \gamma - \eta) v_{dq}^{0}. \tag{A9}$$

Plugging Equation (A10) into Equation (21) and noting that the system is control affine, we see that [23]:

$$\begin{aligned}\frac{dV(k+1)}{dt} &= \frac{\partial V(k+1)}{\partial i_{dq}^{k+1}}(Ai_{dq}^{k+1} + Bu_{dq}^{*} + E) = \frac{\partial V(k+1)}{\partial i_{dq}^{k+1}}(Ai_{dq}^{k+1} + B(\gamma v_{dq}^{i} + \eta v_{dq}^{j} + (1-\gamma-\eta)v_{dq}^{0}) + E) \\ &= \gamma \frac{\partial V(k+1)}{\partial i_{dq}^{k+1}}(Ai_{dq}^{k+1} + Bv_{dq}^{i}) + \eta \frac{\partial V(k+1)}{\partial i_{dq}^{k+1}}(Ai_{dq}^{k+1} + Bv_{dq}^{j}) \\ &+ (1-\gamma-\eta)\frac{\partial V(k+1)}{\partial i_{dq}^{k+1}}(Ai_{dq}^{k+1} + Bv_{dq}^{0}) + \frac{\partial V(k+1)}{\partial i_{dq}^{k+1}}E\end{aligned} \tag{A10}$$

Because γ, η, and $(1 - \gamma - \eta)$ are all nonnegative, the following inequalities hold [12]:

$$\begin{cases} \frac{\partial V(k+1)}{\partial i_{dq}^{k+1}}(Ai_{dq}^{k+1} + Bv_{dq}^{i}) \leq 0 \\ \frac{\partial V(k+1)}{\partial i_{dq}^{k+1}}(Ai_{dq}^{k+1} + Bv_{dq}^{j}) \leq 0 \\ \frac{\partial V(k+1)}{\partial i_{dq}^{k+1}}(Ai_{dq}^{k+1} + Bv_{dq}^{0}) \leq 0 \end{cases}, \tag{A11}$$

which completes the proof in Lemma 3. The theorem also guarantees the stability of the proposed control scheme. □

References

1. Yang, J.F.; Hu, Y.W. Optimal direct torque control of permanent magnet synchronous motor. *Proc. Chin. Soc. Electr. Eng.* **2011**, *31*, 109–115.
2. Depenbrock, M. Direct self-control (DSC) of inverter fed induction machine. *IEEE Trans. Power Electron.* **1988**, *3*, 420–429. [CrossRef]
3. Uddin, M.N.; Zou, H.; Azevedo, F. Online loss-minimization-based adaptive flux observer for direct torque and flux control of PMSM drive. *IEEE Trans. Ind. Appl.* **2016**, *52*, 425–431. [CrossRef]
4. Swierczynski, D.; Kazmierkowski, M.P. Direct torque control of permanent magnet synchronous motor (PMSM) using space vector modulation (DTC-SVM)-simulation and experimental results. In Proceedings of the Conference of the IEEE 2002 28th Annual Conference of the Industrial Electronics Society, Sevilla, Spain, 5–8 November 2002.

5. Tang, L.; Zhong, L.; Rahman, M.; Hu, Y. A novel direct torque controlled interior permanent magnet synchronous machine drive with low ripple in flux and torque and fixed switching frequency. *IEEE Trans. Power Electron.* **2004**, *19*, 346–354. [CrossRef]
6. Cho, Y.; Lee, K.B.; Song, J.H.; Lee, Y. Torque-ripple minimization and fast dynamic scheme for torque predictive control of permanent magnet synchronous motors. *IEEE Trans. Power Electron.* **2015**, *30*, 2182–2190. [CrossRef]
7. Vafaie, M.H.; Dehkordi, B.M.; Moallem, P.; Kiyoumarsi, A. A new predictive direct torque control method for improving both steady-state and transient-state operations of the PMSM. *IEEE Trans. Power Electron.* **2016**, *31*, 3738–3753. [CrossRef]
8. Zhang, Y.; Zhu, J. Direct torque control of permanent magnet synchronous motor with reduced torque ripple and commutation frequency. *IEEE Trans. Power Electron.* **2011**, *26*, 235–248. [CrossRef]
9. Rodriguez, J.; Kazmierkowski, M.P.; Espinoza, J.R.; Zanchetta, P.; Abu-Rub, H.; Young, H.A.; Rojas, C.A. State of the art of finite control set model predictive control in power electronics. *IEEE Trans. Ind. Inform.* **2012**, *9*, 1003–1016. [CrossRef]
10. Rodriguez, J.; Pontt, J.; Silva, C.; Salgado, M.; Rees, S.; Ammann, U.; Lezana, P.; Huerta, R.; Cortes, P. Predictive control of three-phase inverter. *Electron. Lett.* **2004**, *40*, 561–562. [CrossRef]
11. Preindl, M.; Bolognani, S. Model predictive direct torque control with finite control set for PMSM drive systems, part 1: Maximum torque per ampere operation. *IEEE Trans. Ind. Inform.* **2013**, *9*, 1912–1921. [CrossRef]
12. Liu, Q.; Hameyer, K. A finite control set model predictive direct torque control for the PMSM with MTPA operation and torque ripple minimization. In Proceedings of the 2015 IEEE International Electric Machines & Drives Conference (IEMDC), Coeur d'Alene, ID, USA, 10–13 May 2015; pp. 804–810.
13. Geyer, T.; Papafotiou, G.; Morari, M. Model predictive direct torque control; part I: Concept, algorithm, and analysis. *IEEE Trans. Ind. Electron.* **2009**, *56*, 1894–1905. [CrossRef]
14. Preindl, M.; Schaltz, E.; Thogersen, P. Switching frequency reduction using model predictive direct current control for high-power voltage source inverters. *IEEE Trans. Ind. Electron.* **2010**, *58*, 2826–2835. [CrossRef]
15. Mayne, D.Q.; Rawlings, J.B.; Rao, C.V.; Scokaert, P.O.M. Constrained model predictive control: Stability and optimality. *Automatica* **2000**, *36*, 789–814. [CrossRef]
16. Alam, K.S.; Akter, M.P.; Xiao, D.; Zhang, D.; Rahman, M.F. Asymptotically Stable Predictive Control of a Grid-Connected Converter Based on Discrete Space Vector Modulation. *IEEE Trans. Ind. Inform.* **2018**, *15*, 2775–2785. [CrossRef]
17. Aguilera, R.P.; Quevedo, D.E. On stability and performance of finite control set MPC for power converters. In Proceedings of the Workshop Predictive Control Electrical Drives Power Electronics, Munich, Germany, 14–15 October 2011; pp. 55–62.
18. Lee, J.S.; Lorenz, R.D.; Valenzuela, M.A. Time-optimal and loss minimizing deadbeat-direct torque and flux control for interior permanent magnet synchronous machines. *IEEE Trans. Ind. Appl.* **2014**, *50*, 1880–1890. [CrossRef]
19. Xu, W.; Lorenz, R.D. Dynamic loss minimization using improved deadbeat-direct torque and flux control for interior permanent-magnet synchronous machines. *IEEE Trans. Ind. Appl.* **2014**, *50*, 1053–1065. [CrossRef]
20. Chen, Y.; Sun, D.; Lin, B.; Ching, T.W.; Li, W. Dead-beat direct torque and flux control based on sliding-mode stator flux observer for PMSM in electric vehicles. In Proceedings of the 41st Annual Conference of the IEEE Industrial Electronics Society, Yokohama, Japan, 9–12 November 2015; pp. 2270–2275.
21. Barcaro, M.; Fornasiero, E.; Bianchi, N.; Bolognani, S. Design procedure of IPM motor drive for railway traction. In Proceedings of the IEEE International Electric Machines & Drives Conference (IEMDC), Niagara Falls, ON, Canada, 15–18 May 2011; pp. 983–988.
22. Bianchi, N.; Bolognani, S.; Ruzojcic, B. Design of a 1000 HP permanent magnet synchronous motor for ship propulsion. In Proceedings of the 13th European Conference on Power Electronics and Applications (EPE'09), Barcelona, Spain, 8–10 September 2009; pp. 1–8.
23. Prior, G.; Krstic, M. Quantized-input control Lyapunov approach for permanent magnet synchronous motor drives. *IEEE Trans. Control Syst. Technol.* **2013**, *21*, 1784–1794. [CrossRef]

8

Adaptive Robust Control System for Axial Flux Permanent Magnet Synchronous Motor of Electric Medium Bus Based on Torque Optimal Distribution Method

Shuang Wang, Jianfei Zhao *, Tingzhang Liu and Minqi Hua

School of Mechatronic Engineering and Automation, Shanghai University, Baoshan District, Shanghai 200444, China; wang-shuang@shu.edu.cn (S.W.); liutzhcom@oa.shu.edu.cn (T.L.); huaminqi@shu.edu.cn (M.H.)

* Correspondence: jfzhao@shu.edu.cn.

Abstract: In this paper, an adaptive robust drive control system for an axial flux permanent magnet synchronous motor of an electric medium-sized bus based on the optimal torque distribution method is studied. The drive control system is mainly divided into two parts. First, a torque distribution method is proposed. The optimal torque distribution method based on particle swarm optimization algorithm is used to increase the high efficiency interval of the system and apply it to the energy feedback braking. Secondly, in order to reduce the nonlinear disturbance of the system and improve the accuracy of the unified control, this paper models and studies the vector system based on adaptive robust control. Finally, the whole drive control system is modeled, simulated and experimented. The simulation and experimental results show that the torque distribution method proposed in this paper can effectively increase the high-efficiency running time of the electric medium bus, and improve the shortcomings of insufficient mileage of the electric medium-sized bus. The use of a current controller based on adaptive robust control improves the control accuracy of the drive system and can effectively suppress the disturbances generated by it.

Keywords: adaptive robust control; AFPMSM; energy feedback; particle swarm optimization; torque optimal distribution method

---

1. Introduction

The passenger capacity of medium bus is generally 9–20, which is suitable for small and medium-size cities. With the increasing awareness of greenhouse gas emissions, the emergence of electric buses can meet the call for energy conservation and emission reduction [1]. Compared with the internal combustion engine powered medium bus, the electric medium bus have many advantages. For example, they have low vibration noise, simple structure, high power transmission efficiency, easy vehicle layout, and excellent power performance [2,3]. The selection of pure electric medium bus drive motors must meet the vehicle's dynamic requirements, such as maximum speed, acceleration performance, and maximum grade [4]. Among a variety of vehicle drive motors, permanent magnet synchronous motors (PMSMs) have many applications [5,6]. Among them, the axial flux permanent magnet synchronous motors (AFPMSMs) have the advantages of low speed, large torque and high energy density, and are more suitable for use in an electric medium bus with a larger passenger capacity. Double stator-single rotor AFPMSM (two-disc AFPMSM) has better heat dissipation and larger rated torque. Therefore, the two-disc AFPMSM is selected as the driving motor for electric medium bus in this paper.

The two-disc AFPMSM can be equivalent to two PMSMs connected coaxially, and for electric medium-sized buses, there is only one given torque, so this paper involves the problem of multi-motor torque distribution. At present, the commonly used torque distribution method is mainly applied to distributed drive vehicles. Compared with the conventional central direct drive electric vehicles, the drive motors of the various wheels of the distributed drive electric vehicle can be independently controlled, and the torque of each wheel can be distributed in any proportion within its capability range. The energy can be controlled by properly distributing the wheel torque so that the motor works as much as possible in the high efficiency range. The author of [7] proposed a multi-objective optimization method that considers system efficiency and safety for torque distribution. The authors of [8] mainly use torque distribution to enable micro electric vehicles to improve powertrain efficiency. The authors of [9] used the optimal vehicle state estimation method for directional tire torque distribution. In this paper, a torque optimal allocation method is proposed for the purpose of efficiency optimization. The particle swarm optimization algorithm is used to optimize the torque distribution mathematical model to obtain the optimal torque distribution solution. In addition, regenerative braking is one of the most effective ways to extend the durability of electric vehicles [10–12]. In order to further increase the cruising range, this paper applies the previously described optimal torque distribution method to the braking situation.

In addition, due to the large number of passengers in medium bus, some researchers have studied the safety and stability of driving. Authors of [13–16] studied the safety-structure from the structure of the medium bus, among which authors of [13,14] focused on studying the strength of conventional bus structures under operating conditions, authors of [15,16] studied the crashworthiness under rollover accident.

In addition, the research on control systems is mainly divided into two categories, motor design and optimization and motor control. The research on AFPMSM mainly focuses on the optimization of the motor model. The authors of [17] used the combined solution of Maxwell's equations and magnetic equivalent circuits to model the AFPMSM analytically. The authors of [18] used an auxiliary multi-objective optimization algorithm to optimize the design of AFPMSM with dual rotor and single stator. This paper studies the anti-interference and current tracking capabilities of the driving system of medium-sized buses from the perspective of drive control. During the operation of the electric medium bus, due to the complicated operating environment, it will encounter various nonlinear disturbances. Adaptive robust control is used to overcome the effects of nonlinear disturbances, thereby improving tracking accuracy of the current loop. In the study of adaptive robust control, the authors of [19] used adaptive synthesis robust control strategies based on μ synthesis to resist the interference of high-frequency dynamic problems generated by the motor structure mode on linear motor control. The authors of [20] used neural networks to learn adaptive robust controllers to resist interference from unknown factors. The author of [21] used an adaptive robust controller based on extended disturbance observer to improve the control accuracy of linear motors. In this paper, adaptive robust control is applied to the drive control system of AFPMSM for anti-disturbance control.

This paper mainly studies the drive control system of electric medium bus. From the above, the drive control system is mainly divided into two parts. The first part mainly studies the torque distribution method, with the system's highest working efficiency as the distribution target. The second part mainly studies the motor control part. In order to reduce the waveform ripple and improve the system control accuracy, this paper models and studies the adaptive robust control vector system. Finally, the above methods are simulated and the motor experiments and loading experiments are performed, and the results are summarized.

2. Two-Disc AFPMSM Mathematical Model

Compared with the traditional AFPMSM, the AFPMSMs with multi-disc structure improve the overall efficiency of the motor by adjusting the number of stators and rotors running [22]. Therefore, this paper takes the AFPMSM of double-stator-single-rotor structure as the research object, and establishes its mathematical model as the theoretical basis for deducing its control strategy. The structure of the dual-stator-single-rotor AFPMSM (also called double-disc AFPMSM) is shown in Figure 1.

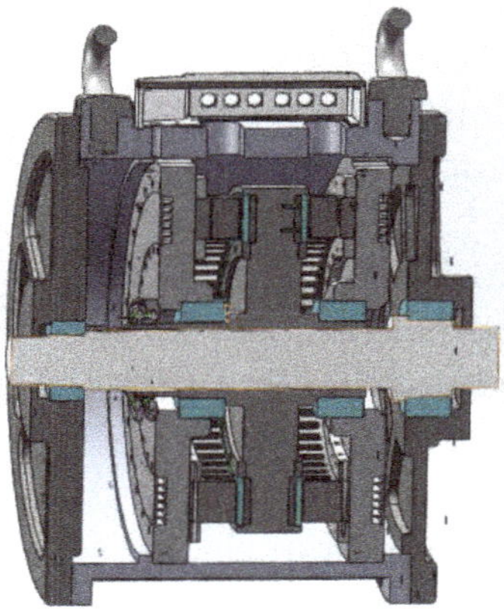

Figure 1. Internal view of double-disc axial flux permanent magnet synchronous motor (AFPMSM).

In order to distinguish the two stators of the AFPMSM, they are respectively defined as the stator 1 and the stator 2. The simplified PMSMs are the motor 1 and the motor 2. For the motor 1 and the motor 2, a d-q axis rotating coordinate system is established, which is d_1-q_1 and d_2-q_2, and the coordinate system rotation speeds are ω_1 and ω_2, and the rotation directions are the same. Since the double disc AFPMSM shares one rotor and is coaxially connected, two mathematical models of the d-q axis rotating coordinate system can be established at the same time. The mathematical model of the AFPMSM is as follows.

Voltage equation:

$$\begin{cases} u_{d1} = R_{s1}i_{d1} + L_{d1}\frac{di_{d1}}{dt} - \omega_1 L_{q1}i_{q1} \\ u_{q1} = R_{s1}i_{q1} + L_{q1}\frac{di_{q1}}{dt} + \omega_1 L_{d1}i_{d1} + \omega_1\psi_f \\ u_{d2} = R_{s2}i_{d2} + L_{d2}\frac{di_{d2}}{dt} - \omega_2 L_{q2}i_{q2} \\ u_{q2} = R_{s2}i_{q2} + L_{q2}\frac{di_{q2}}{dt} + \omega_2 L_{d2}i_{d2} + \omega_2\psi_f \end{cases}, \tag{1}$$

Magnetic chain equation:

$$\begin{cases} \psi_{d1} = L_{d1}i_{d1} + \psi_f \\ \psi_{q1} = L_{q1}i_{q1} \\ \psi_{d2} = L_{d2}i_{d2} + \psi_f \\ \psi_{q2} = L_{q2}i_{q2} \end{cases}, \tag{2}$$

Torque equation:

$$\begin{cases} T_{e1} = \frac{3}{2}p[\psi_f i_{q1} + (L_{d1} - L_{q1})i_{d1}i_{q1}] \\ T_{e2} = \frac{3}{2}p[\psi_f i_{q2} + (L_{d2} - L_{q2})i_{d2}i_{q2}] \end{cases}, \tag{3}$$

Equation of motion:

$$\begin{cases} T_{e1} - T_{L1} - B\omega_1 = \frac{J}{p}\frac{d\omega_1}{dt} \\ T_{e2} - T_{L2} - B\omega_2 = \frac{J}{p}\frac{d\omega_2}{dt} \end{cases}, \tag{4}$$

In Equations (1)–(4), R_{s1} and R_{s2} are two stator resistances respectively, and u_{d1}, u_{q1}, u_{d2}, and u_{q2} are d-q axis components of the winding voltage vectors of the stator 1 and the stator 2, i_{d1}, i_{q1}, and i_{d2}. i_{q2} is the d-q axis component of the winding current vectors of the stator 1 and stator 2, L_{d1}, L_{q1}, L_{d2}, L_{q2} are the d-q axis components of the winding inductances of the stator 1 and stator 2, ψ_{d1}, ψ_{q1}, ψ_{d2}, ψ_{q2} are the d-q axis components of the winding flux of stator 1 and stator 2, T_{e1} and T_{e2} are the electromagnetic torques of the stator 1 and the stator 2, T_{L1} and T_{L2} are the load torques of the stator 1 and the stator 2, J is the moment of inertia, B is the viscosity coefficient, and T_L is the load torque.

Since the two stators share one rotor, it can be approximated that the two motor modules are coaxially connected, so $\omega_{r1} = \omega_{r2} = \omega_r$ can be obtained. According to the conclusion of coaxial connection, two equivalent motors can now be analyzed under the same d-q reference coordinate system. The two stators are structurally identical and symmetrical, therefore the stator resistances R_{S1} and R_{S2} are equal. According to the uniform air gap of the motor, it can be obtained that the direct-axis inductance and the cross-axis inductance of the two motors are equal. This article uses a hidden-pole motor, so it is also concluded that the inductance of the AC and DC axes is equal. At this time, the electromagnetic torque equation can be rewritten as:

$$T_e = T_{e1} + T_{e2} = \frac{3}{2} p \psi_f \left(i_{q1} + i_{q2}\right), \tag{5}$$

The equation of motion is:

$$T_e - T_L - B\omega_r = \frac{J}{p}\frac{d\omega_r}{dt}, \tag{6}$$

3. Torque Optimal Distribution Method

3.1. Dual Stator AFPMSM Drive System Topolgy

The traditional electric medium bus has only one motor drive system, and the vehicle manager only corresponds to one motor controller [23], and the two are connected by controller area network (CAN) communication. Unlike conventional two-motor electric vehicle drive systems, the dual-station AFPMSM needs to control two sets of stator windings. Although they are driven by separate inverters, the same motor controller can be used, so that the vehicle manager and the motor controller can be directly connected via controller area network (CAN) communication. In order to study the torque distribution method more conveniently, a torque distributor is added between the vehicle manager and the motor controller, and the torque is distributed to the motor through the torque distributor. The topology of the double-disc AFPMSM drive system is shown in Figure 2.

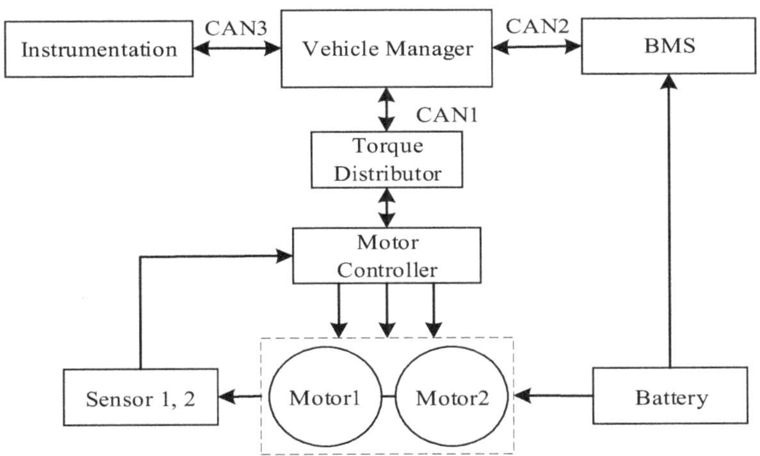

Figure 2. Topology of double-disc AFPMSM drive system for electric medium bus.

The general electric vehicle drive system only has a given torque T_m. The driving motor in this paper can be regarded as two motors after equivalent. Therefore, T_m needs to be allocated. The common method is to evenly distribute torque. In order to save battery power, an optimal torque distribution strategy based on particle swarm optimization is proposed, so that the system can operate in a high efficiency range. The system control block diagram is shown in Figure 3. According to the optimal torque distribution method, the distributed torques of the two motors are obtained, and then the motor control is performed.

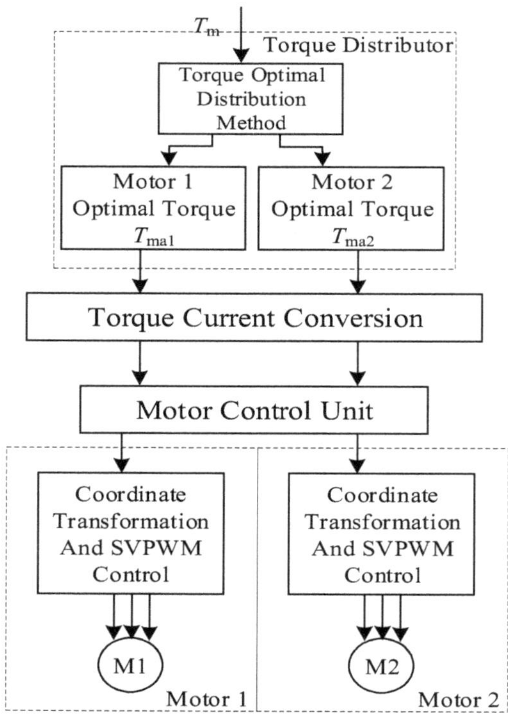

Figure 3. Control block diagram of double-disc AFPMSM drive system for electric medium bus.

3.2. *Optimal Torque Distribution Control Method Based on Particle Swarm Optimization in Driving State*

The torque optimal distribution strategy based on particle swarm optimization (PSO) is modeled as follows. Taking the double-disc AFPMSM for electric medium bus studied in this paper as an example, the total output torque is T, and the range is [0, 500 Nm]. The output torques of the two motor modules are T_1 and T_2, respectively, and the range is [0, 250 Nm], then:

$$T = T_1 + T_2, \tag{7}$$

Assuming that the mechanical angular velocity of the motor is ω, the output power of the two sets of motor modules is $T_1\omega$ and $T_2\omega$, the input power is P_1 and P_2, and the operating efficiency is η_1 and η_2. Let $T_1 = a_1T$, $T_2 = a_2T$ where $a_1 + a_2 = 1$, $a_1, a_2 \in [0, 1]$. Then the input power of the two motor is:

$$P_1 = \frac{T_1\omega}{\eta_1} = \frac{a_1}{\eta_1}T_1\omega, \tag{8}$$

$$P_2 = \frac{T_2\omega}{\eta_2} = \frac{a_2}{\eta_2}T_2\omega, \tag{9}$$

The total output power of the Motor Module is:

$$P_o = T_1\omega + T_2\omega = T\omega, \tag{10}$$

The total input power is:

$$P_i = P_1 + P_2 = (\frac{a_1}{\eta_1} + \frac{a_2}{\eta_2})T\omega, \tag{11}$$

The total efficiency of the motor module is:

$$\eta = \frac{P_o}{P_i} = \frac{T\omega}{(\frac{a_1}{\eta_1} + \frac{a_2}{\eta_2})T\omega} = \frac{1}{\frac{a_1}{\eta_1} + \frac{a_2}{\eta_2}}, \tag{12}$$

Assume:

$$f(a_1) = \frac{a_1}{\eta_1}, f(a_2) = \frac{a_2}{\eta_2}, \tag{13}$$

It can be known from Equation (10) that the output power is constant during the operation of the electric medium bus and it is only necessary to reduce the total input power to improve the system efficiency. If the maximum value of η is required, that is, the minimum value of $f(a_1) + f(a_2)$ is obtained.

Let

$$A = f(a_1) + f(a_2), \tag{14}$$

Then the problem translates into how the two sets of motor modules are assigned torque ratios for a given torque so that the value of A is minimized.

For Equation (14), when one of the torque distribution ratios a_1 or a_2 is determined, the value of A can be determined. However, the speed and torque at a certain moment are not involved in the Equation (14). In the optimization using the particle swarm optimization algorithm, the optimal distribution must be obtained based on the total given torque and speed. Therefore, the three-dimensional model of $f(a_1)$, the total torque command T, and the current rotational speed n can be obtained by data fitting. Since the two sets of motor modules are identical, the efficiency values are the same under different speeds and torques, so only the total torque command T is required to be the x-axis, and the rotational speed n is the y-axis. The value of $f(a_1)$ is calculated as the z-axis for all torque distribution ratios and corresponding efficiencies at different speeds and torques. The three-dimensional model is shown in Figure 4.When the torque distribution system inputs the torque and the rotational speed at any time, any value of $f(a_1)$ will have a certain value of $f(a_2)$ corresponding to it on the z-axis, so that the value of A under all torque distribution ratios can be calculated. The optimization of the PSO algorithm is to find the smallest one of all fitness functions in the three-dimensional stereogram model, and output the corresponding ratio of a_1 and a_2 to achieve the optimal torque distribution.

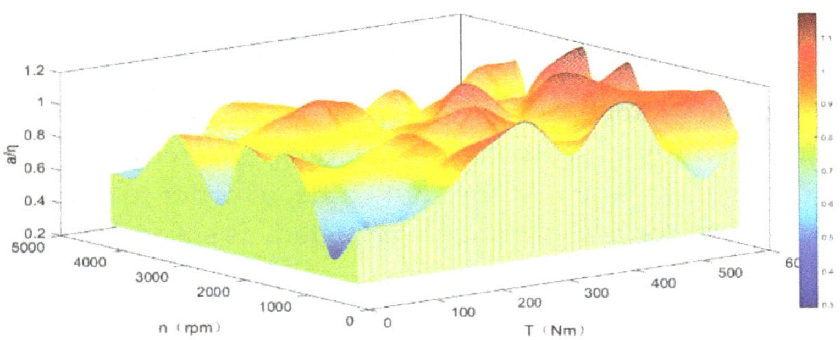

Figure 4. Fitting 3D model under driving state.

Through the analysis above, $A = f(a_1) + f(a_2)$ can be used as fitness function, so the fitness function is designed as follows:

$$\min A = f(a_1) + f(a_2), \tag{15}$$

In a search space of a D-dimensional parameter, the population size of the particles is *Size*. Each particle represents a candidate solution to the solution space, where the position of the *i-th* ($1 \le i \le Size$) particle in the entire solution space is represented as X_i and the velocity is represented as V_i. The optimal solution generated by the *i-th* particle from the initial to the current iteration number search is the individual extremum p_i, and the current optimal solution of the entire population is *BestS*. *Size* particles are randomly generated, and the position matrix and velocity matrix of the initial population are randomly generated. The learning factors are set as c_1 and c_2, the maximum evolution algebra is G, and g is the current evolutionary algebra. The equation for the velocity and position of a particle in the solution space is as follows:

$$V_i^{kg+1} = w(t) \times V_i^g + c_1 r_1 (p_i^g - X_i^g) + c_2 r_2 (BestS_i^g - X_i^g), \tag{16}$$

$$X_i^{g+1} = X_i^g + V_i^{g+1}, \tag{17}$$

Among them, $g = 1,2, \ldots ,G$, $I = 1,2, \ldots$,Size, r_1 and r_2 are random numbers from 0 to 1; c_1 is a local learning factor, and c_2 is a global learning factor, generally c_2 is larger and $w(t)$ is the inertia weight. The particle swarm optimization algorithm has the advantages of strong local search ability, fast calculation speed, and few parameters. However, during the running process, the particle swarm has strong convergence in the local, and it is easy to ignore all and fall into the local optimal solution [24]. In view of the shortcomings of the particle swarm algorithm, the inertia weight $w(t)$ is added to the velocity term, which represents the ability of the particle to update the velocity, which has a great influence on the convergence and accuracy of the whole algorithm. A larger $w(t)$ can improve the global search ability of the algorithm, while a smaller $w(t)$ can improve the local search ability of the algorithm, so the value of $w(t)$ should be decremented during the iterative process, which allows the particle to strike a balance between its search ability and convergence speed. The value of $w(t)$ is determined according to Equation (18).

$$w(t) = w(t)_{\max} - \frac{w_{\max} - w_{\min}}{k_{\max}} \times k, \tag{18}$$

In the equation, w_{max} represents the initial weight, w_{min} represents the final weight, k represents the current iteration number of particles, and k_{max} represents the maximum iteration number of particles. The particle swarm optimization is easy to converge too early and fall into local optimum, which makes it impossible to obtain global optimum solution. Combining with the requirements of speed control for electric vehicles, this paper improves the particle swarm optimization algorithm. The inertia weight is determined by the exponential decrement method, as defined by Equation (19).

$$w = w_{\max} \left(\frac{w_{\min}}{w_{\max}}\right)^{1/(1+10k/k_{\max})}, \tag{19}$$

In the initial stage, w is larger, and has a strong ability to search in a wide range. In the later stage, w is smaller and has a strong ability to search in a small range, thereby improving the performance of the particle swarm algorithm as a whole. At the same time, in order to avoid premature convergence of the algorithm, the learning factor is dynamically adjusted, as shown in Equation (20):

$$\begin{cases} c_1 = 2 - \sin \frac{k\pi}{k_{\max}} \\ c_2 = 1 + \sin \frac{k\pi}{k_{\max}} \end{cases}, \tag{20}$$

In the early stage of population search, c_1 is larger and c_2 is smaller, which facilitates the particle to learn its own optimal solution and improves the global search ability. In the later stage of population search, c_2 is larger and c_1 is smaller, which facilitates the population to move closer to the global optimal solution and enhances the local optimization performance.

3.3. Energy Feedback Brake Control Based on Optimal Torque Distribution Method

For electric medium bus, in order to further improve the cruising range, the energy feedback brake will be added to the vehicle. It not only saves energy, but also solves the problem that the electric medium bus has a short driving range of one charge, and can also improve the braking performance of the car and reduce the friction loss of the brake pad when the car brakes.

In this paper, the energy feedback brake control is carried out under the condition that the battery is safely charged, the rotation speed is not too low, and the power generation power is in the safe interval. Since the motor provides braking torque in the energy feedback state, the torque optimal control problem of the two sets of motor modules is involved. It is known in the foregoing studies that reasonable torque distribution can improve system efficiency. This conclusion is also applicable in the case of energy feedback. Therefore, in order to further improve the endurance of electric mid-size passenger cars, the optimal torque distribution strategy based on particle swarm optimization is also applied to improve the power generation efficiency of the motor. The basic block diagram of the system incorporating energy feedback is shown in Figure 5. The three-dimensional perspective of the braking mode obtained by the particle swarm optimization algorithm is shown in Figure 6.

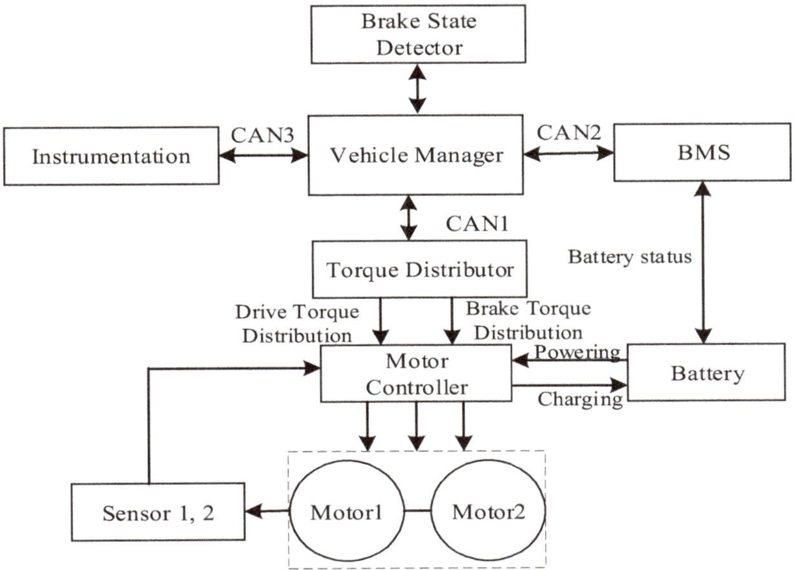

Figure 5. Basic block diagram of the energy feedback system.

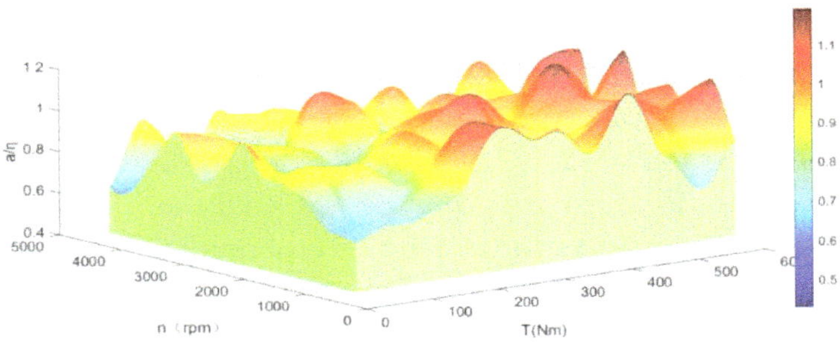

Figure 6. Fitting 3D model in energy feedback state.

4. Electric Medium Passenger Bus Vector Control System Based on Adaptive Robust Current Control

Compared with the application of motors in other aspects, the motor drive system of electric medium-sized buses has higher requirements for the current following ability and control accuracy of the motor. During the operation of electric medium-sized buses, due to the complicated operating environment, it will encounter various non-linear disturbances, resulting in current and torque ripples in the system. Therefore, this paper chooses to study from the perspective of control, and uses adaptive robust control to overcome the effects of nonlinear disturbances, thereby improving the tracking accuracy of the current loop.

According to Equation (1), a voltage model containing non-ideal back-emf can be obtained as shown in Equation (21).

$$L\frac{d}{d_t}i_q = u_q - Ri_q - e_q + \Delta_q, \tag{21}$$

where e_q is the non-ideal back-EMF, and Δq is the sum of the deviation voltages caused by all nonlinear disturbances.

Assume that the non-ideal back EMF model is:

$$e_q = \overline{S}_e K_e, \tag{22}$$

where in $\overline{S}_e$ represents a wave function containing a fundamental wave and a 6th harmonic, K_e represents a back EMF coefficient matrix. Their expressions are shown in Equations (23) and (24), respectively.

$$\overline{S}_e = \frac{3}{2}\omega_e[1\ \cos(6\sigma_e)], \tag{23}$$

$$K_e = \left[K_{q1}\ K_{q6}\right]^T, \tag{24}$$

Bringing Equation (22) into Equation (21), the following equation of state can be obtained:

$$L\frac{d}{d_t}i_q = u_q - Ri_q - \overline{S}_e K_e + \Delta_q, \tag{25}$$

To establish a standard adaptive robust control model, let $\dot{x} = \frac{d}{d_t}i_q$, let $\gamma^{\mathrm{T}} = S_e$ and $\sigma = K_e$. Define the amount of virtual control $\overline{u}$ as:

$$\overline{u} = u_q - Ri_q, \tag{26}$$

where u_q is the actual output of the controller and Ri_q is calculated from the known amount and the feedback amount.

Substituting the above assumption and Equation (26) into Equation (25), Equation (26) is obtained as shown below.

$$L\dot{x} = \overline{u} + \gamma^T\sigma + \Delta_q, \tag{27}$$

The adaptive robust current controller designed in this paper makes the output of the double-disc AFPMSM model overcome the effects of non-ideal back EMF and other nonlinear disturbances. The tracking error with the expected value x_d is as small as possible.

It can be seen from the equation of state, Equation (27) that the previously designed double-disc AFPMSM model clearly includes parameter uncertainties and nonlinear disturbances. Adaptive robust control can compensate the uncertainty in the system through the design of adaptive law, and synthesize the robust control law to overcome the influence of nonlinear disturbance, so it is suitable for the design of PMSM current controller. According to Equations (27) and (28) can be obtained.

$$\overline{u} = L\dot{x} - \gamma^T\sigma - \Delta_q, \tag{28}$$

The control law form of the adaptive robust controller is as shown in Equation (29).

$$\overline{u} = u_a + u_r, \tag{29}$$

where u_a is the compensation term for adaptive control, which can be expressed as:

$$u_a = L\dot{x}_d - \gamma^T\hat{\sigma}, \tag{30}$$

where $\dot{x}_d$ is the differential of the expected value of the q-axis current and $\hat{\sigma}$ is the estimated value of the unknown parameter σ.

u_{r} is a robust control term and can be expressed as:

$$u_r = u_{r1} + u_{r2}, \tag{31}$$

In Equation (31), u_{r1} is a linear proportional feedback term and $u_{r1} = -k_{r1}z$, k_{r1} is a proportional coefficient, and $z = x - x_d$ represents a tracking error. According to the control law represented by Equations (28) and (29), the dynamic equation of the system tracking error is:

$$L\dot{z} - u_{r1} = u_{r2} - \left[\gamma^T\widetilde{\sigma} - \Delta_q\right], \tag{32}$$

where $\widetilde{\sigma}$ is the parameter estimation bias and $\widetilde{\sigma} = \hat{\sigma} - \sigma$, u_{r2} is the robust control term. According to the robust control principle, the design requirements are:

$$\begin{cases} zu_{r2} \leq 0 \\ z\left\{u_{r2} - \left[\gamma^T\widetilde{\sigma} - \Delta_q\right]\right\} \leq \varepsilon \end{cases}, \tag{33}$$

In the above equation, ε represents any positive integer, the first condition can be guaranteed to be naturally dissipated, and the second condition indicates that u_{r2} can suppress nonlinear disturbance, modeling error and estimation error of adaptive parameters. There are many ways to select u_{r2} that meet the conditions [19]. The most common one is:

$$\begin{cases} u_{r2} = -\frac{1}{4\varepsilon}h^2 z \\ h = \left|\gamma^T\right|\left|\sigma_{\max} - \sigma_{\min}\right| + \Delta_{q\max} \end{cases}, \tag{34}$$

It can be known from the control law in Equation (29) and the tracking error dynamic in Equation (32) that the tracking performance of the adaptive robust controller depends on the design of the robust control term u_r. Since the adaptive law design is synthesized by tracking error, the parameter projection method is used to modify the adaptive law [20]. Therefore, the adaptive law of adaptive robust control is expressed as:

$$\dot{\hat{\sigma}} = \mathrm{Pr}oj(\Gamma\gamma z), \tag{35}$$

where Γ is a diagonal adaptive law matrix, and $\mathrm{Pr}oj(\lambda)$ is a projection operator, which can be expressed as:

$$\mathrm{Pr}oj(\lambda) = \begin{cases} 0, \text{if} \begin{cases} \hat{\sigma} = \sigma_{\min} \text{ and } \lambda < 0 \\ \hat{\sigma} = \sigma_{\max} \text{and } \lambda > 0 \end{cases} \\ \lambda, \text{others} \end{cases}, \tag{36}$$

When using the control law of Equation (29) and the adaptive law of Equation (35), the adaptive robust current control block diagram is shown in Figure 7.

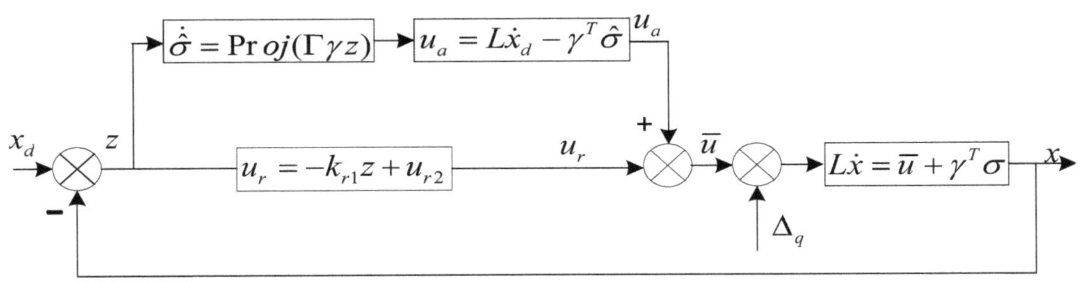

Figure 7. Adaptive robust current control block diagram.

5. Simulation and Experimental Results

5.1. Drive System Simulation

A two-disc AFPMSM simulation system is set up, and the current controller adopts adaptive robust control method and PI control, respectively. Finally, compare the current tracking performance of the two controllers. The current controller adopts the control strategy of $i_d = 0$, the d-axis current adopts PI controller, the q-axis current adopts adaptive robust controller, the given current i_q* is 150 A, and the i_q current response waveform is shown in Figure 8. The three-phase current waveform is shown in Figure 9.

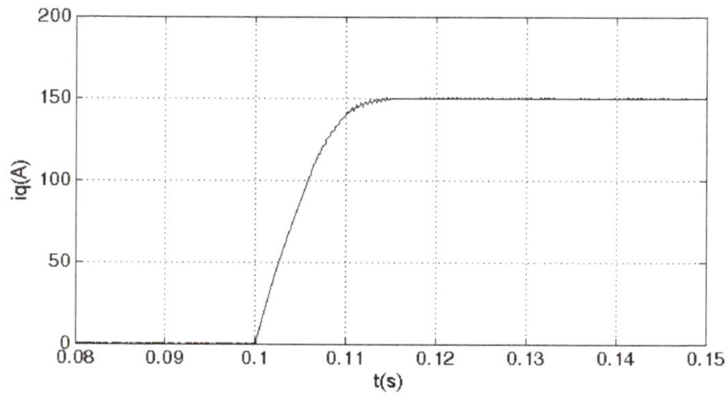

Figure 8. I_q current response waveform based on adaptive robust current controller.

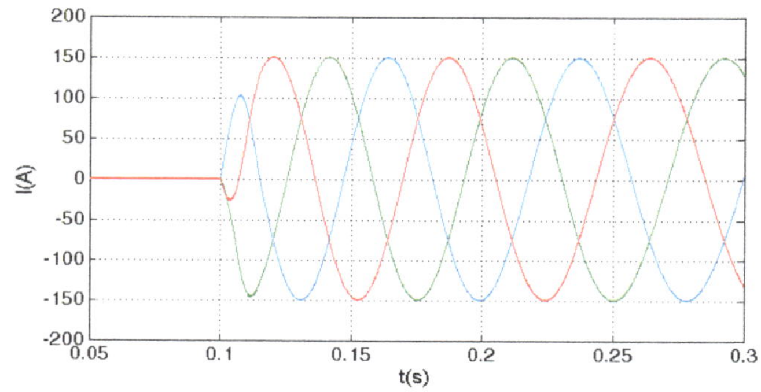

Figure 9. Three-phase current waveform based on adaptive robust current controller.

When the given current i_q* is also 150 A, and the d-axis and q-axis currents all use the PI controller, the i_q current response waveform is shown in Figure 10, and the three-phase current waveform is shown in Figure 11.

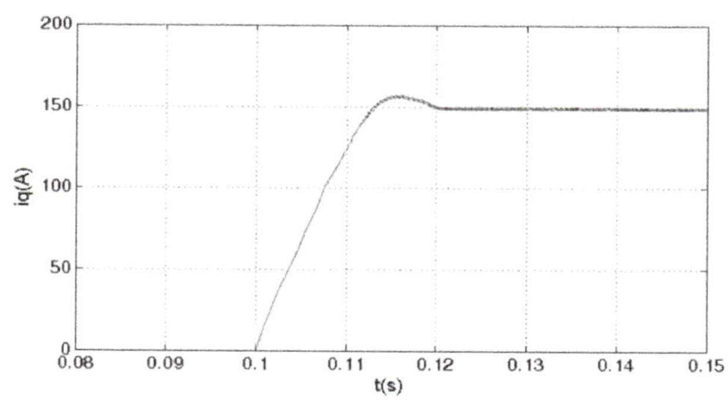

Figure 10. I_q current response waveform based on PI current controller.

It can be seen from the simulation waveform that the control system based on adaptive robust current controller has a response time of about 14 ms from 0 A to 150 A, and the current has almost no overshoot, and the current fluctuation is small at steady state. The control system based on the PI current controller has a response time of approximately 21 ms from 0 A to 150 A and a current overshoot of approximately 6%. The simulation results show that the adaptive robust current controller designed in this paper has better current control performance.

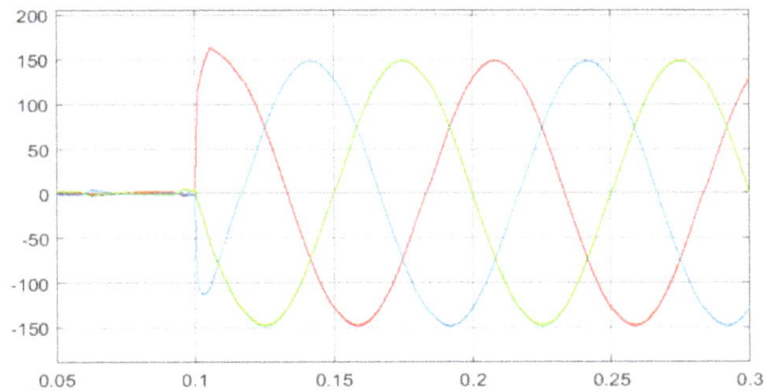

Figure 11. Three-phase current waveform based on PI current controller.

5.2. Experimental Platform Verification Experiment

Figure 12 shows the AFPMSM tow experimental platform. Firstly, the drive test experiment is carried out, and the AFPMSM is used as the drive motor to load. The AC asynchronous motor works in the fixed speed mode, the rotation speed is 1000 rpm. After the rotation speed is stable, the given torque is 150 Nm. The A and B phase current waveforms are shown in Figure 13. The oscilloscope waveform amplitude is about 1.5 V. Then the torque tracking experiment was carried out, and the torque was abruptly changed from 0 Nm to 120 Nm. The torque waveform is shown in Figure 14.

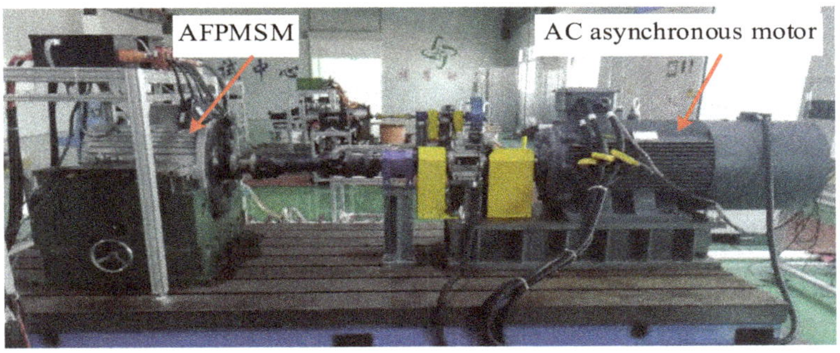

Figure 12. AFPMSM tow experimental platform.

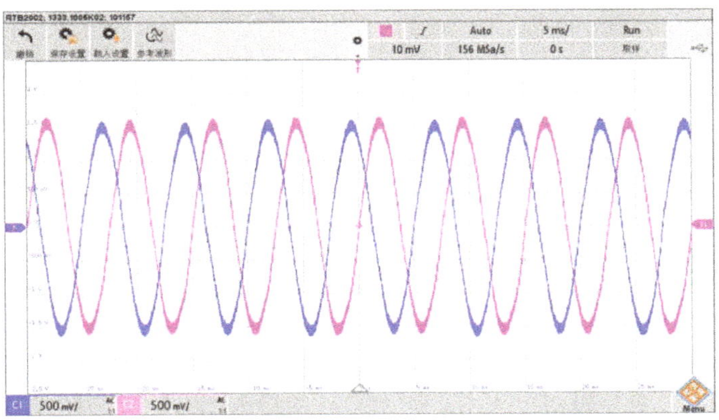

Figure 13. A and B phase current waveforms when the given torque is 150 Nm.

The motor drive system efficiency experiments were carried out under the torque average distribution and the torque optimal distribution control strategy based on PSO. The maximum speed of the motor is 4500 rpm. During the experiment, the motor speed is from 500 rpm to 4500 rpm, and one speed value is taken every 100 rpm for a total of 41 speed values. At every speed value, the output torque is increased from 0 Nm to 600 Nm, and a torque value is selected every 30 Nm. Twenty torque values correspond to 820 efficiency test values. The output torque T, the rotational speed n, the DC-side input voltage U, and the current I of each efficiency point are acquired by a sensor and a storage recording instrument. After calculating the efficiency of the motor drive system, the speed, output torque and efficiency values are finally imported into MATLAB to generate a motor efficiency map. The efficiency map generated by the torque average distribution and the torque optimal distribution control strategy based on the particle swarm optimization algorithm is shown in Figures 15 and 16.

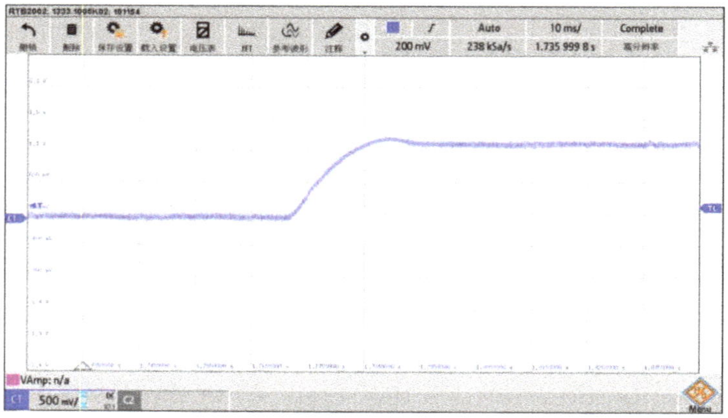

Figure 14. Drive system torque tracking waveform.

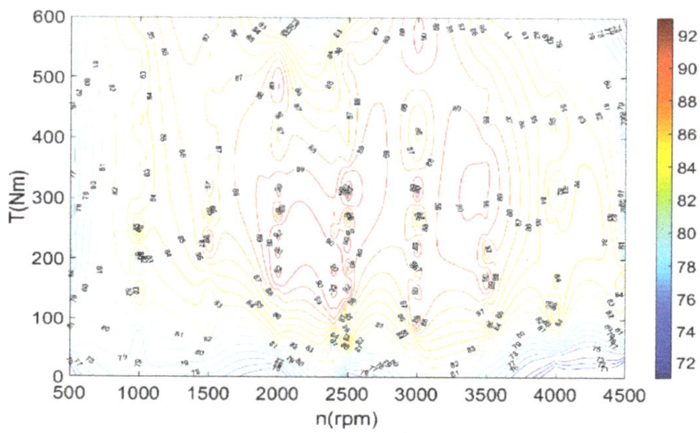

Figure 15. System efficiency map based on traditional torque average distribution strategy.

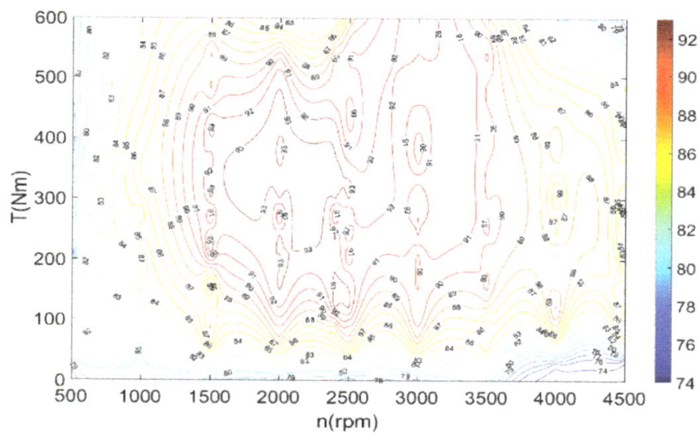

Figure 16. System efficiency map of torque optimal distribution control strategy based on particle swarm optimization.

After comparison, the optimal torque distribution control strategy based on particle swarm optimization algorithm can significantly increase the high efficiency range of AFPMSM. The system efficiency increases by about 15% in the interval of 85% or more, and the system efficiency increases by 20% in the interval of 90% or more. The correctness of the proposed optimal torque distribution control strategy based on particle swarm optimization is verified.

As in the driving state, the efficiency map generated by the torque average distribution and the torque optimal distribution control strategy based on the particle swarm optimization algorithm in the braking situation is as shown in Figures 17 and 18.

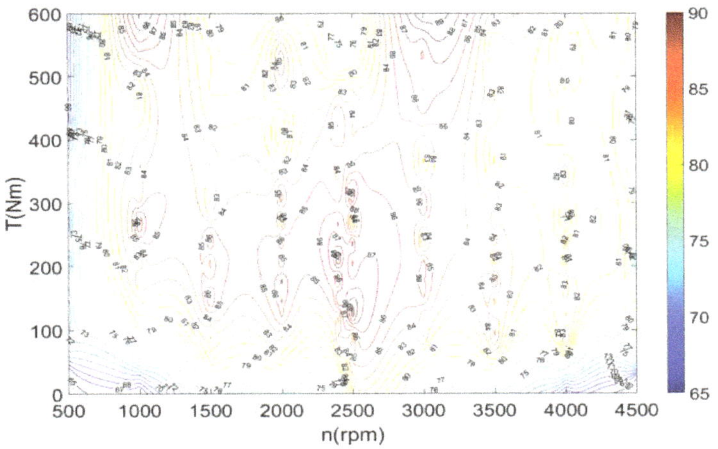

Figure 17. System efficiency map based on traditional torque average allocation strategy in energy feedback state.

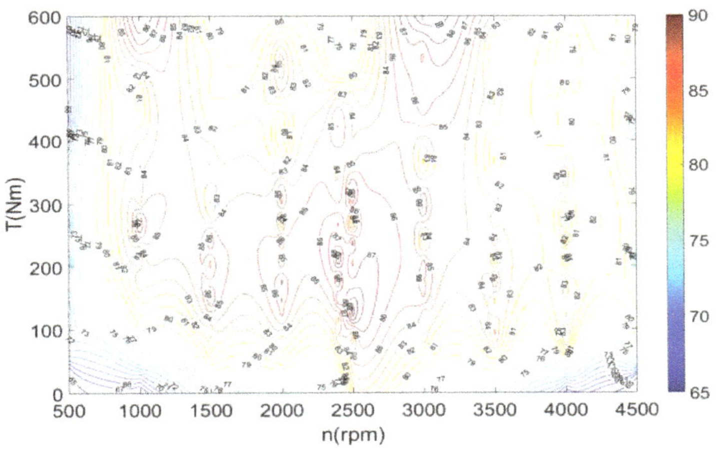

Figure 18. System efficiency map of torque optimal distribution control strategy based on particle swarm optimization in energy feedback state.

After comparison, the optimal torque distribution control strategy based on particle swarm optimization algorithm can significantly increase the high efficiency range of AFPMSM in energy feedback state. The system efficiency is increased by about 25% in the interval above 85%, and the system efficiency increases by 10% in the interval of 90% or more, which verifies the correctness of the optimal torque distribution control strategy in the energy feedback state.

5.3. Vehicle Experiment

As shown in Figure 19, the experimental vehicle is used to simulate the electric medium bus by means of load. The electric mid-size bus simulated in this experiment has an empty load of about 3000 kg. The battery specifications are shown in Table 1. The internal structure of the experimental vehicle is shown in Figure 20. The endurance capability based on the torque average distribution and the optimal torque distribution control strategy based on the particle swarm optimization algorithm is tested on the urban road. The load conditions are no-load and 700 kg (about 10 people). With 300 km as the target cruising range, the data in the table indicates the percentage of completion. The experimental results are shown in Table 2.

Figure 19. Experimental vehicle.

Table 1. The battery specifications.

The Battery Specifications	Specific Parameters
Battery Type	Ternary Polymer Lithium Battery
Battery Capacity	80 kWh
Battery Rated Voltage	384 V
State of Charge	10–100%

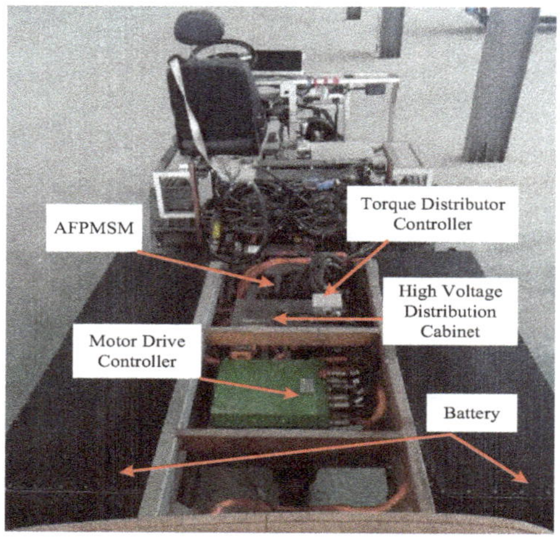

Figure 20. Internal structure of the experimental vehicle.

Table 2. Electric medium-sized passenger car endurance experiment.

Experimental Condition	Torque-Based Average Distribution Method	Optimal Torque Distribution Method
No load	89.67%	97.67%
700 kg load	77.33%	86.00%

Table 2 showed that the optimal torque distribution control strategy based on PSO algorithm can improve the cruising range of 8% under no-load conditions and increase the cruising range by 8.67% under 700 kg load. Therefore, the optimal torque distribution method designed in this paper can effectively improve the system efficiency and improve the endurance.

6. Conclusions

This paper mainly studies the drive control system of electric medium bus. The selected motor is a dual-stator single-rotor AFPMSM, which can be equivalent to two PMSM connected coaxially, therefore the drive control system is divided into two parts: torque distribution and motor control. The first part is the study of the torque distribution method that maximizes the efficiency of the dual-motor system. Simulation and experimental results show that the optimal torque distribution method proposed in this paper can effectively improve system efficiency and endurance. The second part is the motor control part. Simulation and experimental results show that the system based on adaptive robust current controller proposed in this paper has better anti-interference ability and stability than the system based on PI current controller. At the same time, this article also lays the foundation for the subsequent research of multiple modular dual-stator single-rotor combined motor systems, and provides some useful methods and ideas for the drive control system of AFPMSM.

Author Contributions: Conceptualization, S.W., J.Z. and T.L.; methodology, S.W. and J.Z.; software, S.W. and M.H.; validation, J.Z. and T.L.; formal analysis, J.Z. and S.W.; investigation, S.W. and M.H.; resources, J.Z. and T.L.; data curation, S.W. and M.H.; writing—Original draft preparation, J.Z., S.W. and M.H.; writing—Review and editing, J.Z., S.W. and T.L.; visualization, M.H. supervision, J.Z. and S.W.; project administration, J.Z.; funding acquisition, S.W.

References

1. Wang, X.; González, J.A. Assessing feasibility of electric buses in small and medium-sized communities. *Int. J. Sustain. Transp.* **2013**, *7*, 431–448. [CrossRef]
2. Vepsäläinen, K.J.; Tammi, K. Stochastic driving cycle synthesis for analyzing the energy consumption of a battery electric bus. *IEEE Access* **2018**, *6*, 55586–55598.
3. Li, L.; Yang, C.; Zhang, Y.L.; Song, J. Correctional DP-Based Energy Management Strategy of Plug-In Hybrid Electric Bus for City-Bus Route. *IEEE Trans. Veh. Technol.* **2015**, *64*, 2792–2803. [CrossRef]
4. Wang, W.; Zhang, Z.; Shi, J.; Lin, C.; Gao, Y. Optimization of a dual-motor coupled powertrain energy management strategy for a battery electric bus based on dynamic programming method. *IEEE Access* **2018**, *6*, 32899–32909. [CrossRef]
5. Kommuri, S.K.; Defoort, M.; Karimi, H.R.; Veluvolu, K.C. A Robust Observer-Based Sensor Fault-Tolerant Control for PMSM in Electric Vehicles. *IEEE Trans. Ind. Electron.* **2016**, *63*, 7671–7681. [CrossRef]
6. Sant, A.V.; Khadkikar, V.; Xiao, W.; Zeineldin, H.H. Four-Axis Vector-Controlled Dual-Rotor PMSM for Plug-in Electric Vehicles. *IEEE Trans. Ind. Electron.* **2015**, *62*, 3202–3212. [CrossRef]
7. Huang, J.; Liu, Y.; Liu, M.; Cao, M.; Yan, Q. Multi-Objective Optimization Control of Distributed Electric Drive Vehicles Based on Optimal Torque Distribution. *IEEE Access* **2019**, *7*, 16377–16394. [CrossRef]
8. Yuan, X.; Wang, J.; Colombage, K. Torque distribution strategy for a front and rear wheel driven electric vehicle. In Proceedings of the 6th IET International Conference on Power Electronics, Machines and Drives, Bristol, UK, 27–29 March 2012; pp. 1–6.
9. Chen, L.; Chen, T.; Xu, X.; Jiang, H.; Sun, X. Multi-objective coordination control strategy of distributed drive electric vehicle by orientated tire force distribution method. *IEEE Access* **2018**, *6*, 69559–69574. [CrossRef]
10. Xu, G.; Xu, K.; Zheng, C.; Zhang, X.; Zahid, T. Fully Electrified Regenerative Braking Control for Deep Energy Recovery and Safety Maintaining of Electric Vehicles. *IEEE Trans. Veh. Technol.* **2016**, *65*, 1186–1198. [CrossRef]
11. Li, W.; Du, H.; Li, W. Driver intention based coordinate control of regenerative and plugging braking for electric vehicles with in-wheel PMSMs. *IET Intell. Transp. Syst.* **2018**, *12*, 1300–1311. [CrossRef]
12. Heydari, S.; Fajri, P.; Rasheduzzaman, M.; Sabzehgar, R. Maximizing Regenerative Braking Energy Recovery of Electric Vehicles through Dynamic Low-Speed Cutoff Point Detection. *IEEE Trans. Transp. Electrif.* **2019**, *5*, 262–270. [CrossRef]
13. Jain, R.; Tandon, P.; Vasantha, K. Optimization methodology for beam gauges of the bus body for weight reduction. *Appl. Comput. Mech.* **2014**, *8*, 47–62.
14. Croccolo, D.; Agostinis, M.D.; Vincenzi, N. Structural Analysis of an Articulated Urban Bus Chassis via Finite Element Method: A Methodology Applied to a Case Study. *J. Mech. Eng.* **2011**, *57*, 799–809. [CrossRef]
15. Kongwat, S.; Jongpradist, P.; Kamnerdtong, T. Optimization of Bus Body based on Structural Stiffness and Rollover Constraints. In Proceedings of the Asian Congress of Structural and Multidisciplinary Optimization, Nagasaki, Japan, 22–26 May 2016; pp. 22–26.
16. Kunakron-ong, P.; Ruangjirakit, K.; Jongpradist, P. Design and analysis of electric bus structure in compliance with ECE safety regulations. In Proceedings of the 2017 2nd IEEE International Conference on Intelligent Transportation Engineering (ICITE), Singapore, 1–3 September 2017; pp. 25–29.

17. Hemeida, A.; Sergeant, P. Analytical modeling of surface PMSM using a combined solution of Maxwell–s equations and magnetic equivalent circuit. *IEEE Trans. Magn.* **2014**, *50*, 1–13. [CrossRef]
18. Lim, D.K.; Cho, Y.S.; Ro, J.S.; Jung, S.Y.; Jung, H.K. Optimal design of an axial flux permanent magnet synchronous motor for the electric bicycle. *IEEE Trans. Magn.* **2015**, *52*, 1–4. [CrossRef]
19. Chen, N.Z.; Yao, B.; Wang, Q. μ-Synthesis-Based Adaptive Robust Control of Linear Motor Driven Stages with High-Frequency Dynamics: A Case Study. *IEEE/ASME Trans. Mechatron.* **2015**, *20*, 1482–1490. [CrossRef]
20. Wang, Z.; Hu, C.; Zhu, Y.; He, S.; Yang, K.; Zhang, M. Neural Network Learning Adaptive Robust Control of an Industrial Linear Motor-Driven Stage with Disturbance Rejection Ability. *IEEE Trans. Ind. Inform.* **2017**, *13*, 2172–2183. [CrossRef]
21. Yang, Y.; Wang, Y.; Jia, P. Adaptive robust control with extended disturbance observer for motion control of DC motors. *Electron. Lett.* **2015**, *51*, 1761–1763. [CrossRef]
22. Zhao, J.; Hua, M.; Liu, T. Collaborative Optimization and Fault Tolerant Control Method for Multi-disc Permanent Magnet Synchronous Motors for Electric Vehicles. *Proc. CSEE* **2019**, *39*, 386–394.
23. Gan, C.; Jin, N.; Sun, Q.; Kong, W.; Hu, Y.; Tolbert, L.M. Multiport bidirectional SRM drives for solar-assisted hybrid electric bus powertrain with flexible driving and self-charging functions. *IEEE Trans. Power Electron.* **2018**, *33*, 8231–8245. [CrossRef]
24. Bonyadi, M.R.; Michalewicz, Z. Analysis of Stability, Local Convergence, and Transformation Sensitivity of a Variant of the Particle Swarm Optimization Algorithm. *IEEE Trans. Evol. Comput.* **2016**, *20*, 370–385. [CrossRef]

9

Comparison of an Off-Line Optimized Firing Angle Modulation and Torque Sharing Functions for Switched Reluctance Motor Control

Peter Bober and Želmíra Ferková *

Faculty of Electrical Engineering and Informatics, Technical University of Kosice, 04200 Kosice, Slovakia; peter.bober@tuke.sk

* Correspondence: zelmira.ferkova@tuke.sk

Abstract: In this paper, a comparison of the simple firing angle modulation method (FAM) and the more advanced torque sharing function (TSF)-based control of switched reluctance motor (SRM) is presented. The off-line procedure to tailor and optimize the parameters of chosen methods for off-the-shelf SRM is explained. Objective functions for optimization are motor efficiency, torque ripple, and integral square error. The off-line optimization uses a finite element method (FEM) model of the SRM. The model was verified by measurement on the SRM. Simulation results showed that FAM has comparable efficiency to TSF, but has a much higher value of torque ripple. The presented off-line procedure can be used for single or multi-objective optimization.

Keywords: switched reluctance motor; torque sharing functions; finite element method; firing angle modulation; torque ripple; efficiency; optimization

---

1. Introduction

The principle of Switched Reluctance Motor (SRM) is known from the 1830s when the first attempts to build a usable machine were made by many constructors and inventors [1,2]. One of them was Robert Davidson who, in 1838, developed a reluctance motor to power an electric locomotive at the Edinburgh-Glasgow railway [3]. The early attempts suffer from inadequate mechanical switches and poor electromagnetic and mechanical design. The SRM waited 150 years for the theoretical foundations to be developed and new power electronics devices to appear. A work of Lawrenson et al. [4] lays general foundations for the practical design of a family of switched reluctance motors in 1980. Another thirty years were needed to achieve progress in the field of power electronics and powerful microcontrollers. Currently, many companies are producing SRM in the world, and the worldwide market is expected to grow by roughly 5.2% over the next five years [5]. The construction of SRM can be tailored to broad application areas. Report [5] showed that 22.95% of the SRM market demand is in the Automobile Industry, 19.43% in the Appliance Industry, and 39.28% in Industrial Machinery in 2016.

Typical SRM construction has salient poles on both the stator and the rotor. The windings are only on the stator, and phase electric current is unidirectional. Reluctance torque produced by the change in inductance as a function of the angle creates a rotary movement. The advantages resulting from this simple construction are low cost, robustness, high efficiency, and wide speed range up to 150,000 rpm [6]. Each energized phase winding generates torque for a certain rotor angle interval. Therefore, SRM control depends heavily on rotor position information to turn-on and turn-off the phase current at the chosen angle. Torque ripples that occur during the transition from one winding to another are the major drawback of SRMs. The main goal of ongoing research is a torque ripple reduction and efficiency improvement by the optimization of the structure and magnetic design of the motor [6–10], and by adopting advanced motor control [9,11–17].

An overview of SRM control strategies is presented in [9]. The methods are grouped into (1) current and angle modulation, (2) average torque control (ATC) and direct torque control (DTC), (3) torque sharing function (TSF)-based control, (4) feedback linearization control (FBL), (5) iterative learning control (ILC), and (6) intelligent control (IC).

Current and angle modulation control methods are focused on defining turn-on and turn-off angles to lower torque ripple. In addition, the current controller can maintain constant current or follow a current profile specified in look-up tables. ACT and DTC use estimated average or instant torque in the control loop. TSFs define the torque reference for each phase so that their sum is equal to the output torque reference. FBL control algorithm compensates nonlinear characteristics of the motor by utilizing state feedback in the closed-loop system. ILC learns iteratively from the difference between required and actual phase current and motor torque. IC uses self-learning and the adaptive ability of fuzzy logic, genetic algorithms, and neural networks to optimize current profiles either on-line or off-line.

The majority of described SRM control methods, especially in the automotive industry, are complex and require high computational power. Authors in [14] use 1 GHz CPU and FPGA running at 100 MHz. The EPM570 Intel CPLD—Complex Programmable Logic Devices—with a frequency of 150–300 MHz and dsPIC30f6010A MPU is used in [15]. High-dynamic four-quadrant switched reluctance drive based on DITC published in [16] uses DSP/FPGA rapid-prototyping platform. This computational power is needed for torque and speed estimation, look-up tables, vibration and noise suppression, and precise current control, where high sampling frequency for low inductances is required. In the coming years, the price for computing power will fall. Nevertheless, the growing SRM market has the potential for simple and cheap drives where the hardware cannot support complex control methods. It can be assumed that not all SRMs available on the market will be designed optimally. The question then arises as to what characteristics the drive with such a motor and a simple control method will have.

Two control methods applied to the commercially available switched reluctance motor were chosen to investigate in this research. The first method is a conventional control with Firing Angle Modulation (FAM) and constant current profile similar to [18,19]. This method uses on-line turn-on and turn-off angle calculation and simple hysteresis current controller with a low demand on computing power. No look-up tables nor torque estimation is needed.

The second method is more complex TSF-based control that uses special functions to distribute torque between the outgoing and incoming motor phase. According to the evaluation in [20], the cubic, sinusoidal, and exponential torque sharing functions result in similar minimum torque ripples, which are much smaller than the torque ripple produced by the linear torque sharing function. A minimum effective rate-of-change of flux linkage was used in [11,21] as an objective function that should be minimized by selected TSF to extend the torque-speed range. Research results in [11,13,20–23] show that different objective functions lead to different TSF. Therefore, multi-objective optimization with weighted criterion [11,13,19] or the Pareto-optimal approach [22] are used.

Many published works [6–8,11,13,19–23] use two to four objective functions to compare or optimize control methods or the structural design of the motor. This reduction of problem formulation is necessary to handle the complexity of a full industrial design process and can be considered as one stage in the multi-stage optimization process. Authors in [10] address this complexity and present a two-step procedure for a motor design where different performance measures (such as manufacturing cost and iron weight) are included along with efficiency and torque ripple. Here some general optimization frameworks, e.g., ARTAP published in [24], are helpful since various domain-specific numerical solvers are needed for this type of motor design task.

The main aim of this paper is to present a comparison of two selected methods that control the off-the-shelf switched reluctance motor. The procedure to tailor and optimize the parameters of the chosen methods is explained. Objective functions for optimization are motor efficiency, torque ripple, and integral square error between instant and average torque. The offline optimization uses the finite element method (FEM) model of the SRM. The model is verified by comparing simulation results to the motor measurement.

2. Modeling and Control of SRM

2.1. Switched Reluctance Motor and Power Converter

Control methods are tailored and optimized for the SRM that is commercially available on the market. It is a small 200 W three-phase 12/8 pole switched reluctance motor. Rated voltage is 120 V, and the maximum phase current is 6 A. The geometry of the motor is shown in Figure 1a. Each stator phase winding consists of four coils connected in a series-parallel manner. The width of the stator and rotor pole is almost the same. Therefore, inductance profile has a sharp peak in an aligned position, and transition from a motor to generator mode is fast.

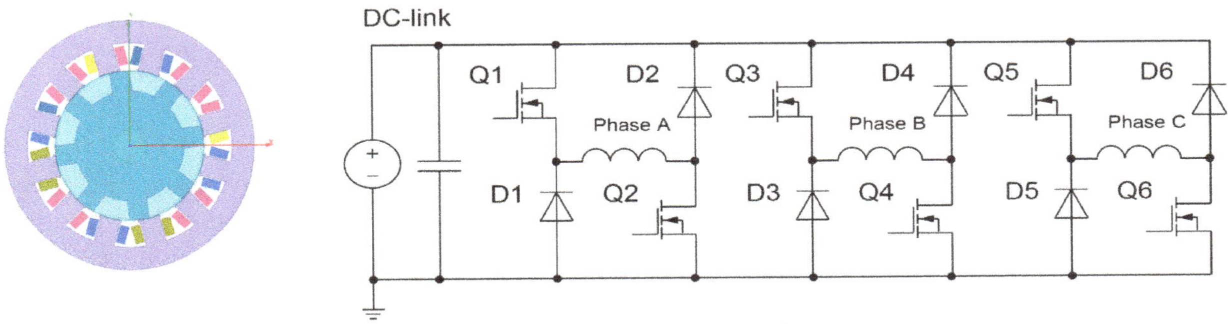

(a) (b)

Figure 1. (**a**) Geometry of the switched reluctance motor; (**b**) standard asymmetric H-bridge converter topology.

The experimental setup uses a power converter that was built for measurement purposes. It has a standard asymmetric H-bridge topology to drive three phases of SRM (Figure 1b). A stabilized power supply supplies a DC-link of the power converter. The voltage can vary from 30 V to 120 V.

2.2. FEM Model

The FEM model of SRM was built in ANSYS Maxwell according to the dimensions of the motor and electrical measurement. When creating the model, a compromise solution was chosen between the mesh size and the computational time. The total number of elements is 11,110 in the 2D model. The mesh is denser in the air gap. The additional leakage inductance models the effect of the end winding leakage flux. The magnetic vector potential is zero on the outer circumference of the yoke. The comparison of the simulation results with the electrical measurement led to the selection of the B-H characteristic corresponding to the electrical steel M400.

The SRM model is driven by three-phase asymmetric H-bridge created in ANSYS Simplorer. The control algorithms are also programmed in ANSYS Simplorer. Figure 2 shows magnetic field distribution in the FEM model of the studied motor for aligned rotor position 22.5° (a) and for 7.2° (b). Figure 2c shows simulated motor characteristics for a maximum phase current of 6 A and 120 V DC-link voltage. The FEM model verification is described in Section 3.2.

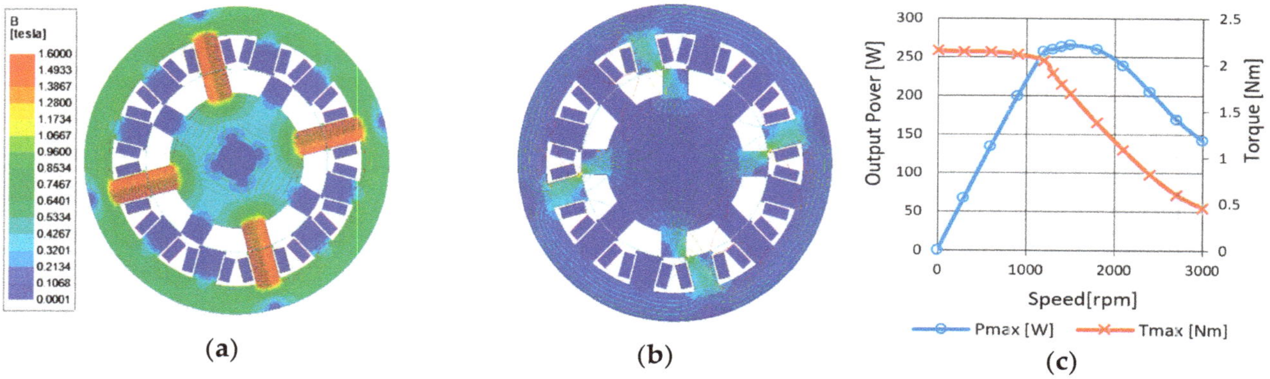

Figure 2. Magnetic field distribution in the finite element method (FEM) model, phase current 6 A: (**a**) aligned position, rotor angle 22.5 degrees; (**b**) rotor angle 7.2 degrees; (**c**) Power–Speed and Torque–Speed characteristics.

2.3. *SRM Control Methods*

The switched reluctance motor is controlled by two methods. The first method is a Firing Angle Modulation, and the second one is a control method with Torque Sharing Functions.

2.3.1. Firing Angle Modulation

The firing angle modulation method is quite straightforward and can be easily implemented. The asymmetric H-bridge converter allows controlling the electric current in each stator phase winding independently by using a simple hysteresis controller. Each motor phase creates torque only in a limited angle interval, and the phases alternate one after the other. Therefore, the current is switched on at the angle θ_{on} and switched off at the angle θ_{off}. The hysteresis current controller controls the current level inside the angle interval θ_{on} and θ_{off}. Figure 3a shows simulated current waveforms of one phase as a function of a rotor angle for different speeds. The shape of the phase current cannot be fully controlled at an available DC-link voltage as it depends on the inductance profile and sampling frequency of the hysteresis controller. The current rising is fast because the phase inductance is low. At the end of the current impulse, the inductance is highest as the stator and rotor poles are aligned. Therefore, the electric current decay is slow, and it can create negative torque, especially at high speeds. Average motor torque corresponds to the current reference I_{ref} and, of course, to the θ_{on} and θ_{off} values. The value of I_{ref} is an output of the speed controller. Figure 3b shows the block diagram of the firing angle modulation method. The procedure to calculate firing angles θ_{on} and θ_{off} is described in Section 2.3.2.

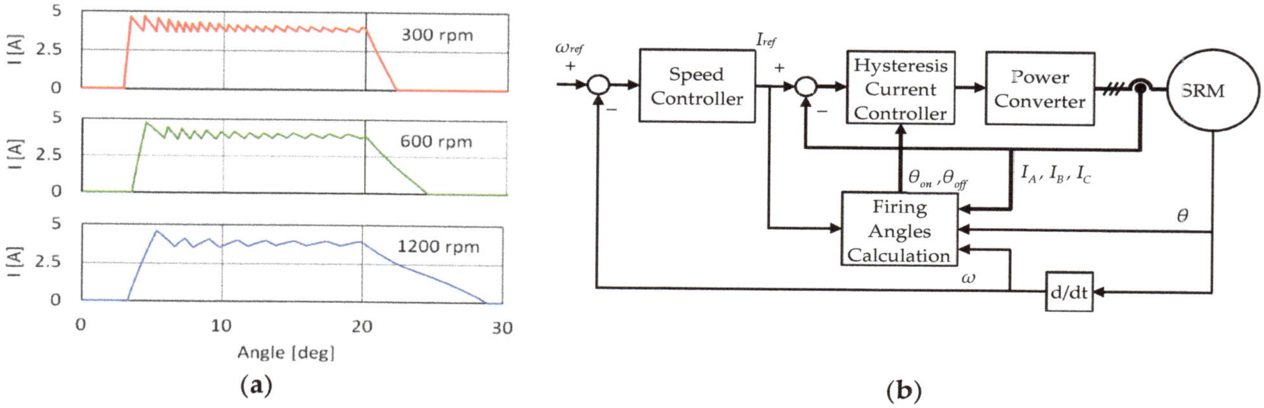

Figure 3. (**a**) Current waveform at speed 300, 600, and 900 rpm, 120 V; (**b**) block diagram of the firing angle modulation method.

2.3.2. Firing Angles Calculation

The phase current generates torque in a rotor angle region, where the phase inductance is changing due to a rotor position change. The electric current rise time is defined by the inductance and the applied voltage. Therefore, the turn-on angle θ_{on} shall precede the increase in inductance to start to generate the desired torque from the very beginning. If the current reaches the specified level too early, it creates a copper loss only. If the current does not reach the specified level at a defined angle, the torque is less than it could be. An early turn-off angle θ_{off} causes torque to drop. However, a late turn-off of the current causes a negative torque to be generated.

A calculation of the turn-on angle θ_{on} is given in [16,17] based on the assumption, that the inductance profile has constant minimum inductance, then the inductance starts to rise at angle θ_m. The value of the current should reach the reference value at this point, and θ_{on} can be calculated as follows:

$$\theta_{on} = \theta_m - 6L_u I_{ref} \frac{n}{V_{DC}}, \tag{1}$$

where θ_{on} is turn-on angle (degrees), θ_m is rotor angle (degrees) where the inductance begins to rise, L_u is the inductance in unaligned position, I_{ref} is a reference current from speed controller, n is rotor speed (rpm), and V_{DC} is the supply voltage. This equation neglects the actual shape of the inductance profile. Therefore, a set of simulation experiments was planned and performed to obtain a more precise formula for turn-on angle calculation. An extrapolation of collected data leads to the formula:

$$\theta_{on} = 7 - \left(0.463 + 8.88e^{-4}n + 1.2e^{-6}n^2\right)\left(\frac{I_{ref0}}{I_{max}}\right)^2, \tag{2}$$

where I_{ref0} is a reference current from speed controller at angle 0°, and I_{max} is a maximum reference current.

The formula for calculation of turn-off angle is:

$$\theta_{off} = 22.5 - \frac{0.2I_{15}2\pi\, n}{60V_{DC}}, \tag{3}$$

where I_{15} is a phase winding current at angle 15°, and V_{DC} is a voltage applied on winding inductance at turn-off time. This formula was derived from the linear approximation of the current fall time. Equations (2) and (3) give angle values for first phase A. Angle values for phases B and C are shifted by 15° and 30°, respectively. Formulas (2) and (3) are valid only for the given SRM and used power converter with a DC-link voltage of 120 V.

2.3.3. Torque Sharing Functions

The motor torque created by consecutive phases is overlapping. Torque sharing function defines a required torque profile for each energized phase that the total motor torque has a small ripple. The shape of TSF can be chosen arbitrarily or is calculated according to selected criteria. Figure 4 shows sinusoidal TSF defined on the interval (0, θ_p) as follows [10]:

$$TSF(\theta) = \begin{cases} 0, & (0 \le \theta < \theta_{on}) \\ f_{up}(\theta), & (\theta_{on} \le \theta < \theta_{on} + \theta_{ov}) \\ T_c & \left(\theta_{on} + \theta_{ov} \le \theta < \theta_{off}\right) \\ f_{dn}(\theta), & \left(\theta_{off} \le \theta < \theta_{off} + \theta_{ov}\right) \\ 0, & \left(\theta_{off} + \theta_{ov} \le \theta < \theta_p\right) \end{cases}, \tag{4}$$

$$f_{up}(\theta) = \frac{T_c}{2}\left[1 - \cos\left(\frac{\theta - \theta_{on}}{\theta_{ov}}\pi\right)\right], \tag{5}$$

$$f_{dn}(\theta) = T_c - f_{up}\left(\theta + \theta_{ov} - \theta_{off}\right), \tag{6}$$

where θ is the rotor angle, $\theta_p = 2\pi/8$ is the rotor pole pitch, θ_{on} is the turn-on angle, θ_{ov} is the overlap angle, $\theta_{off} = \theta_{on} + 2\pi/24$ is the turn-on angle of the next phase, T_c is the torque command, $f_{up}(\theta)$ is the rising, and $f_{dn}(\theta)$ the declining part of the TSF shape.

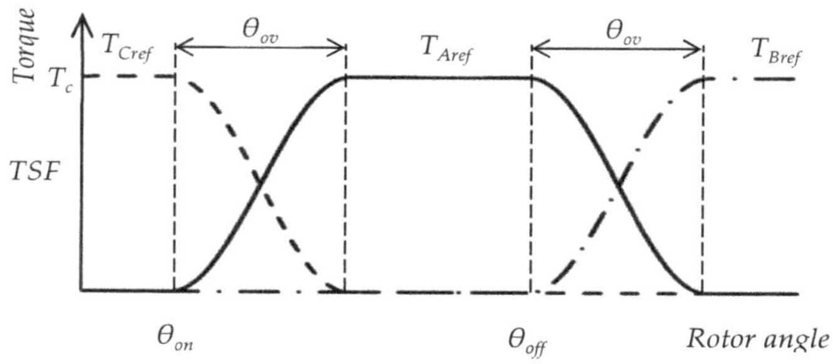

Figure 4. Sinusoidal torque sharing function.

The block diagram of used TSF-based control is shown in Figure 5. The torque command T_c from the *Speed Controller* is translated into three torque profiles T_{Aref}, T_{Bref}, and T_{Cref} for each motor phase in the *TSF* block using Equations (4)–(6). The *Torque to Current* block translates the torque profile to current reference utilizing a look-up table obtained from the FEM model. A *Hysteresis Current Controller* is used to control phase currents. Optimal TSF parameters θ_{on} and θ_{ov} are calculated by functions $f_1(T_c,\omega)$ and $f_2(T_c,\omega)$ from torque command T_c and actual rotor speed ω. The next section explains how to construct functions f_1 and f_2.

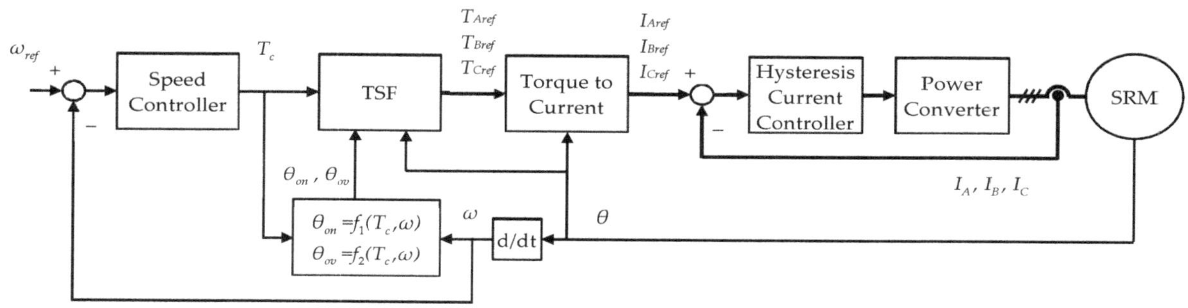

Figure 5. Block diagram of the switched reluctance motor (SRM) control using torque sharing function (TSF) with optimized parameters θ_{on} and θ_{ov}.

2.3.4. Optimization of TSF

The objective function needs to be defined to formulate an optimization problem. Three criteria are selected in this research to find optimal values of parameters θ_{on} and θ_{ov}:

- Efficiency η

$$\eta = \frac{Mechanical\ output\ power}{Electrical\ input\ power} \tag{7}$$

- Relative torque ripple T_{rip}

$$T_{rip} = \frac{T_{max} - T_{min}}{T_{avg}} \tag{8}$$

  where T_{max}, T_{min}, and T_{avg} are the maximum torque, the minimum torque, and average torque, respectively.
- Integral square error criterion ISE

$$ISE = \sqrt{\frac{1}{t_2 - t_1}\int_{t_1}^{t_2}\left(T - T_{avg}\right)^2 dt} \tag{9}$$

where T, and T_{avg} are the instantaneous torque and average torque, respectively, measured in a time-interval (t_1, t_2). This objective function is a measure of overall torque oscillation.

Simulation experiments calculate the numerical value of an objective function. One simulation run is needed for each individual combination of parameters θ_{on} and θ_{ov}. Calculations for all combinations in all operating points are time-consuming. Therefore, a workaround is used: the shape of the objective function is replaced by the interpolation function calculated from a limited set of combinations of θ_{on} and θ_{ov} values. Then, the extreme of the interpolated function determines the optimal parameters for one operating point defined by torque and speed. The procedure described above is repeated to obtain a sufficient number of function points of $f_1(T_c,\omega)$ and $f_2(T_c,\omega)$. Functions obtained in this way are approximations of the optimal solution. Results from optimization are presented in Section 3.3.1.

The described off-line optimization procedure is an alternative to other optimization methods, such as the genetic algorithm. Its advantage is that, unlike genetic algorithms, it does not need new FEM computation when the objective function changes. Instead, the new values of the objective functions are calculated from the interpolated functions for which the computation time is significantly shorter. Once the interpolated functions have been enumerated, no further FEM computation is needed to apply different types of multi-objective optimization.

It must be said that the ability of the current controller to follow the reference current derived from the TSF is limited, especially at high speed. Therefore, the phase torque may not track the TSF. However, described off-line optimization involves the influence of the current controller to the actual phase torque. There is a possibility to reduce the torque ripple further using a more advanced current controller. Authors in [17] propose a predictive pulse width modulation (PWM) current control method that serves this purpose and a modified proportional–derivative (PD) controller that lowers the transient response overshoot is described in [25].

3. Results

3.1. Experimental Setup

The experimental setup consists of 200 W three-phase 12/8 pole switched reluctance motor connected to an induction machine that serves as load. Torque sensor *KISTLER 4520A20* with *CoMo torque evaluation instrument 4700BP0UA* was used to measure the torque, speed, and mechanical power on the shaft. *Infratek 106A Power Analyzer* measured input power. *STM32F303RET6* microcontroller, with the control loop cycle time 50 µs, controlled the asymmetric H-bridge power converter.

3.2. FEM Model Verification

The FEM model of SRM was verified by comparing measured and calculated motor characteristics. Figure 6a shows measured and calculated inductance profiles for currents 1.5 A, 6 A, and 9 A [26]. The more significant deviation occurs at the current of 9 A, but this is not a problem as the operating currents are up to 6 A. Figure 6b shows measured and calculated torque profiles for phase currents 2 A, 4 A, and 6 A. Phase currents from the measurement and FEM model are shown in Figure 7a. The behavior of the FEM hysteresis current controller slightly differs from the real one because it depends on the controller sampling, hysteresis-band, and the current sensor response time. Figure 7b shows the measured and calculated efficiency of the drive. Mechanical and power converter losses are included. Despite the above differences, the authors believe that the match between model and measurement is sufficient to allow meaningful conclusions to be drawn from the simulation results.

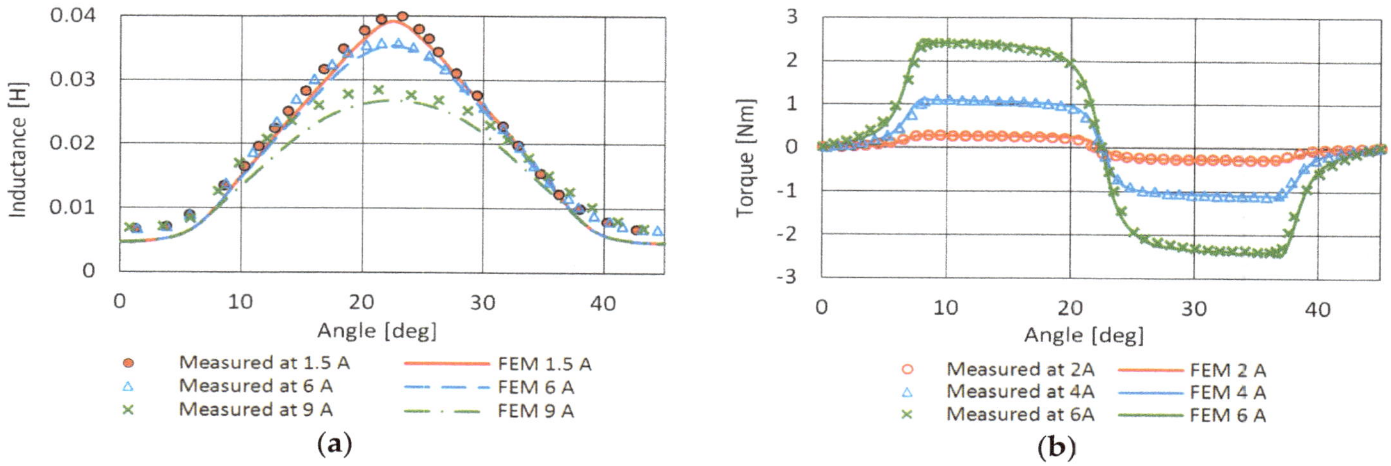

Figure 6. (**a**) Comparison of measured and calculated inductance profiles; (**b**) measured and calculated torque profile.

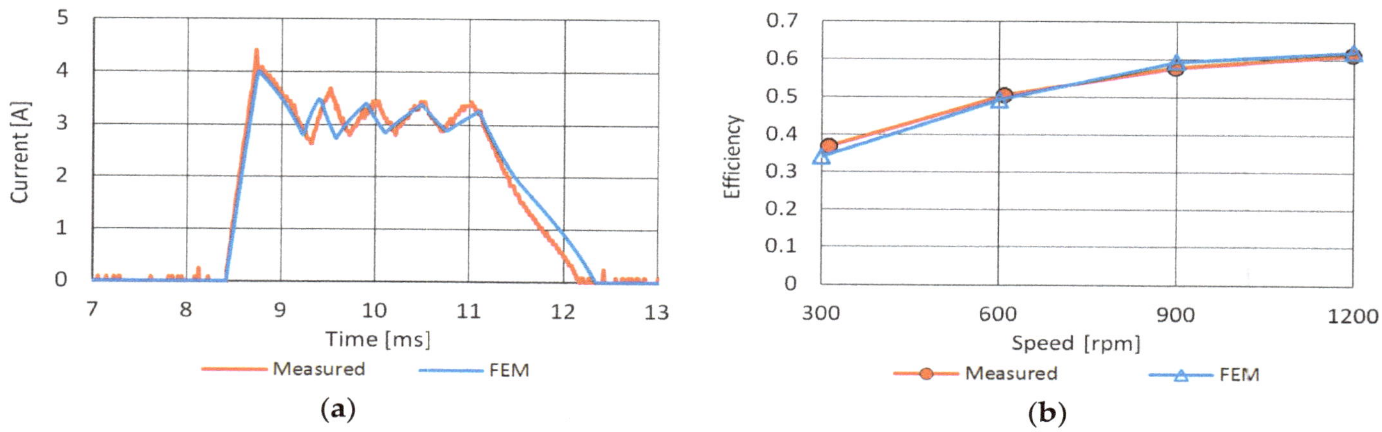

Figure 7. (**a**) Comparison of measured and calculated phase current, $\theta_{on} = 0°$, $\theta_{off} = 19.25°$, speed is 1200 rpm, DC-link voltage is 100 V, load torque is 1 Nm; (**b**) measured and calculated motor efficiency for constant angles $\theta_{on} = 0°$, $\theta_{off} = 17.6°$, mechanical and power converter losses are included.

3.3. Simulation Results

3.3.1. Calculation of Optimal TSF Parameters

Results of preliminary simulation experiments were used to determine the range and number of values for speed, torque, θ_{on}, and θ_{ov} parameters. Then, simulation runs for each parameter combination were calculated, and the values of selected objective functions were recorded. The Matlab piece-wise interpolating polynomials function uses recorded values to calculate interpolation function for each objective. The filled contour graph in Figure 8 visualizes the motor efficiency, relative torque ripple, and integral square error criterion for two operational points: torque 1 Nm, speed 600 rpm (Figure 8a), and 1200 rpm (Figure 8b).

The data tips show values of θ_{on} and θ_{ov} parameters for the specified operating point that are optimal according to the corresponding objective function. Figure 9 shows optimal values of θ_{on} and θ_{ov} as a function of speed separately in two graphs for speeds 300, 600, 900, and 1200 rpm. Simulated motor torque for TSF-based control with optimal parameter settings according to three objective functions is shown in Figure 10a for speed 600 rpm, and in Figure 10b for speed 1200 rpm.

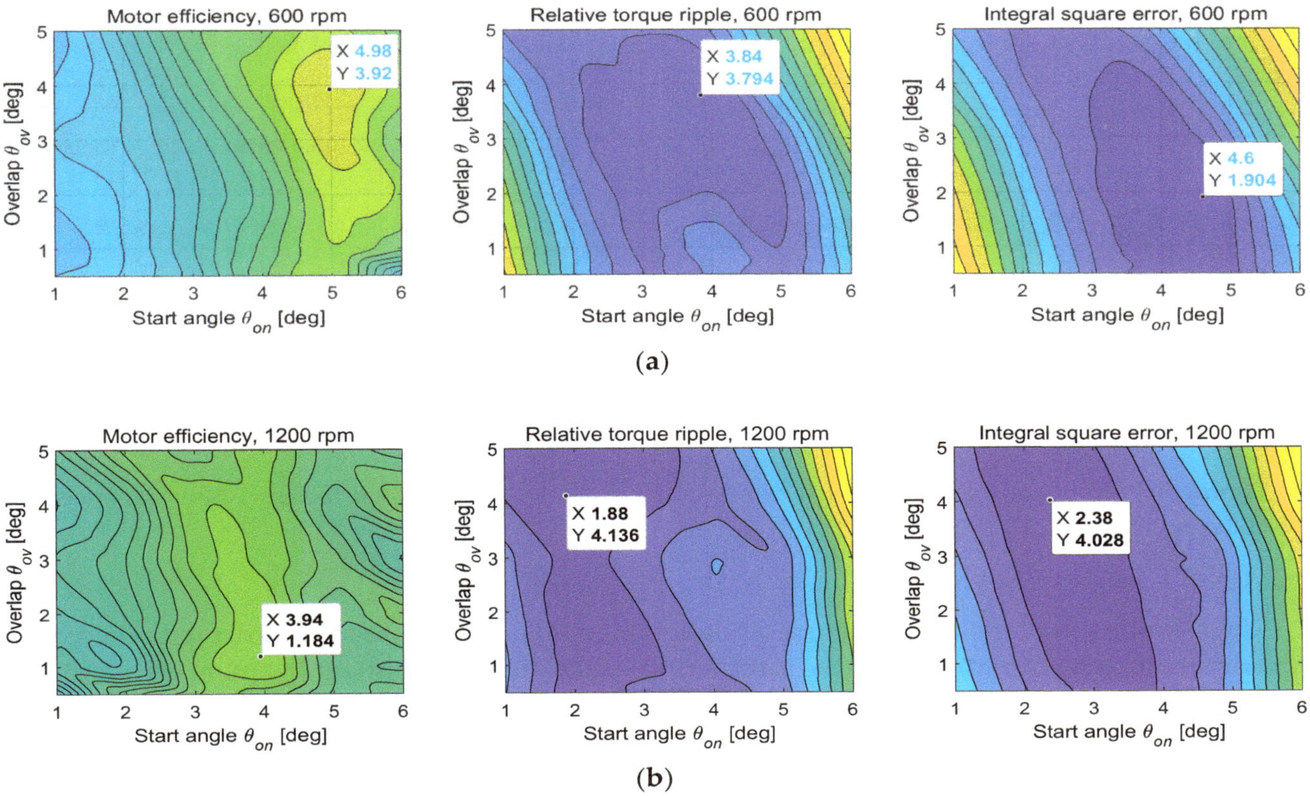

Figure 8. Values of the objective functions at load torque 1 Nm: (**a**) speed 600 rpm; (**b**) speed 1200 rpm.

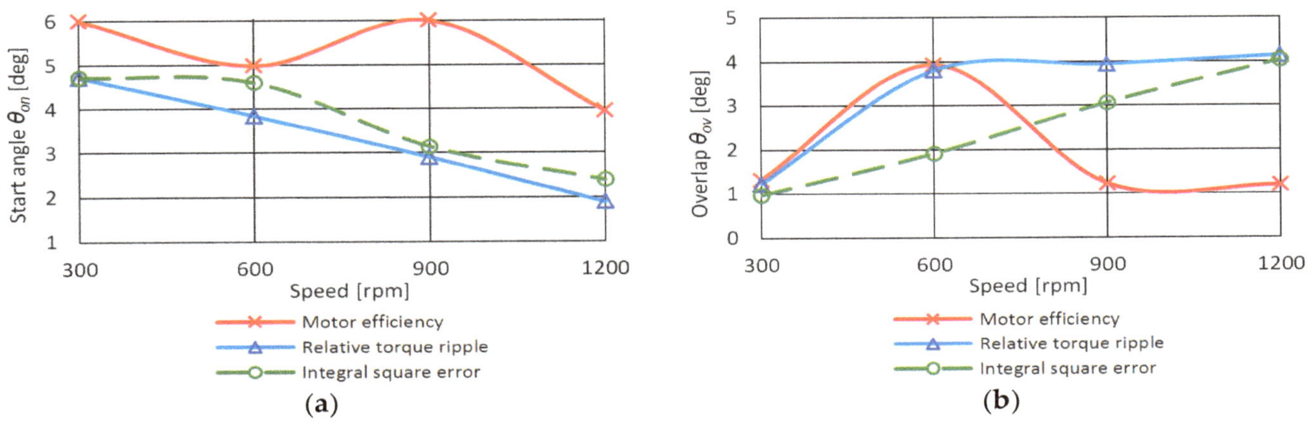

Figure 9. Optimal parameter values at load torque 1 Nm: (**a**) start angle θ_{on}; (**b**) overlap θ_{ov}.

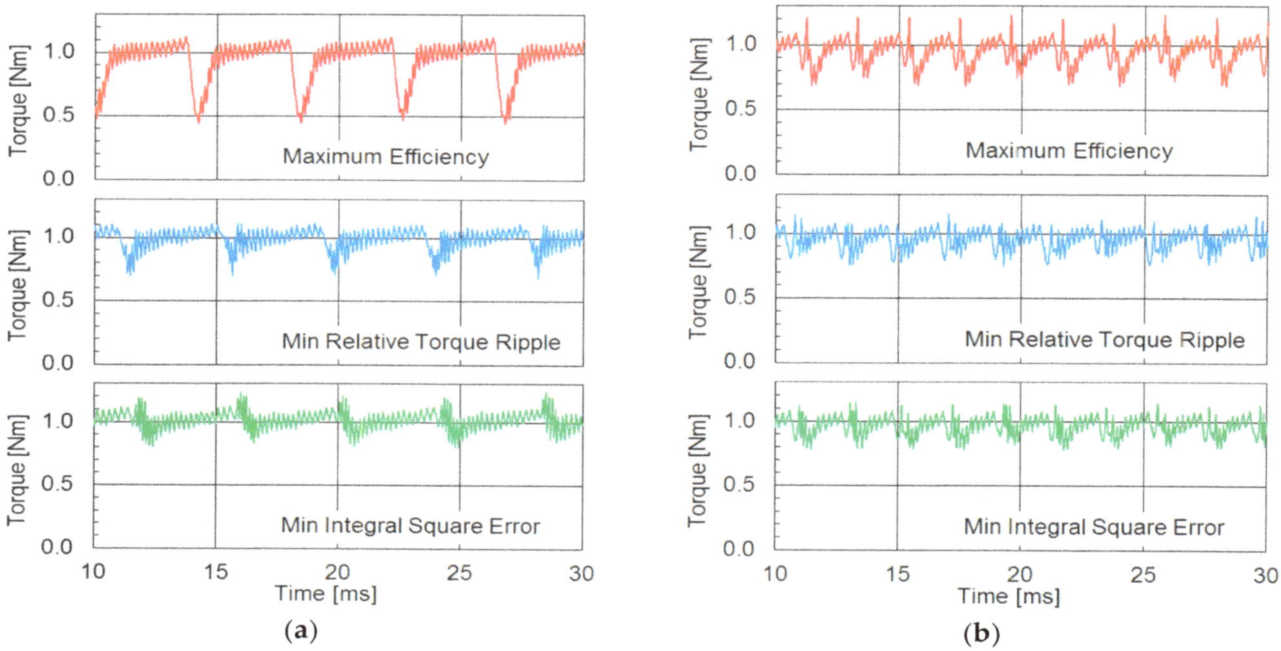

Figure 10. Simulated motor torque for optimal parameter values at load torque 1 Nm: (**a**) speed 600 rpm; (**b**) speed 1200 rpm.

3.3.2. Comparison of FAM and TSF-based control

The comparison of the firing angle modulation method and TSF-based control method according to maximal efficiency, minimal relative torque ripple, and minimum integral square error criteria is presented in Figure 11. Each graph shows the value of the corresponding objective function for four cases. The first three cases are TSF motor control methods optimized according to the defined objective functions. The fourth case is the FAM method, where θ_{on} and θ_{off} are calculated according to Equations (2) and (3), respectively.

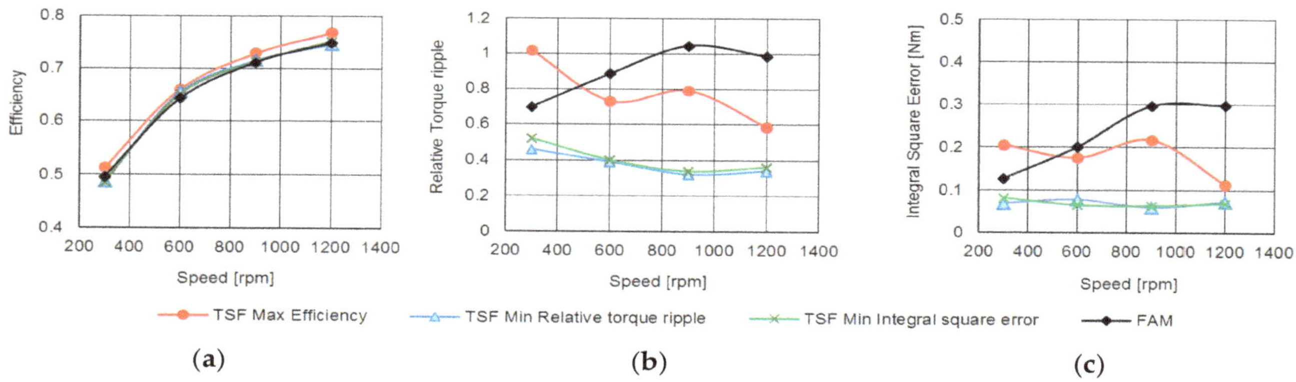

Figure 11. Comparison of the firing angle modulation (FAM) method and TSF at load torque 1 Nm: (**a**) efficiency; (**b**) relative torque ripple; (**c**) integral square error.

Figure 11a shows that the control with TSF optimized for efficiency has the highest efficiency value, but the differences between the other control methods are small, and the efficiency of the FAM is only 1.7% lower (Figure 11a). As shown in Figure 11b, the control with TSF optimized for relative torque ripple, and TSF optimized for integral square error has a small value of relative torque ripple, unlike the FAM and TSF, optimized for efficiency, that has a large value of relative torque ripple. Similar results are displayed in Figure 11c for the integral square error criterion.

Inconsistency in the calculated data appeared in Figure 11c. The value of integral square error criterion (ISE) should be the smallest one for TSF optimized for ISE. However, TSF optimized for relative torque ripple has the lowest value of ISE for 300 and 900 rpm. This unexpected finding shows that the accuracy of a numerical solution is limited.

4. Discussion and Conclusions

This paper presents tailoring and a comparison of two methods to control a commercially available small switched reluctance motor, whose construction is not optimized for the minimization of torque ripple. Furthermore, a new approach to off-line evaluation of objective functions for motor control optimization that reduce computational time is presented.

The first control method is a firing angle modulation, which can be easily implemented on a low-cost microcontroller if the price of the drive is critical for a certain market segment. The second one is a control method with Torque Sharing Functions, which is more complex and uses torque-angle-current look-up tables calculated in advance by FEM.

Objective functions of the control methods optimization are maximal efficiency, minimal relative torque ripple, and minimum integral square error criterion.

The research shows that FAM has the same overall efficiency as TFS optimized for the torque ripple and TSF optimized for integral square error, but FAM has a high torque ripple. The results of simulation experiments show that the presented off-line optimization procedure can find optimal parameters matching the selected objective. However, it is not possible to achieve the lowest torque ripples and the highest efficiency at the same time. This is in line with the results of other authors [11,19,20,22]. Multi-objective optimization should be used to solve the problem. Herein, the results from the presented off-line optimization procedure can serve this purpose. Once the set of simulation runs is executed and the values of each objective function are calculated, another multi-objective optimization, e.g., Pareto-optimal approach, can be applied.

To conclude, the efficiency of FAM control is comparable with a more advanced TSF method, and it makes sense to use it in cost-effective drives where the high torque ripple does not affect the device, e.g., pumps and fans. More advanced SRM control must be used if low torque ripple is required. The presented off-line optimization procedure can be used for single or multi-objective optimization.

The main contribution of this paper is the quantitative comparison of two control methods for SRM that can help in decision making in the process of designing a commercial drive for existing off-the-shelf switched reluctance motors. The second contribution is the explanation of how to tailor and optimize the parameters of control methods using FEM. Presented off-line calculations with the objective function interpolation reduce the computational time. Once the objective functions have been enumerated, no further FEM computation is needed to apply different types of multi-objective optimization.

In future research, the attention will be given to on-line optimization where the objective function value is obtained from the direct measurement of efficiency, vibration, or acoustic noise.

Author Contributions: Conceptualization, P.B. and Ž.F.; methodology, P.B. and Ž.F.; software ANSYS Maxwell and Simplorer, Ž.F.; software Matlab, P.B.; validation, P.B. and Ž.F.; formal analysis, Ž.F.; investigation, P.B. and Ž.F.; writing P.B.; funding acquisition, Ž.F. All authors have read and agreed to the published version of the manuscript.

Abbreviations and Symbols

ATC	Average Torque Control
CPLD	Complex Programmable Logic Devices
DITC	Direct Instantaneous Torque Control
DSP	Digital Signal Processor
DTC	Direct Torque Control

ISE	Integral Square Error criterion
FAM	Firing Angle Modulation method
FBL	Feedback Linearization control
FEM	Finite Element Method
FPGA	Field-Programmable Gate Array
IC	Intelligent Control
ILC	Iterative Learning Control
MPU	Micro Processor Unit
rpm	revolution per minute [r/min]
SRM	Switched Reluctance Motor
TSF	Torque Sharing Function
$f_1(T_c,\omega)$	The function for optimal value of θ_{on} parameter of TSF
$f_2(T_c,\omega)$	The function for optimal value of θ_{ov} parameter of TSF
$f_{dn}(\theta)$	The declining part of the TSF shape
$f_{up}(\theta)$	The rising part of the TSF shape
I_{15}	The phase current at angle 15°
I_A, I_B, I_C	The instantaneous current in phase A, B, and C
I_{Aref}, I_{Bref}, I_{Cref}	The reference current for phase A, B, and C
I_{max}	The maximum current reference
I_{ref}	The reference current
I_{ref0}	The reference current at angle 0°
L_u	The inductance in unaligned position
n	The rotor speed in [rpm]
T_{Aref}, T_{Bref}, T_{Cref}	The torque profile for motor phase A, B, and C
T_{avg}	The average torque
T_c	The torque command
T_{max}	The maximum torque on the interval
T_{min}	The minimum torque on the interval
T_{rip}	The relative torque ripple
V_{DC}	The supply voltage
η	The efficiency
θ	The rotor angle
θ_m	The rotor angle where the inductance begins to rise
θ_{off}	The turn-off angle
θ_{on}	The turn-on angle
θ_{ov}	The overlap angle
θ_p	The rotor pole pitch
ω	The rotor angular speed
ω_{ref}	The rotor angular speed reference

References

1. Santo, A.E.; Calado, M.R.; Cabrita, C.M. Sliding Mode Position Controller for a Linear Switched Reluctance Actuator. In *Sliding Mode Control*; IntechOpen: London, UK, 2011; pp. 181–202. [CrossRef]
2. Ahn, J.-W.; Lukman, G.F.L. Switched Reluctance Motor: Research Trends and Overview. *China Electrotech. Soc. Trans. Electr. Mach. Syst.* **2019**, *2*, 339–347. [CrossRef]
3. Jarvis, R. Davidson's locomotive: How did he do it? *Eng. Sci. Educ. J.* **1996**, *5*, 281–288. [CrossRef]
4. Lawrenson, P.; Stephenson, J.; Blenkinsop, P.; Corda, J.; Fulton, N. Variable-speed switched reluctance motors. *IEE Proc. B Electr. Power Appl.* **1980**, *127*, 253. [CrossRef]
5. More, A. Switched Reluctance Motors Market 2019 Industry Size by Global Major Companies Profile, Competitive Landscape and Key Regions 2025. 2019. Available online: https://www.theexpresswire.com/pressrelease/Switched-Reluctance-Motors-Market-2019-Industry-Size-by-Global-Major-Companies-Profile-Competitive-Landscape-and-Key-Regions-2025-360-Research-Report_10273615 (accessed on 21 March 2020).

6. Kozuka, S.; Tanabe, N.; Asama, J.; Chiba, A. Basic characteristics of 150,000r/min switched reluctance motor drive. In Proceedings of the IEEE PES General Meeting, Pittsburgh, PA, USA, 20–24 July 2008; pp. 1–4. [CrossRef]
7. Miller, T. Optimal design of switched reluctance motors. *IEEE Trans. Ind. Electron.* **2002**, *49*, 15–27. [CrossRef]
8. Belhadi, M.; Krebs, G.; Marchand, C.; Hannoun, H.; Mininger, X. Geometrical optimization of SRM on operating mode for automotive application. *Electr. Eng.* **2017**, *100*, 303–310. [CrossRef]
9. Gan, C.; Wu, J.; Sun, Q.; Kong, W.; Li, H.; Hu, Y. A Review on Machine Topologies and Control Techniques for Low-Noise Switched Reluctance Motors in Electric Vehicle Applications. *IEEE Access* **2018**, *6*, 31430–31443. [CrossRef]
10. Lin, C.-H.; Hwang, C.-C. High Performances Design of a Six-Phase Synchronous Reluctance Motor Using Multi-Objective Optimization with Altered Bee Colony Optimization and Taguchi Method. *Energies* **2018**, *11*, 2716. [CrossRef]
11. Xue, X.; Cheng, K.-W.E.; Ho, S.L. Optimization and Evaluation of Torque-Sharing Functions for Torque Ripple Minimization in Switched Reluctance Motor Drives. *IEEE Trans. Power Electron.* **2009**, *24*, 2076–2090. [CrossRef]
12. Evangeline, J.S.; Kumar, S.S.; Jayakumar, J. Torque modeling of Switched Reluctance Motor using LSSVM-DE. *Neurocomputing* **2016**, *211*, 117–128. [CrossRef]
13. Ye, W.; Ma, Q.; Zhang, P. Improvement of the Torque-Speed Performance and Drive Efficiency in an SRM Using an Optimal Torque Sharing Function. *Appl. Sci.* **2018**, *8*, 720. [CrossRef]
14. Zhang, M.; Bahri, I.; Mininger, X.; Vlad, C.; Xie, H.; Berthelot, E. A New Control Method for Vibration and Noise Suppression in Switched Reluctance Machines. *Energies* **2019**, *12*, 1554. [CrossRef]
15. Cheng, H.; Chen, H.; Yang, Z. Average torque control of switched reluctance machine drives for electric vehicles. *IET Electr. Power Appl.* **2015**, *9*, 459–468. [CrossRef]
16. Fuengwarodsakul, N.H.; Menne, M.; Inderka, R.; De Doncker, R. High-Dynamic Four-Quadrant Switched Reluctance Drive Based on DITC. *IEEE Trans. Ind. Appl.* **2005**, *41*, 1232–1242. [CrossRef]
17. Cai, H.; Wang, H.; Li, M.; Shen, S.; Feng, Y.; Zheng, J. Torque Ripple Reduction for Switched Reluctance Motor with Optimized PWM Control Strategy. *Energies* **2018**, *11*, 3215. [CrossRef]
18. Bose, B.K.; Miller, T.J.E.; Szczesny, P.M.; Bicknell, W.H. Microcomputer Control of Switched Reluctance Motor. *IEEE Trans. Ind. Appl.* **1986**, *22*, 708–715. [CrossRef]
19. Hamouda, M.; Szamel, L. *Reduced Torque Ripple based on a Simplified Structure Average Torque Control of Switched Reluctance Motor for Electric Vehicles*; Institute of Electrical and Electronics Engineers (IEEE): Piscataway, NJ, USA, 2018.
20. Xue, X.D.; Cheng, K.W.E.; Cheung, N.C. Evaluation of torque sharing functions for torque ripple minimization of switched reluctance motor drives in electric vehicles. In Proceedings of the 2008 Australasian Universities Power Engineering Conference, Sydney, NSW, Australia, 14–17 December 2008; pp. 1–6.
21. Ye, J.; Bilgin, B.; Emadi, A. An Extended-Speed Low-Ripple Torque Control of Switched Reluctance Motor Drives. *IEEE Trans. Power Electron.* **2014**, *30*, 1457–1470. [CrossRef]
22. Li, H.; Bilgin, B.; Emadi, A. An Improved Torque Sharing Function for Torque Ripple Reduction in Switched Reluctance Machines. *IEEE Trans. Power Electron.* **2018**, *34*, 1635–1644. [CrossRef]
23. Ro, H.-S.; Lee, K.-G.; Lee, J.-S.; Jeong, H.-G.; Lee, K.-B. Torque Ripple Minimization Scheme Using Torque Sharing Function Based Fuzzy Logic Control for a Switched Reluctance Motor. *J. Electr. Eng. Technol.* **2015**, *10*, 118–127. [CrossRef]
24. Panek, D.; Orosz, T.; Karban, P. Artap: Robust Design Optimization Framework for Engineering Applications. In Proceedings of the 2019 Third International Conference on Intelligent Computing in Data Sciences (ICDS), Marrakech, Morocco, 28–30 October 2019; pp. 1–6.
25. Sovicka, P.; Rafajdus, P.; Vavrus, V. Switched reluctance motor drive with low-speed performance improvement. *Electr. Eng.* **2019**, *102*, 27–41. [CrossRef]
26. Ferkova, Z.; Suchy, L.; Cernohorsky, J. Measurement of switched reluctance motor parameters. In Proceedings of the 2017 19th International Conference on Electrical Drives and Power Electronics (EDPE), Dubrovnik, Croatia, 4–6 October 2017; pp. 287–290. [CrossRef]

10

A Super-Twisting Sliding-Mode Stator Flux Observer for Sensorless Direct Torque and Flux Control of IPMSM

Junlei Chen, Shuo Chen, Xiang Wu, Guojun Tan * and Jianqi Hao

School of Electrical and Power Engineering, China University of Mining and Technology, Xuzhou 221116, China
* Correspondence: gjtan@cumt.edu.cn.

Abstract: The scheme based on direct torque and flux control (DTFC) as well as active flux is a good choice for the interior permanent magnet synchronous motor (IPMSM) sensorless control. The precision of the stator flux observation is essential for this scheme. However, the performance of traditional observers like pure integrator and the low-pass filter (LPF) is severely deteriorated by disturbances, especially dc offset. Recently, a sliding-mode stator flux observer (SMFO) was proposed to reduce the dc offset in the estimated stator flux. However, it cannot eliminate the dc offset totally and will cause the chattering problem. To solve these problems, a novel super-twisting sliding-mode stator flux observer (STSMFO) is proposed in this paper. Compared with SMFO, STSMFO can reduce the chattering and fully eliminate the dc offset without any amplitude and phase compensation. Then, the precision of the stator flux and rotor position can be greatly improved over a wide speed region. The detailed mathematical analysis has been given for comparing it with another three traditional observers. The numerical simulations and experimental testing with an IPMSM drive platform have been implemented to verify the capability of the proposed sensorless scheme.

Keywords: interior permanent magnet synchronous motor (IPMSM); active flux; sensorless control; stator flux observation; super-twisting sliding-mode stator flux observer (STSMFO)

---

1. Introduction

Interior permanent magnet synchronous motor (IPMSM) has been utilized in wide industrial fields because of its advantages like high torque density, fast response, and low torque ripple [1–3]. Direct torque control (DTC) is one of the most popular strategies for ac machines for its fast torque response and strong robustness [4,5]. However, the traditional DTC relies on the hysteresis comparators and the switching table, which will result in the unfixed switching frequency and the inaccurate compensation of the flux magnitude and the torque error. Furthermore, the ripple will be caused by the hysteresis comparators in the stator flux, and ripple can deteriorate the performance of the IPMSM drive [6]. To overcome this problem, the space vector modulation (SVM) is introduced into the DTC algorithm and the DTC-SVM is proposed [7]. The reference voltage is selected via three variables which are torque error, reference stator flux amplitude, and feedback flux vector, respectively. The introduction of the SVM can fix the switching frequency and reduce the ripple. However, the robustness is poor due to the single torque closed-loop structure of the DTC-SVM. In [8], the direct torque and flux control (DTFC) is proposed to enhance the robustness of the DTC-SVM. Compared with the DTC-SVM, one extra proportional-integral (PI) regulation is utilized to control flux and it can realize the flux to track the flux reference more precisely.

For the sake of realizing the high-performance control of IPMSM, the accurate position information is essential. However, position sensors not only decrease the reliability of the IPMSM system but also increase the cost. Therefore, research of the sensorless strategies for PMSMs have been paid much attention. In general, sensorless methods are divided into two categories which are based on signal injection and machine model according to speed region. At low speeds, the signal injection is generally adopted to realize rotor estimation [9–11]. However, it can cause the extra loss and torque ripple. With the speed increasing to the medium domain, model-based schemes are widely utilized to observe the position [12–16]. Due to its ease of implementation, plenty of research is focused on the designation of the back-emf observer like the extended Kalman filter [12] and the sliding-mode observer (SMO) [13]. However, the capability of the back-emf-based sensorless scheme is poor at low speeds because of the non-negligible noise. In [17], the active flux concept is proposed to transform salient-pole motors into virtual nonsalient-pole motors. The active flux is independent of speed and the direction aligns with rotor direction. Therefore, the active flux-based sensorless scheme can be readily implemented. Even at low speeds, the sensorless control can be carried out without signal injection [18]. Moreover, the stator flux can be observed by the DTFC and the corresponding robustness is enhanced due to the regulation of the DTFC. Therefore, the active flux-based sensorless scheme is suitable for the DTFC-based IPMSM drive system

Based on the IPMSM model, the stator flux is the integral of the back-emf and it is critical for flux observers to consider the influence of disturbances, especially dc offset [19]. When the pure integrator is adopted as a flux observer, the flux dc offset will increase with time, eventually resulting in the saturation of the observer [20]. In [21], a disturbance observer is designed to eliminate the dc component, but it is difficult to implement due to its complex structure. In [22], the dc component of the back-emf is filtered out by a low-pass filter (LPF). Actually, the LPF is a pure integrator cascaded with a high-pass filter. However, the amplitude attenuation and the phase delay are inevitable because of the introduction of the LPF. Therefore, the accurate compensation is essential for the sensorless control scheme. In [23], the SMFO is proposed to observe the stator flux without compensation, which is robust against disturbances and can reduce the dc offset through its compensation term. However, the dc disturbance rejection capability of the SMFO is limited and the chattering problem of the traditional SMFO is inevitable. It is worth mentioning that even a little dc offset can cause serious speed ripples. In [24], super-twisting algorithm (STA) is proposed to decrease the chattering of the SMO without reducing the robustness and it has been widely utilized in observers and controllers [25–28].

This paper proposed a novel STSMFO for the active flux-based IPMSM sensorless scheme. The stator flux observer based on STA is designed to overcome the disadvantages of the traditional observer, including the saturation effect, the position compensation, and flux dc offset. The corresponding Lyapunov stability of the STSMFO is proved based on references [29,30]. Then, the precision of the flux and position estimation can be greatly enhanced over a wide speed region. Moreover, the proposed sensorless method is better than most model-based sensorless strategies at low speeds due to the fact that it can be successfully utilized at low speeds without signal injection and compensation. Finally, the capability of the scheme based on the novel STSMFO is confirmed through numerical simulation and experimental testing with an IPMSM drive platform.

This paper is organized as follows: Section 2 gives the IPMSM model and the concept of DTFC. Section 3 introduces the active flux-based sensorless strategy and the mathematical analysis of three traditional observers. Section 4 introduces the proposed STSMFO and its analysis. The evaluation of the simulation and the experiment is given in Section 5. Section 6 draws the conclusions.

2. Direct Torque and Flux Control

Figure 1 shows the relationship between the stator flux-oriented coordinate (x-y), rotating reference coordinate (d-q), and stationary reference coordinate (α-β). The IPMSM stator flux and voltage equations in the d-q axis are presented as:

$$\begin{cases} u_d = R_s i_d + \frac{d\psi_d}{dt} - \omega_e \psi_q \\ u_q = R_s i_q + \frac{d\psi_q}{dt} + \omega_e \psi_d \end{cases} \tag{1}$$

$$\begin{cases} \psi_q = L_q i_q + \psi_f \\ \psi_q = L_q i_q \end{cases} \quad (2)$$

where, u_{dq}, i_{dq}, and ψ_{dq} are the stator voltages, currents, and flux linkages in the d-q axis, respectively. R_S is the resistance, L_{dq} are the stator inductances, ω_e is the rotor electrical angular velocity, ψ_f is the rotor flux linkage, δ is the load angle, θ_e is the rotor electrical position, and θ_s is the stator flux angle.

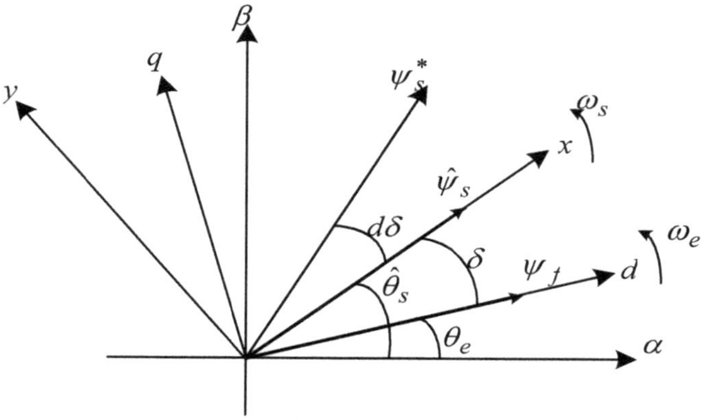

Figure 1. The relationship between various reference frames.

Transforming (1) to an x-y reference frame and combining the relationship between load angle and torque [31], the voltage equation is found as:

$$\begin{cases} u_x = R_s i_x + \frac{d|\psi_s|}{dt} \\ u_y = R_s i_y + \omega_s|\psi_s| = R_s i_y + \omega_e|\psi_s| + \frac{1}{K}\cdot\frac{dT_e}{dt} \end{cases} \quad (3)$$

where, $K = \frac{3p_n}{2L_dL_q}\left[\psi_f L_q cos\delta - |\psi_s|(L_q - L_d)cos2\delta\right]$, p_n is the number of pole pairs, ψ_s is the stator flux vector, and u_{xy} and i_{xy} are the stator voltages and currents in the x-y frame, respectively.

The IPMSM stator flux and torque equations in the α-β axis can be given as:

$$\begin{cases} \psi_\alpha = \int(u_\alpha - R_s i_\alpha)dt \\ \psi_\beta = \int(u_\beta - R_s i_\beta)dt \end{cases} \quad (4)$$

$$|\psi_s| = \sqrt{\psi_\alpha^2 + \psi_\beta^2} \quad (5)$$

$$T_e = \frac{3}{2}p_n(\psi_\alpha i_\beta - \psi_\beta i_\alpha) \quad (6)$$

where, $u_{\alpha\beta}$, $i_{\alpha\beta}$, and $\psi_{\alpha\beta}$ are the α-β axis stator voltages, currents, and flux linkages respectively, and T_e is torque.

3. Active Flux-Based Sensorless Control

To observe the rotor position of IPMSMs easily, the active flux is proposed to transform IPMSMs into virtual nonsalient-pole machines. The d-q axis active flux equation in reference [17] is given as:

$$\begin{cases} \psi_{dAF} = \psi_d - L_q i_d = \psi_f + (L_q - L_d)\cdot i_d \\ \psi_{qAF} = \psi_q - L_q i_q = 0 \end{cases} \quad (7)$$

where, ψ_{dqAF} are the active fluxes in the d-q axis. Equation (7) shows that the active flux aligns with the rotor direction, therefore active flux in the α-β frame results in:

$$\begin{cases} \psi_{\alpha AF} = \psi_\alpha - L_q i_\alpha = \psi_{dAF} cos\theta_e \\ \psi_{\beta AF} = \psi_\beta - L_q i_\beta = \psi_{dAF} sin\theta_e \end{cases} \tag{8}$$

where, $\psi_{\alpha\beta AF}$ are the active fluxes in the α-β axis. Equation (8) shows that the rotor position can be observed via the active flux vector. One way to extract it is a phase-locked loop. Moreover, it is worth mentioning that the active flux is obtained on the basis of stator flux. Therefore, the accuracy of the rotor position observation directly relies on the estimation accuracy of stator flux. In practice, a dc component in back-emf is inevitable and it is the main reason for the deteriorating performance of the flux observer. Considering the dc component, the stator flux equation is presented as:

$$\psi_s = \int (u_s - R_s i_s + A_0) dt = \int e_s dt \tag{9}$$

where u_s and i_s are the stator voltage and current, A_0 is the dc component, and e_s is the back-emf. For the convenience of analysis, e_s is expressed as:

$$e_s = A_0 + A_1 sin(\omega_1 t + \varphi_1) \tag{10}$$

where, $A_1 sin(\omega_1 t + \varphi_1)$ is the fundamental component. A_1, ω_1, and φ_1 are amplitude, angular frequency, and initial angel of the fundamental component, respectively. Taking the Laplace transformation of back-emf, it can be found as:

$$E_s(s) = \frac{A_0}{s} + A_1 \frac{s \cdot sin\varphi_1 + \omega_1 cos\varphi_1}{s^2 + \omega_1^2} \tag{11}$$

where, $E_s(s)$ is the Laplacian form of e_s and s is the Laplacian operator.

3.1. Pure Integrator-Based Observer

The Laplace transform of pure integrator is given as:

$$\psi_{s\_I}(s) = \frac{E_s(s)}{s} \tag{12}$$

where, $\psi_{s\_I}(s)$ is the Laplacian form of ψ_{s_I}, and ψ_{s_I} is the pure integrator-observed stator flux. Taking (11) into (12) then results in [19]:

$$\psi_{s\_I} = A_0 t + \frac{A_1 cos\varphi_1}{\omega_1} + \frac{A_1}{\omega_1} sin\left(\omega_1 t + \varphi_1 - \frac{\pi}{2}\right) \tag{13}$$

As seen, two disturbance terms exist in the estimated stator flux by pure integrator. One is a component increasing with time linearly, which is caused by a dc component in the back-emf and it can eventually result in the saturation of the integrator. Another one is a dc offset determined by the fundamental wave and it will change at various initial positions. These two terms can cause serious distortions of the estimated stator flux. Furthermore, the precise rotor position estimation cannot be achieved due to the distorted active flux.

3.2. Low-Pass Filter (LPF)-Based Observer

For the sake of filtering out the dc component in the back-emf, a high-pass filter is used to cascade with pure integrator. The Laplace transform of the observer is found as:

$$\psi_{s\_LPF}(s) = \frac{1}{s} \cdot \frac{s}{s + \omega_c} \cdot E_s(s) = \frac{E_s(s)}{s + \omega_c} \tag{14}$$

which shows that the observer is actually a LPF. In Equation (14), $\psi_{s\_LPF}(s)$ is the Laplace transform of ψ_{s_LPF}, ψ_{s_LPF} and ω_c are the estimated stator flux and cutoff frequency of LPF, respectively. ω_c is always selected much lower than ω_1 because it is, in effect, the cutoff frequency of the high-pass filter. Taking (11) into (14) then results in [19]:

$$\psi_{s\_LPF} = \frac{A_0}{\omega_c} - \frac{A_0}{\omega_c}e^{-\omega_c t} + \frac{A_1 cos(\varphi_1 + \theta_1)}{\sqrt{\omega_1^2 + \omega_c^2}}e^{-\omega_c t} + \frac{A_1}{\sqrt{\omega_1^2 + \omega_c^2}} sin\left(\omega_1 t + \varphi_1 - \frac{\pi}{2} + \theta_1\right) \quad (15)$$

where, $\theta_1 = tan^{-1}(\omega_c/\omega_1)$ is the phase delay caused by the LPF. Compared with Equation (13), the component causing integrator saturation is removed. The two main distorted terms decrease exponentially to zero with time. Although the dc offset still cannot be eliminated totally, it is already decreased greatly by LPF. However, the phase delay θ_1 and amplitude attenuation $A_1/\sqrt{\omega_1^2 + \omega_c^2}$ of the fundamental wave is inevitable because of the introduction of the LPF. Therefore, the estimated rotor position by LPF needs precise compensation to ensure good performance of the sensorless control.

3.3. First-Order Sliding-Mode Stator Flux Observer (SMFO)

Obviously, no matter whether a pure integrator-based observer or LPF-based observer is used, it cannot achieve high-performance sensorless control. Consequently, the SMFO is proposed to observe the stator flux. Figure 2 shows the diagram of the SMFO.

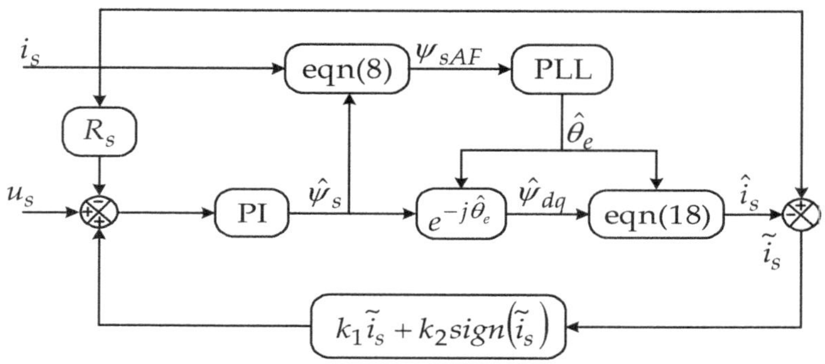

Figure 2. The diagram of the first-order sliding-mode stator flux observer (SMFO).

The SMFO, considering both the current and the voltage model of IPMSM, is designed. Moreover, the former is essentially utilized to compensate the latter. The stator flux and current are defined as the state variable and the output, respectively. On the basis of the IPMSM model, equations of the SMFO result in:

$$\frac{d\psi_s}{dt} = u_s - R_s i_s + k_1\tilde{i}_s + k_2 sign(\tilde{i}_s) \quad (16)$$

$$\begin{bmatrix} \hat{\psi}_d \\ \hat{\psi}_q \end{bmatrix} = \begin{bmatrix} cos\hat{\theta}_e & sin\hat{\theta}_e \\ -sin\hat{\theta}_e & cos\hat{\theta}_e \end{bmatrix} \begin{bmatrix} \hat{\psi}_\alpha \\ \hat{\psi}_\beta \end{bmatrix} \quad (17)$$

$$\begin{bmatrix} \hat{i}_\alpha \\ \hat{i}_\beta \end{bmatrix} = \begin{bmatrix} cos\hat{\theta}_e & -sin\hat{\theta}_e \\ sin\hat{\theta}_e & cos\hat{\theta}_e \end{bmatrix} \begin{bmatrix} \frac{\hat{\psi}_d - \psi_f}{L_d} \\ \frac{\hat{\psi}_q}{L_q} \end{bmatrix} \quad (18)$$

where, $\tilde{i}_s = i_s - \hat{i}_s$ is the stator current estimation error, "^" is the estimated value, and sign () is the sign function. k_1 and k_2 are linear and nonlinear gains of the SMFO which affect the dynamic performance and robustness, respectively. The Lyapunov stability has been proved in reference [8]. Considering the dc offset, (16) can be found as:

$$\frac{d\psi_s}{dt} = u_s - R_s i_s + A_0 + k_1\tilde{i}_s + k_2 sign(\tilde{i}_s) \quad (19)$$

Actually, the SMFO is used to observe the dc offset according to the current estimation error. Further, the estimated dc offset is fed back to the back-emf to realize the accurate estimation of the stator flux. However, the SMFO cannot remove flux dc offset totally. In the steady state, $A_0 + k_1\widetilde{i_s} + k_2 sign(\widetilde{i_s}) = 0$ and the effect of the dc component on the flux estimation is eliminated. However, in the process of reaching a steady state, the estimated flux dc offset is already accumulated. The estimated flux dc offset can be given as:

$$\psi_{s0} = \int_0^{t_1} (A_0 + k_1\widetilde{i_s} + k_2 sign(\widetilde{i_s}))dt \tag{20}$$

where, ψ_{s0} is the flux dc offset, and t_1 is the time to steady state. In general, ψ_{s0} can be omitted because A_0 and t_1 are small, but if new dc disturbances interfere with the system, it will cause the accumulation of flux estimation dc offset and eventually have a non-negligible impact on the stator flux observation.

4. Super-Twisting Sliding-Mode Stator Flux Observer

The estimation precision of the stator flux dramatically affects the capability of the DTFC-based active flux sensorless scheme. The first-order SMFO can observe the stator flux without phase shift and amplitude attenuation. However, the conventional first-order SMFO is still sensitive to dc disturbances and suffers from the chattering problem. To overcome the limitations of the first-order SMFO, the well-known STA was utilized to design an observer for estimating stator flux in this paper. With the STA, the chattering can be reduced and the dc offset can be eliminated effectively.

4.1. Super-Twisting Algorithm

To improve the chattering problem of the traditional SMO, the super-twisting algorithm was proposed in reference [24]. The equation of STA considering perturbation terms can be presented as:

$$\begin{cases} \frac{dx_1}{dt} = -k_1|\widetilde{x_1}|^{0.5} sign(\widetilde{x_1}) + x_2 + \rho_1(x_1, t) \\ \frac{dx_2}{dt} = -k_2 sign(\widetilde{x_1}) + \rho_2(x_2, t) \end{cases} \tag{21}$$

where, x_i, $\widetilde{x_i}(\widetilde{x_i} = \hat{x}_i - x_i)$, k_i and ρ_i are state variables, state variables estimation error, gains of the STA, and perturbation terms, respectively. The Lyapunov stability has been proved in references [29,30]. According to [29,30], the observer is stable when sliding-mode gains and the perturbation terms satisfy (22):

$$\begin{cases} k_1 > 2\delta_1, k_2 > k_1 \frac{5\delta_1 k_1 + 4\delta_1^2}{2(k_1 - 2\delta_1)} \\ |\rho_1| \le \delta_1 |x_1|^{0.5}, \rho_2 = 0 \end{cases} \tag{22}$$

where, δ_1 is a positive constant.

4.2. Super-Twisting Sliding-Mode Stator Flux Observer

To solve problems of chattering and the dc offset in the traditional SMFO, a stator flux observer is proposed based on STA. Substituting $x_1 = \hat{\psi}_s$ into (21), then Equation (21) can be rewritten as:

$$\frac{d\hat{\psi}_s}{dt} = -k_1|\widetilde{\psi_s}|^{0.5} sign(\widetilde{\psi_s}) - k_2 \int sign(\widetilde{\psi_s})dt + \rho_1(\hat{\psi}_s, t) \tag{23}$$

where, $\rho_1(\hat{\psi}_s, t)$ is:

$$\rho_1(\hat{\psi}_s, t) = u_s - R_s i_s + A_0 \tag{24}$$

Taking (24) into (22) results in:

$$|u_s - R_s i_s + A_0| \le \delta_1 |\psi_s|^{0.5} \tag{25}$$

which shows that the system can converge in finite time to sliding surface for a large enough δ_1. By subtracting (23) from (4), the flux error equation is found as:

$$\frac{d\widetilde{\psi}_s}{dt} = k_1\left|\widetilde{\psi}_s\right|^{0.5} sign\left(\widetilde{\psi}_s\right) + k_2 \int sign\left(\widetilde{\psi}_s\right)dt - A_0 \tag{26}$$

At the starting point, the stator flux estimation value is different from the actual value due to the existence of dc offset A_0. Then, the sliding-mode terms $k_1\left|\widetilde{\psi}_s\right|^{0.5} sign\left(\widetilde{\psi}_s\right) + k_2 \int sign\left(\widetilde{\psi}_s\right)dt$ are used to estimate the dc component. Furthermore, the estimated dc component is fed back to compensate the back-emf. At a steady state, the flux error is on the sliding surface. Taking $\widetilde{\psi}_s = 0$ into (26) results in:

$$A_0 = k_1\left|\widetilde{\psi}_s\right|^{0.5} sign\left(\widetilde{\psi}_s\right) + k_2 \int sign\left(\widetilde{\psi}_s\right)dt \tag{27}$$

which shows that the effect of the dc component has been removed. In addition, in the process of reaching the steady state, the accumulated flux estimation dc offset can be given as:

$$\psi_{s0\_STA} = \int\left(A_0 - k_1\left|\widetilde{\psi}_s\right|^{0.5} sign\left(\widetilde{\psi}_s\right) - k_2 \int sign\left(\widetilde{\psi}_s\right)dt\right)dt \tag{28}$$

As we all know, the integral term $k_2 \int sign\left(\widetilde{\psi}_s\right)dt$ will remain until the ψ_{s0_STA} is zero. Therefore, compared with the SMFO, the chattering caused by the sign function can be reduced greatly and the estimated flux dc offset can be eliminated completely due to the integral term and a large gain k_2 of the STSMFO. In addition, it is worth mentioning that the stator flux amplitude reference value and estimated flux angle are utilized to replace the actual stator flux vector. The diagram of the STSMFO is shown in Figure 3.

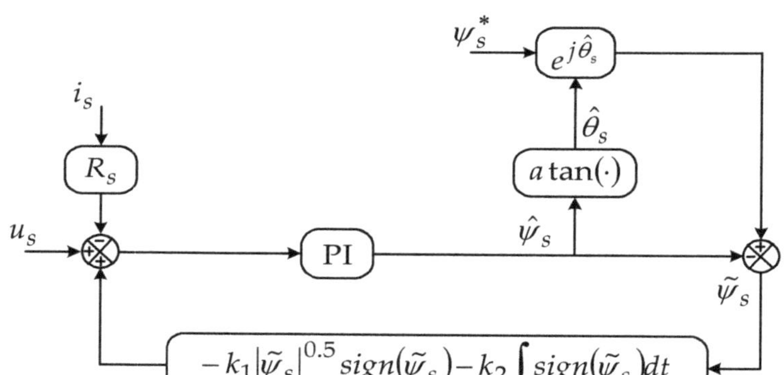

Figure 3. The diagram of the super-twisting sliding-mode stator flux observer (STSMFO).

In conclusion, two disturbance terms by pure integrator can cause serious distortions of the estimated stator flux. The LPF can eliminate the dc offset at the cost of amplitude attenuation and phase shift. The SMFO can reduce the dc offset greatly without any influence on the fundamental wave. However, the SMFO will accumulate the small disturbance and cause chattering. Finally, the proposed STSMFO can remove the dc offset completely and smooth the estimated stator flux.

5. Evaluation via Simulation and Experiment

The IPMSM control scheme is given in Figure 4. Both the simulation and the experiment based on Figure 4 were carried out to verify the capability of the STSMFO. The parameters of the IPMSM are listed in Table 1.

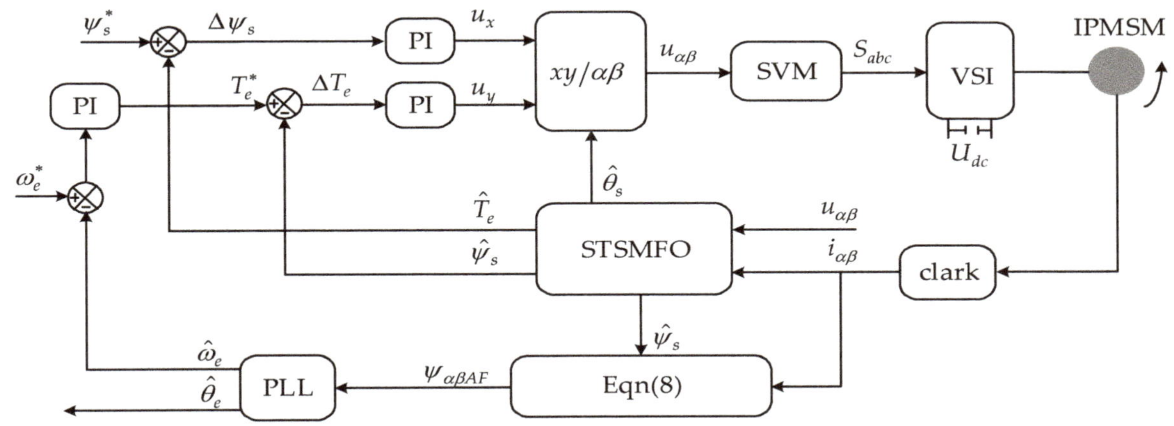

Figure 4. The proposed interior permanent magnet synchronous motor (IPMSM) sensorless scheme.

Table 1. IPMSM Parameters.

Parameter	Value
rated speed	3000 rpm
rated power	60 kW
rated voltage	380 V
rated current	130 A
rotational inertia	0.2 kg·m^2
p_n	4
R_s	0.1 Ω
L_d/L_q	0.95/2.05 mH
ψ_f	0.225 Wb

5.1. Simulation Results

MATLAB/Simulink software was utilized to numerically demonstrate the capability of the STSMFO. The simulation results of α-axis estimated stator fluxes at 300 rpm and 1500 rpm are given in Figure 5.

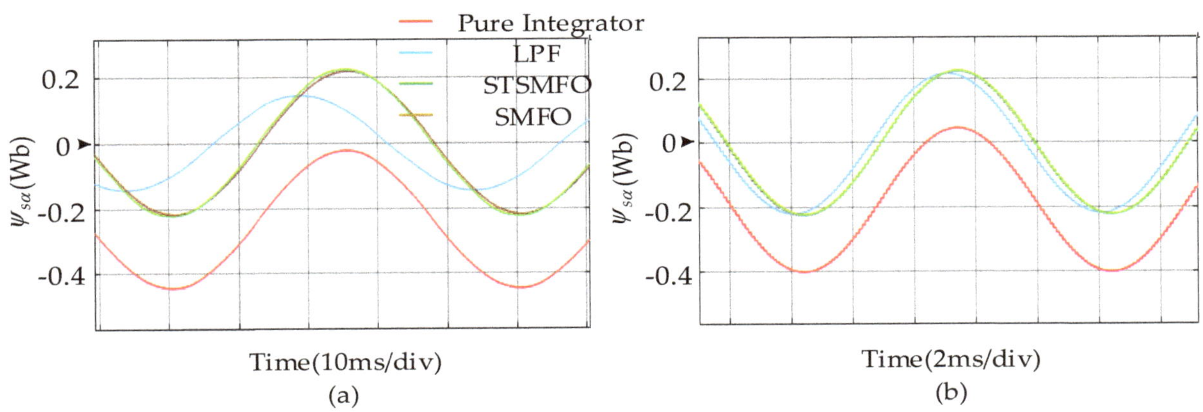

Figure 5. Simulation results of α-axis estimated stator fluxes: (**a**) The stator fluxes at 300 rpm, (**b**) The stator fluxes at 1500 rpm.

The pure integrator estimated stator flux deviates from the ideal value due to the dc offset. Figure 5a,b shows that the LPF will delay the phase and attenuate the amplitude of flux estimation. Moreover, as previously analyzed, the problems have been improved with the increase of the frequency of the fundamental wave. The simulation result of the rotor position observation at 300 rpm is shown in Figure 6 which shows that the pure integrator cannot estimate the rotor position correctly. In addition, the estimation error of the STSMFO is 0.01 rad which is lower than 0.06 rad of the SMFO and much lower than 0.85 rad of the LPF.

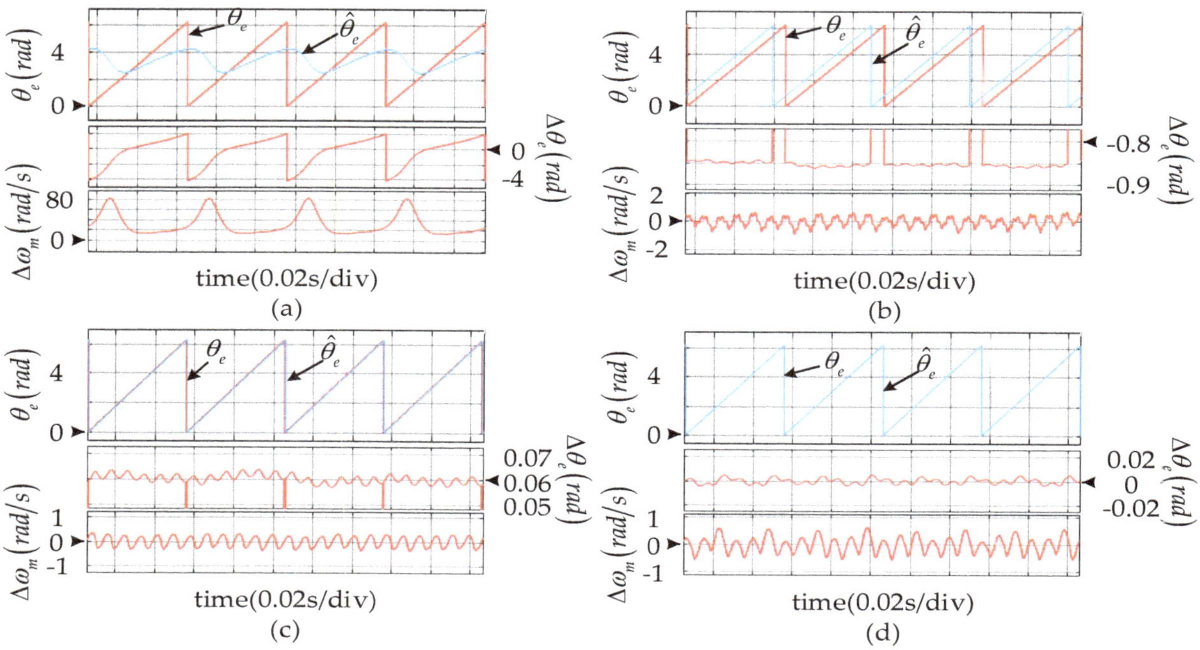

Figure 6. The simulation result of the position observation at 300 rpm: (**a**) Integrator, (**b**) low-pass filter (LPF), (**c**) The SMFO, (**d**) The STSMFO.

Although both the SMFO and the STSMFO can precisely estimate the stator flux, the SMFO cannot totally eliminate the dc offset. To prove it further, Figure 7 plots the simulation results of the flux estimation under the 9 V dc offset (added to the $u_{s\alpha}$). From the figure, it is clear that the SMFO cannot eliminate the dc offset, while the STSMFO removes the dc offset completely due to the integral term of the STA. Furthermore, the flux dc offset ψ_{s0} is inverse in proportion to the SMFO gains. However, there is a tradeoff between the ψ_{s0} and the SMFO gains because the larger SMFO gains will deteriorate the chattering problem.

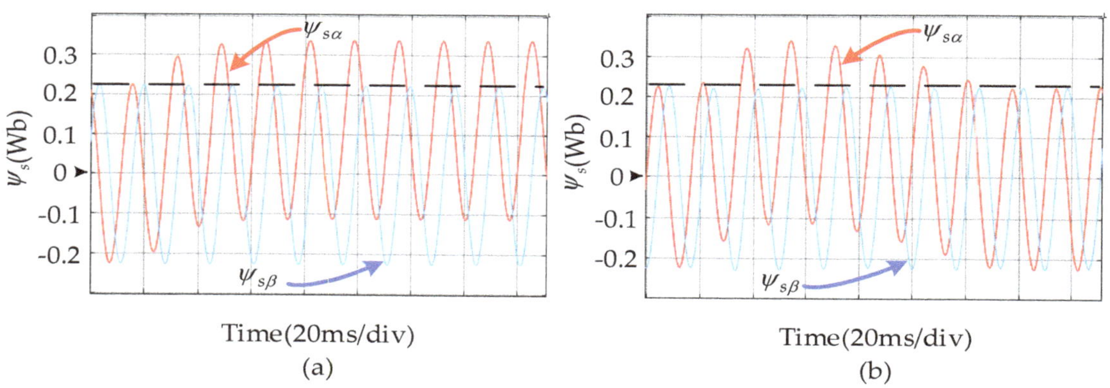

Figure 7. Simulation results of estimated stator fluxes under the 9 V dc offset: (**a**) The SMFO, (**b**) The STSMFO.

5.2. Experimental Results

Experiments on a 2-level IPMSM drive system [32] were subsequently carried out to further verify the effectiveness of the proposed observer. The composition of the IPMSM test platform is given in Figure 8, where a DSP TMS320F2812 was chosen as the MCU and output signals of a D/A chip (TLV5610) were displayed by an oscilloscope to monitor system variables in real-time. The three-phase

PWM inverter [33] is composed of IGBTs (FF650R17IE4) and a 540V dc-link voltage was obtained by the PWM rectifier. The sampling and switching periods were set to 100 μs. The sensors of current (CHB-500SG) and voltage (LV25-P) were utilized to measure ab-phase currents and dc-link voltage. The actual rotor position was detected by a rotary decoder (PGA411-Q1) for comparison. The main parameters of the IPMSM were the same as those utilized in the simulation. Moreover, the parameters of the PI regulator in phase locked loop (PLL) were $K_p = 15$ and $K_i = 250$. In addition, the parameters of the STSMFO were $K_1 = 2.5$ and $K_2 = 5000$.

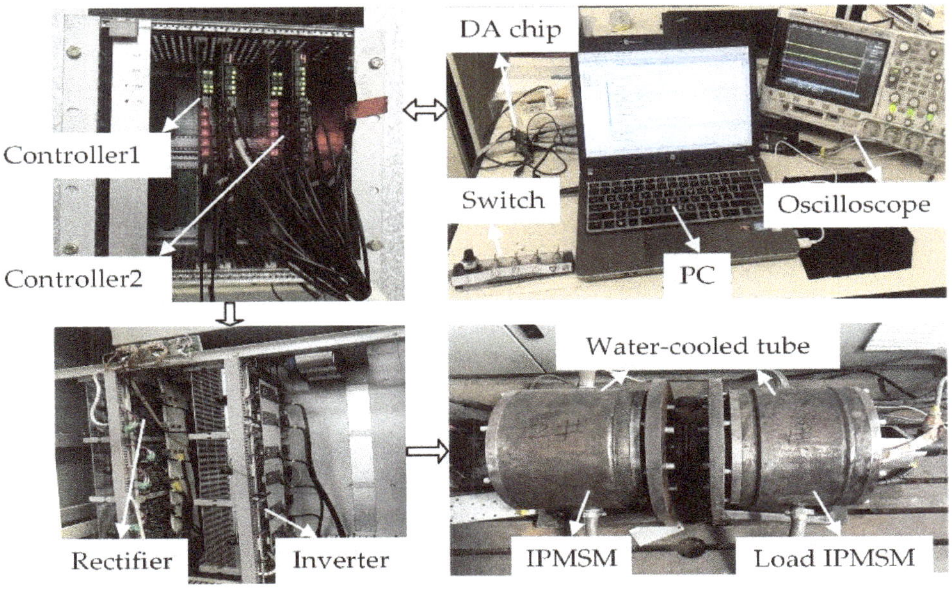

Figure 8. The IPMSM experimental platform.

5.2.1. Steady-State Performance

Figures 9 and 10 show the estimated stator flux trajectories and rotor positions at 300 rpm, respectively. The result in Figure 9a confirms that the pure integrator will introduce a monotonously increasing dc offset in the flux estimation with respect to time. It can really deteriorate the performance of the observer, resulting in the pure integrator not being able to estimate the correct position at all. The error of estimation is approximately equal to 300 rpm. The dc offset can be removed by the LPF, but the amplitude is attenuated from the flux amplitude reference which is 0.225 Wb to 0.17 Wb. Moreover, the LPF also generates a phase delay and causes a position error of about 30°. Although both the SMFO and the STSMFO can estimate the stator flux accurately, the SMFO cannot remove the dc offset totally and results in a larger position error than the STSMFO.

Besides, Figures 11 and 12 plot the estimated stator flux trajectories and rotor positions at 1200 rpm. As same as the situation at 300 rpm, the pure integrator cannot observe the flux correctly and the flux locus still deviates from the ideal locus. As described in Equation (15), the problems of amplitude attenuate, and phase delay caused by the LPF are improved with the increasing speed. Compared with the performance at 300 rpm, the amplitude of the estimated flux increases from 0.175 Wb to 0.205 Wb and the position error decreases to 10°. However, the LPF still cannot meet the requirements of a high-performance sensorless control. The performance of the SMFO and the STSMFO are similar to the situation at 300 rpm. They can estimate the stator flux accurately. Moreover, the position error caused by the STSMFO is still a bit lower than the SMFO.

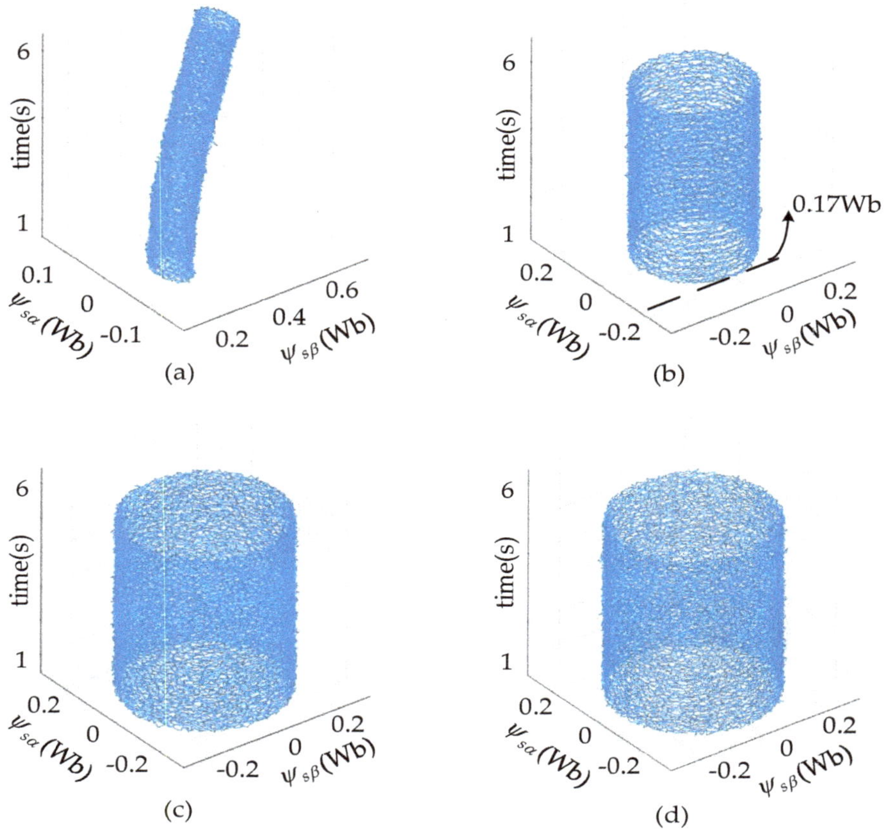

Figure 9. The estimated stator flux loci at 300 rpm: (**a**) Integrator, (**b**) LPF, (**c**) The SMFO, (**d**) The STSMFO.

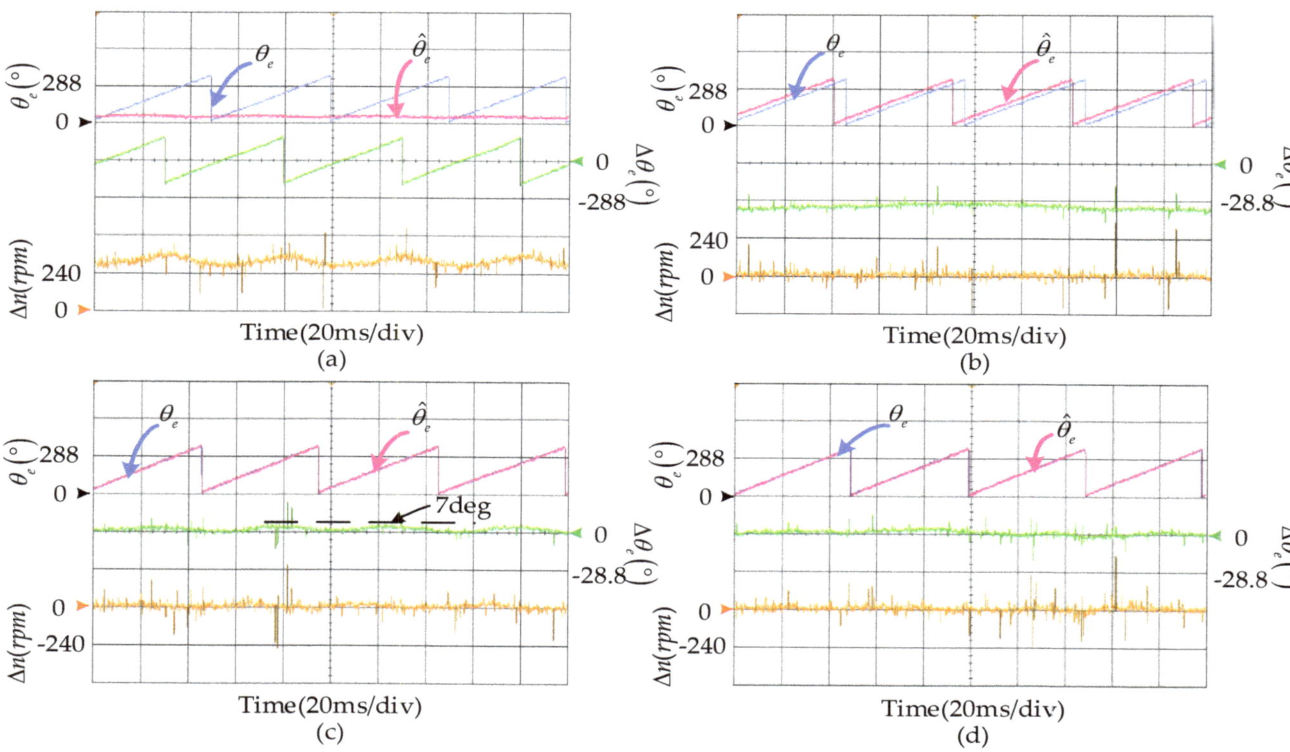

Figure 10. The performance of the position observation at 300 rpm: (**a**) Integrator, (**b**) LPF, (**c**) The SMFO, (**d**) The STSMFO.

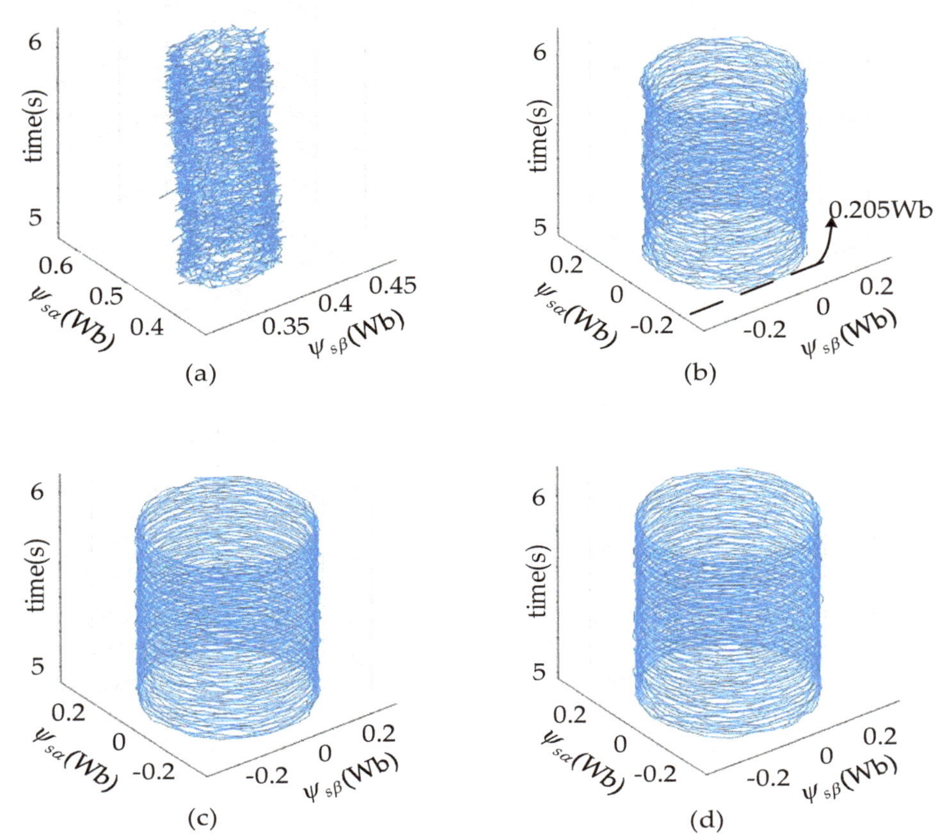

Figure 11. The estimated stator flux loci at 1200 rpm: (**a**) Integrator, (**b**) LPF, (**c**) The SMFO, (**d**) The STSMFO.

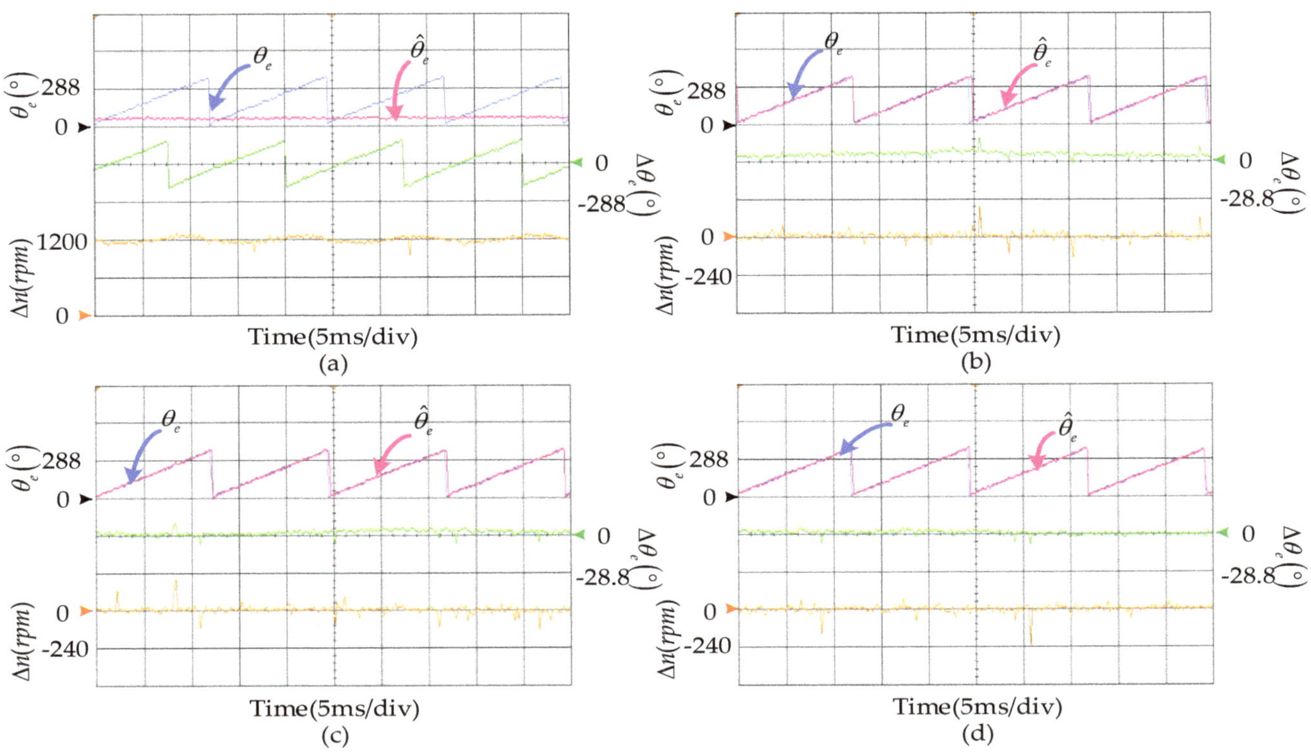

Figure 12. The performance of the position observation at 1200 rpm: (**a**) Integrator, (**b**) LPF, (**c**) The SMFO, (**d**) The STSMFO.

The experimental results of sensorless control which is based on the STSMFO at the 150 rpm are presented in Figure 13. The stator flux and its amplitude are given in Figure 13a. Figure 13b plots the actual and estimated position, estimated speed, and position error. As can be seen in the figure, the stator flux can be observed accurately and the chattering is small. Moreover, the figure shows a good position tracking ability of the STSMFO with a position error being controlled within 7°. Therefore, the STSMFO can be carried out at a low speed region.

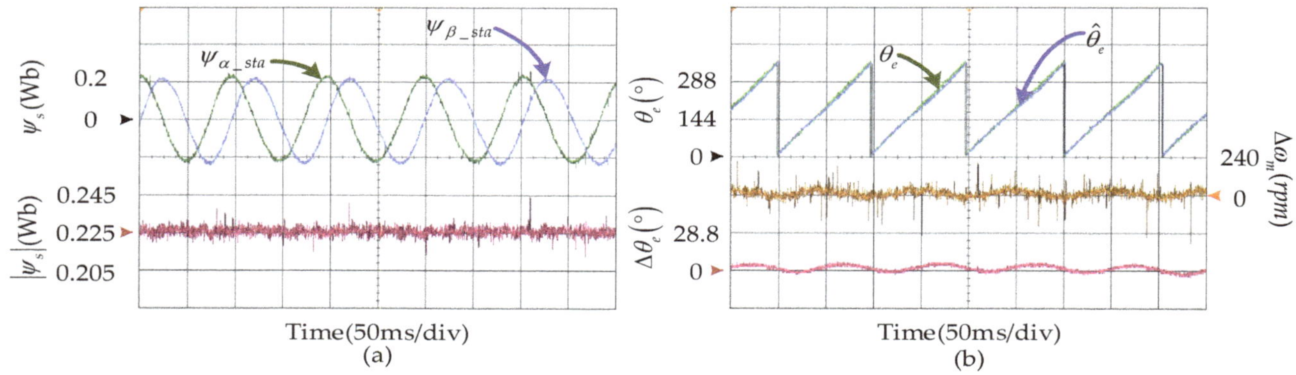

Figure 13. The sensorless drive by the STSMFO at 150 rpm: (**a**) The stator flux estimation, (**b**) The rotor position estimation.

5.2.2. DC Disturbance Rejection

To evaluate the dc disturbance rejection capability of the STSMFO, experiments with added dc disturbance signals are also implemented and the results are illustrated in Figures 14 and 15. The 20 V dc offset is added to the $u_{s\alpha}$ and the estimated fluxes by two observers are presented in the figure. As can be seen, there is a 0.125 Wb flux dc offset on the estimated stator flux by the SMFO since it cannot totally eliminate the dc disturbance. The flux observed by the STSMFO is also subject to dc offset, but it can recover to normal value within two sample times. The corresponding three-dimensional trajectories in Figure 14 illustrate this more intuitively.

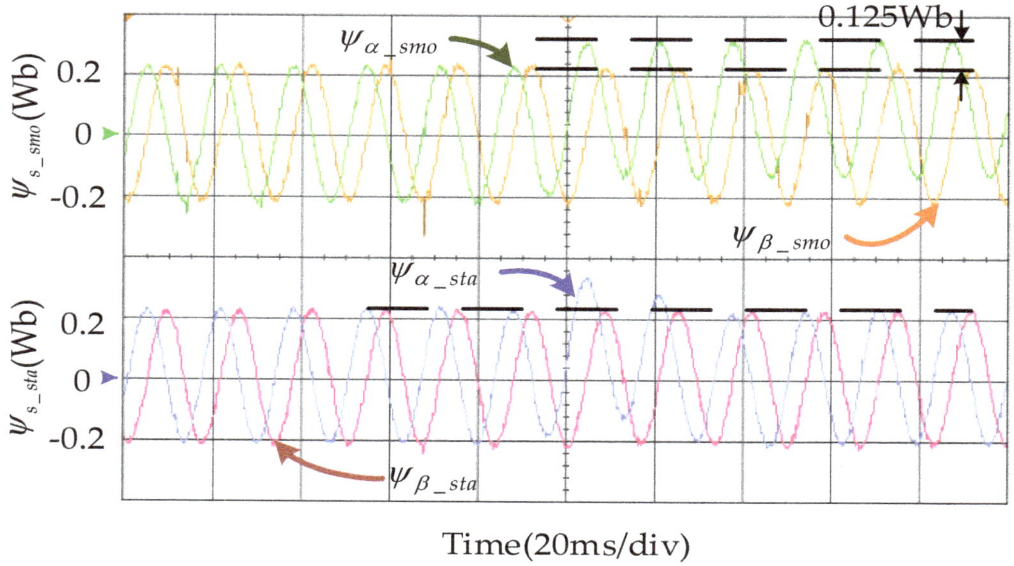

Figure 14. The experimental results of estimated stator fluxes under the 20 V dc offset.

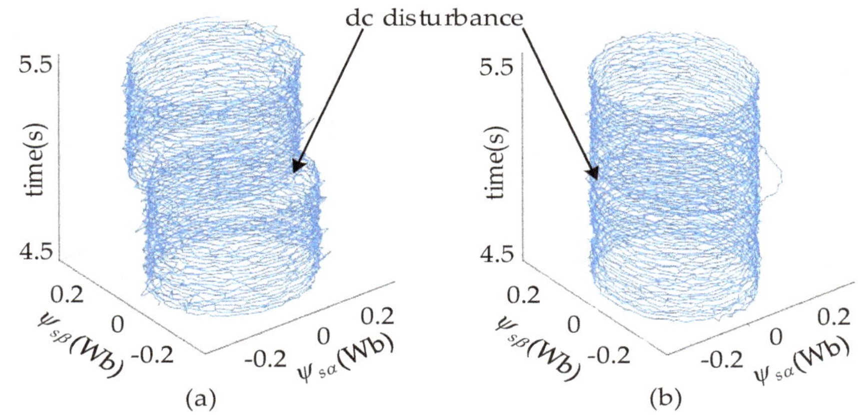

Figure 15. Three-dimensional stator flux trajectories under dc disturbance: (**a**) The SMFO, (**b**) The STSMFO.

The sensorless drive results under the 20V dc offset are illustrated in Figure 16. The picture plots the position error, the stator flux, and its amplitude. Compared with the open loop estimation, the estimated flux dc offset of the closed-loop estimation is smaller due to the flux regulation of the DTFC. However, a small flux dc offset can still cause large chattering in the speed and it can deteriorate the performance of the sensorless control. As for the STSMFO, the flux dc offset can be eliminated within one sample time and the chattering problem of the estimated rotor position can also be solved accordingly. As can be seen, the STSMFO has stronger disturbance rejection performance than the SMFO.

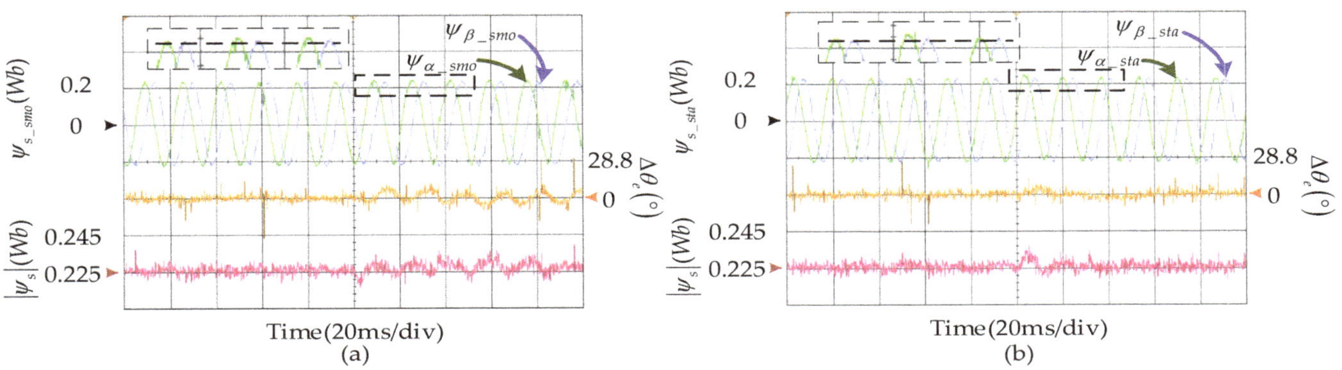

Figure 16. The sensorless drive results under the 20 V dc offset. (**a**) The SMFO, (**b**) The STSMFO.

5.2.3. Dynamic Capability

To further verify the performance of the STSMFO, the result of speed variation between 300 rpm and 1,500 rpm is given in Figure 17, which shows that the observer can accurately estimate the speed during the variable speed and control position error within 8°. Figure 18 plots the experimental result of the proposed sensorless method at 900 rpm under a 30 N·m load disturbance. The estimated speed, estimated position error, estimated speed error, and load torque are all shown in the figure. The results indicate a good speed tracking ability of the STSMFO under a load step, with a rotor position error being about 14.4°.

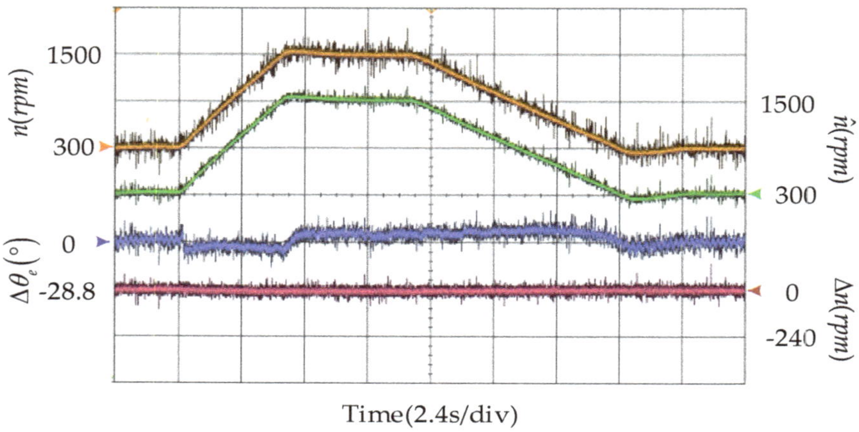

Figure 17. The STSMFO-based sensorless drive during variable speed operation.

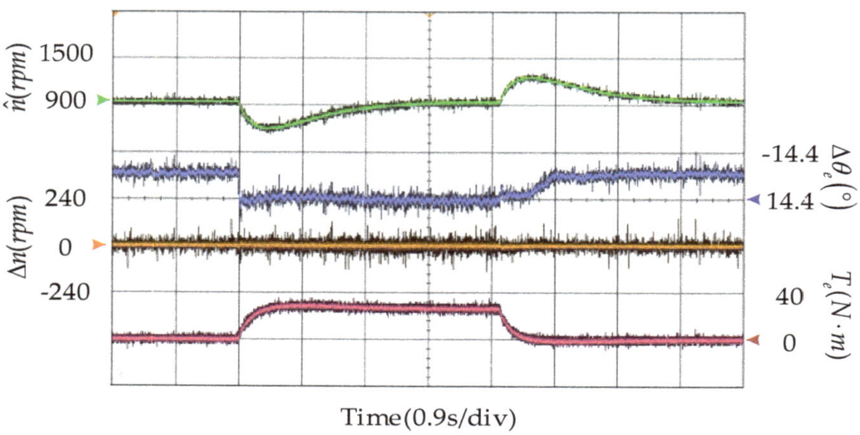

Figure 18. The STSMFO-based sensorless drive with 30 N·m load step at 900 rpm.

To validate the capability of the STSMFO at low speeds, the experimental result of sensorless drive during speed variation form 60 rpm to 300 rpm is presented in Figure 19. The actual and estimated speed, speed, and position estimation error are presented in the figure. It shows that the estimated rotor position error chatters a little around 14° at 60 rpm, but the speed estimation maintains a relatively high tracking accuracy over the whole speed domain, thus ensuring a stable motor operation.

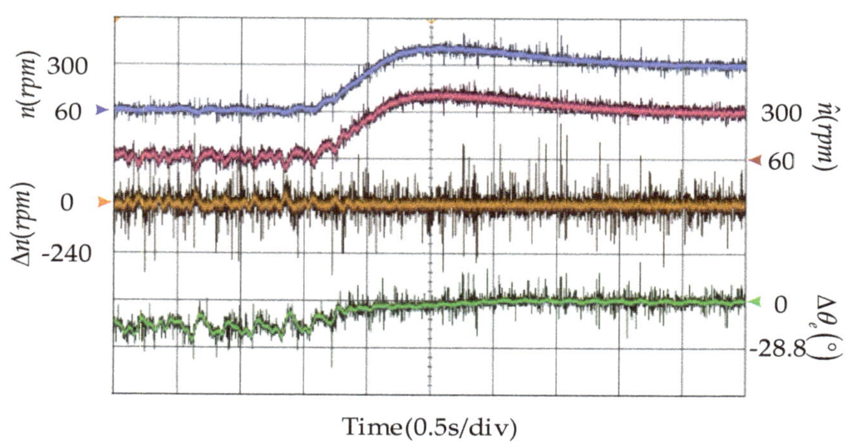

Figure 19. The STASFO-based sensorless drive results at a low-speed region.

6. Conclusions

This paper proposed a novel STSMFO for the active flux-based sensorless scheme. To demonstrate its effectiveness, a detailed mathematical analysis was given for comparing it with traditional observers, i.e., the pure integrator, the LPF, and the SMFO. It shows that the STSMFO can easily solve the problems caused by dc offset, unknown integral initial value, and saturation. Moreover, the robustness of STSMFO is much better than SMFO. Then, the precision of the stator flux estimation can be enhanced greatly without any amplitude and phase compensation. Furthermore, the rotor position can be observed accurately via the active flux over a wide speed domain. The numerical simulations and experimental tests have confirmed the capability of the steady state, the dynamic response, and the disturbance rejection.

Author Contributions: J.C. proposed the new scheme. J.C., S.C., and X.W. performed the simulations, experiments, and data analysis. J.C. wrote the first draft and S.C., X.W., G.T., and J.H. guided and revised the manuscript.

References

1. Li, J.; Huang, X.; Niu, F.; You, C.; Wu, L.; Fang, Y. Prediction Error Analysis of Finite-Control-Set Model Predictive Current Control for IPMSMs. *Energies* **2018**, *11*, 2051. [CrossRef]
2. Lara, J.; Xu, J.; Chandra, A. Effects of Rotor Position Error in the Performance of Field Oriented Controlled PMSM Drives for Electric Vehicle Traction Applications. *IEEE Trans. Ind. Electron.* **2016**, *63*, 4738–4751. [CrossRef]
3. Rahman, M.M.; Uddin, M.N. Third Harmonic Injection Based Nonlinear Control of IPMSM Drive for Wide Speed Range Operation. *IEEE Trans. Ind. Appl.* **2014**, *55*, 3174–3184. [CrossRef]
4. Takahashi, I.; Noguchi, T. A New Quick-Response and High-Efficiency Control Strategy of an Induction Motor. *IEEE Trans. Ind. Appl.* **1986**, *22*, 820–827. [CrossRef]
5. Ouanjli, N.E.; Derouich, A.; Ghzizal, A.E.; Motahhir, S.; Chebabhi, A.; Mourabit, Y.E.; Taoussi, M. Modern improvement techniques of direct torque control for induction motor drives—A review. *Prot. Control Mod. Power Syst.* **2019**, *4*, 136–147. [CrossRef]
6. Choi, Y.; Choi, H.H.; Jung, J. Feedback Linearization Direct Torque Control with Reduced Torque and Flux Ripples for IPMSM Drives. *IEEE Trans. Power Electron.* **2016**, *31*, 3728–3737. [CrossRef]
7. Tang, L.; Zhong, L.; Rahman, M.F.; Hu, Y. A Novel Direct Torque Controlled Interior Permanent Magnet Synchronous Machine Drive with Low Ripple in Flux and Torque and Fixed Switching Frequency. *IEEE Trans. Power Electron.* **2004**, *19*, 346–354. [CrossRef]
8. Foo, G.; Rahman, M.F. Sensorless Direct Torque and Flux-Controlled IPM Synchronous Motor Drive at Very Low Speed Without Signal Injection. *IEEE Trans. Ind. Electron.* **2010**, *57*, 395–403. [CrossRef]
9. Chen, J.; Yang, S.; Tu, J. Comparative Evaluation of a Permanent Magnet Machine Saliency-Based Drive with Sine-Wave and Square-Wave Voltage Injection. *Energies* **2018**, *11*, 2189. [CrossRef]
10. Yousefi-Talouki, A.; Pescetto, P.; Pellegrino, G.; Boldea, I. Combined Active Flux and High Frequency Injection Methods for Sensorless Direct Flux Vector Control of Synchronous Reluctance Machines. *IEEE Trans. Power Electron.* **2018**, *33*, 2447–2457. [CrossRef]
11. Yoon, Y.; Sul, S.; Morimoto, S.; Ide, K. High-Bandwidth Sensorless Algorithm for AC Machines Based on Square-Wave-Type Voltage Injection. *IEEE Trans. Ind. Appl.* **2011**, *47*, 1361–1370. [CrossRef]
12. Shi, Y. Online Identification of Permanent Magnet Flux Based on Extended Kalman Filter for IPMSM Drive with Position Sensorless Control. *IEEE Trans. Ind. Electron.* **2012**, *59*, 4169–4178. [CrossRef]
13. Chen, S.; Zhang, X.; Wu, X.; Tan, G.; Chen, X. Sensorless Control for IPMSM Based on Adaptive Super-Twisting Sliding-Mode Observer and Improved Phase-Locked Loop. *Energies* **2019**, *12*, 1225. [CrossRef]
14. Vieira, R.P.; Gastaldini, C.C.; Azzolin, R.Z.; Gründling, H.A. Sensorless Sliding-Mode Rotor Speed Observer of Induction Machines Based on Magnetizing Current Estimation. *IEEE Trans. Ind. Electron.* **2014**, *61*, 4573–4582. [CrossRef]

15. Fan, Y.; Zhang, L.; Cheng, M.; Chau, K.T. Sensorless SVPWM-FADTC of a New Flux-Modulated Permanent-Magnet Wheel Motor Based on a Wide-Speed Sliding Mode Observer. *IEEE Trans. Ind. Electron.* **2015**, *62*, 3143–3151. [CrossRef]
16. Gadoue, S.M.; Giaouris, D.; Finch, J.W. MRAS Sensorless Vector Control of an Induction Motor Using New Sliding-Mode and Fuzzy-Logic Adaptation Mechanisms. *IEEE Trans. Energy Convers.* **2010**, *25*, 394–402. [CrossRef]
17. Boldea, I.; Paicu, M.C.; Andreescu, G. Active Flux Concept for Motion-Sensorless Unified AC Drives. *IEEE Trans. Power Electron.* **2008**, *23*, 2612–2618. [CrossRef]
18. Boldea, I.; Paicu, M.C.; Andreescu, G.; Blaabjerg, F. "Active Flux" DTFC-SVM Sensorless Control of IPMSM. *IEEE Trans. Energy Convers.* **2009**, *24*, 314–322. [CrossRef]
19. Zhao, R.; Xin, Z.; Loh, P.C.; Blaabjerg, F. A Novel Flux Estimator Based on Multiple Second-Order Generalized Integrators and Frequency-Locked Loop for Induction Motor Drives. *IEEE Trans. Power Electron.* **2017**, *32*, 6286–6296. [CrossRef]
20. Xu, W.; Jiang, Y.; Mu, C.; Blaabjerg, F. Improved Nonlinear Flux Observer-Based Second-Order SOIFO for PMSM Sensorless Control. *IEEE Trans. Power Electron.* **2019**, *34*, 565–579. [CrossRef]
21. Choi, J.; Nam, K.; Bobtsov, A.A.; Pyrkin, A.; Ortega, R. Robust Adaptive Sensorless Control for Permanent-Magnet Synchronous Motors. *IEEE Trans. Power Electron.* **2017**, *32*, 3989–3997. [CrossRef]
22. Lin, T.C.; Zhu, Z.Q.; Liu, J.M. Improved Rotor Position Estimation in Sensorless-Controlled Permanent-Magnet Synchronous Machines Having Asymmetric-EMF with Harmonic Compensation. *IEEE Trans. Ind. Electron.* **2015**, *62*, 6131–6139. [CrossRef]
23. Roberto, M.; Edmundo, B.; Marco, A.A.; Hernández, C. Sensorless Predictive DTC of a Surface-Mounted Permanent-Magnet Synchronous Machine Based on Its Magnetic Anisotrogy. *IEEE Trans. Ind. Electron.* **2013**, *60*, 3016–3024.
24. Levant, A. Sliding order and sliding accuracy in sliding mode control. *Int. J. Control* **1993**, *58*, 1247–1263. [CrossRef]
25. Liang, D.; Li, J.; Qu, R. Sensorless Control of Permanent Magnet Synchronous Machine Based on Second-Order Sliding-Mode Observer with Online Resistance Estimation. *IEEE Trans. Ind. Appl.* **2017**, *53*, 3672–3682. [CrossRef]
26. Liang, D.; Li, J.; Qu, R.; Kong, W. Adaptive Second-Order Sliding-Mode Observer for PMSM Sensorless Control Considering VSI Nonlinearity. *IEEE Trans. Power Electron.* **2018**, *33*, 8994–9004. [CrossRef]
27. Li, Z.; Zhou, S.; Xiao, Y.; Wang, L. Sensorless Vector Control of Permanent Magnet Synchronous Linear Motor Based on Self-adaptive Super-twisting Sliding Mode Controller. *IEEE Access* **2019**, *7*, 44998–45011. [CrossRef]
28. Holakooie, M.H.; Ojaghi, M.; Taheri, A. Modified DTC of a Six-Phase Induction Motor with a Second-Order Sliding-Mode MRAS-Based Speed Estimator. *IEEE Trans. Power Electron.* **2019**, *34*, 600–611. [CrossRef]
29. Moreno, J.A.; Osorio, M. A Lyapunov approach to second-order sliding mode controlers and observer. In Proceedings of the 47th IEEE Conference on Decision and Control, Cancun, Mexico, 9–11 December 2008; pp. 2856–2861.
30. Moreno, J.A.; Osorio, M. Strict Lyapunov Functions for the Super-Twisting Algorithm. *IEEE Trans. Autom. Control* **2012**, *57*, 1035–1040. [CrossRef]
31. Zhong, L.; Rahman, M.F.; Hu, W.; Lim, K.W. Analysis of Direct Torque Control in Permanent Magnet Synchronous Motor Drives. *IEEE Trans. Power Electron.* **1997**, *12*, 528–536. [CrossRef]
32. Wu, X.; Tan, G.; Ye, Z.; Liu, Y.; Xu, S. Optimized Common-Mode Voltage Reduction PWM for Three-Phase Voltage Source Inverters. *IEEE Trans. Power Electron.* **2016**, *31*, 2959–2969. [CrossRef]
33. Xia, Y.; Gou, B.; Xu, Y. Anew ensemble-based classifier for IGBT open-circuit fault diagnosis in three-phase PWM converter. *Prot. Control Mod. Power Syst.* **2018**, *3*, 364–372. [CrossRef]

11

The Novel Rotor Flux Estimation Scheme Based on the Induction Motor Mathematical Model Including Rotor Deep-Bar Effect

Grzegorz Utrata [1,*], Jaroslaw Rolek [2] and Andrzej Kaplon [2]

1 Institute of Environmental Engineering, Czestochowa University of Technology, Brzeźnicka 60a, 42-200 Czestochowa, Poland
2 Department of Industrial Electrical Engineering and Automatic Control, Kielce University of Technology, Tysiąclecia Państwa Polskiego 7, 25-314 Kielce, Poland
* Correspondence: gutrata@is.pcz.pl.

Abstract: During torque transients, rotor electromagnetic parameters of an induction motor (IM) vary due to the rotor deep-bar effect. The accurate representation of rotor electromagnetic parameter variability by an adopted IM mathematical model is crucial for a precise estimation of the rotor flux space vector. An imprecise estimation of the rotor flux phase angle leads to incorrect decoupling of electromagnetic torque control and rotor flux amplitude regulation which in turn, causes deterioration in field-oriented control of IM drives. Variability of rotor electromagnetic parameters resulting from the rotor deep-bar effect can be modeled by the IM mathematical model with rotor multi-loop representation. This paper presents a study leading to define the unique rotor flux space vector on the basis of the IM mathematical model with rotor two-terminal network representation. The novel rotor flux estimation scheme was validated with the laboratory test bench employing the IM of type Sg 132S-4 with two variants of rotor construction: a squirrel-cage rotor and a solid rotor manufactured from magnetic material S235JR. The accuracy verification of the rotor flux estimation was performed in a slip frequency range corresponding to the IM load adjustment range up to 1.30 of the stator rated current. This study proved the correct operation of the developed rotor flux estimation scheme and its robustness against electromagnetic parameter variability resulting from the rotor deep-bar effect in the considered slip frequency range.

Keywords: deep-bar effect; mathematical model; estimation; induction motors; motor drives

---

1. Introduction

The development of advanced control methods of induction motors (IMs), such as direct and indirect field-oriented control [1,2] or direct torque control [3], have contributed to the widespread use of this type of motor in modern drive systems intended for various applications in industry. In the rotor-flux-orientation, the stator phase currents through the Park's transformation are represented by the field- and torque-producing components. In cases when the rotor flux amplitude is stabilized by the field-producing component of the stator current space vector, IM electromagnetic torque is linearly proportional to the torque-producing component [1]. The decoupling of IM electromagnetic torque control and rotor flux amplitude regulation is realized based on the phase angle of the rotor flux space vector. Since direct measurement of the rotor flux is practically not achievable, development of indirect methods for rotor flux space vector estimation is reported in the world literature, especially model-based methods.

In IM field-oriented control, slip frequency is controlled within the set range of values, except for very short torque transients. With slip frequency changes, rotor electromagnetic parameters vary due to the rotor deep-bar effect. For maintaining the high dynamic performance of the IM rotor-flux-oriented control during torque transients, the accurate representation of rotor electromagnetic parameter variability by an adopted IM mathematical model is required.

Inaccurate representation of this variability by the adopted IM mathematical model, which serves as basis for the rotor flux estimation scheme, leads to an erroneous estimation of the rotor flux space vector. In consequence, the erroneous estimation of the vector components results in deterioration of decoupling effectiveness of electromagnetic torque control and rotor flux amplitude regulation, thus deteriorating the overall performance of the IM rotor-flux-oriented control [4–6]. For this reason, the compensation of the influence of the rotor deep-bar effect on the rotor flux estimation accuracy is important for the rotor-flux-oriented control of squirrel-cage IMs, especially the ones where the rotor bar is large enough to incorporate high rotor current.

Until now, estimation schemes for the rotor flux space vector have been elaborated predominantly on the basis of the IM classical mathematical model with rotor single-loop representation with constant parameters. In order to compensate for the influence of the rotor electromagnetic parameter variability on the rotor flux estimation accuracy, the estimation schemes extended by algorithms enabling tracking variability of rotor electromagnetic parameters were proposed [4,6–15]. These schemes work very well with reference to IMs with squirrel-cage rotors, in which the electromagnetic parameters do not show significant variability resulting from the rotor deep-bar effect. The response of the proposed algorithms for variability tracking of rotor electromagnetic parameters may not be fast enough to follow rapid parameter variability during torque transients. These algorithms were mainly intended to model rotor resistance changes associated with temperature variation [4,6–15].

The variability of rotor electromagnetic parameters resulting from the rotor deep-bar effect can be modeled by the IM mathematical model with rotor multi-loop representation [16–25]. Nevertheless, an estimation scheme of the rotor flux space vector which algorithm would be formulated on the basis of such IM mathematical models has not been developed so far. What is more, the authors of these works [19–22] stated that defining the unique rotor flux space vector in the IM mathematical model with rotor multi-loop representation is not possible, and thus they proposed IM airgap-flux or pseudorotor-flux oriented control, developed with the use of the mathematical model of this type.

The results of simulation and experimental studies presented previously [19–22] indicate very good dynamic performance of the vector-controlled squirrel-cage and double-cage IMs. This fact encouraged us to carry on work on the application of the IM mathematical model with rotor multi-loop representation in the IM rotor-flux-oriented control, since such a control strategy has a simpler structure and a more effective decoupling of electromagnetic torque control and rotor flux amplitude regulation than airgap-flux-oriented control [22].

This paper presents a study which leads to development of the rotor flux estimation scheme on the basis of the IM mathematical model with rotor two-terminal network representation. The overall goal of this work was focused on the accuracy verification of the rotor flux estimation in a slip frequency range corresponding to the IM load adjustment range up to 1.30 of the stator rated current. Thus, the considered slip frequency range exceeded the typical operating range of slip frequency for IM field-oriented control. This study aimed to prove the proper modeling of the electromagnetic parameter variability resulting from the rotor deep-bar effect by the novel rotor flux estimation scheme. Due to the assumed concept of the conducted work, the experimental investigations were realized in an open-loop drive system (without speed feedback or slip compensator), at a fixed setpoint of stator voltages and step commands of load torque. The evaluation of operation accuracy of the developed rotor flux estimation scheme was realized indirectly with the use of the registered shaft torque.

The results of the conducted study point out an improvement of the estimation accuracy of the rotor flux space vector obtained by the scheme developed on the basis of the IM mathematical model with rotor two-terminal network representation, in comparison to the accuracy which was gained by the estimation schemes formulated with the use of the IM classical mathematical model. In particular, this applies to the tested IM characterized by the intense rotor skin effect. Consequently, the obtained results confirm the correct operation of the novel rotor flux estimation scheme and its robustness for electromagnetic parameter variability resulting from the rotor deep-bar effect.

2. Mathematical Models of an Induction Motor

One of the fundamental problems associated with the use of the IM classical mathematical model with constant parameters in IM control algorithms is the variability of motor electromagnetic parameters which is conditioned by changes of motor winding temperature, ferromagnetic core saturation, as well as the rotor deep-bar effect [26]. Figure 1a presents the T-type equivalent circuit corresponding to the IM classical mathematical model expressed in the Laplace-domain (p-domain), in which the rotor resistance R_2 and leakage inductance $L_{\sigma 2}$ are represented by parameters varying as a function of slip frequency ω_2. The variability of rotor electromagnetic parameters resulting from the rotor deep-bar effect can be modeled in the rotor equivalent circuit by a two-terminal network with constant parameters [16–25]. The electromagnetic parameters of such an IM mathematical model can be determined based on the p-domain motor inductance:

$$L_1(p) = \frac{\underline{\Psi}_{1r}(p)}{\underline{I}_{1r}(p)} = \omega_b \frac{Z_{ab}(p)}{p} \tag{1}$$

where $\underline{\Psi}_{1r}(p)$ and $\underline{I}_{1r}(p)$ are the Laplace transforms of the stator flux and current space vectors, respectively, p is a complex frequency, $Z_{ab}(p)$ denotes the p-domain impedance between the terminals "a" and "b" of the IM equivalent circuit presented in Figure 1a, ω_b is the base frequency (Appendix A), and the subscript "r" denotes physical quantity space vectors expressed in an orthogonal coordinate system rotating at the shaft angular velocity ω_{sh}.

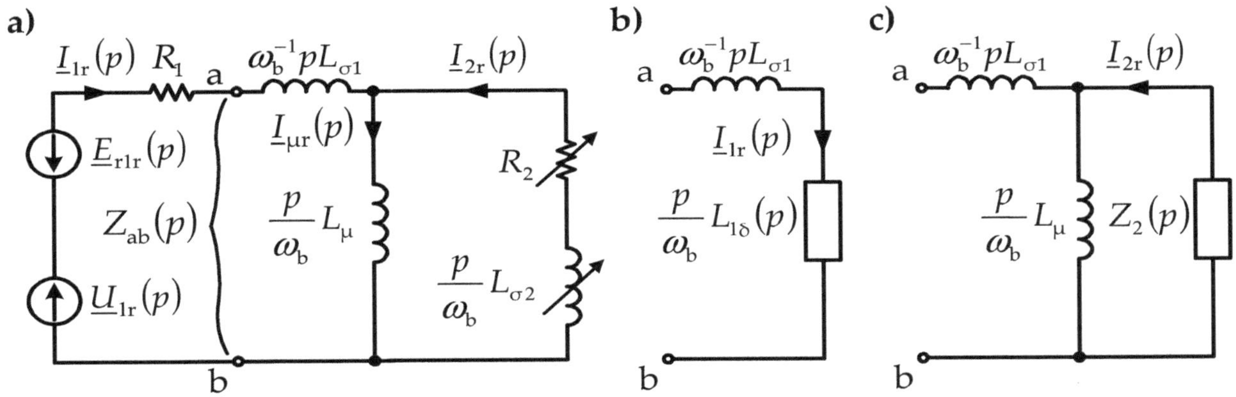

Figure 1. The induction motor (IM) equivalent circuits expressed in the Laplace-domain: (**a**) The IM equivalent circuit with rotor resistance and leakage inductance represented by parameters varying as a function of slip frequency. (**b**) The representation of the p-domain motor inductance by a series connection of the stator leakage inductance and the p-domain inductance associated with the airgap flux. (**c**) The representation of the p-domain inductance associated with the airgap flux by a parallel connection of the magnetizing inductance and the p-domain rotor impedance.

Equation (1), as well as the subsequent equations included in this paper, are expressed in the per-unit (p.u.) system. The base values of the used p.u. system are defined in Appendix A. Moreover, rotor physical quantities and electromagnetic parameters are referred to the stator.

The p-domain motor inductance $L_1(p)$ is a series connection of the stator leakage inductance $L_{\sigma 1}$ and the p-domain inductance associated with the airgap flux $L_{1\delta}(p)$ (Figure 1b):

$$L_1(p) = L_{\sigma 1} + L_{1\delta}(p). \tag{2}$$

The p-domain motor inductance $L_{1\delta}(p)$ can be further represented as a parallel connection of the magnetizing inductance L_μ and the p-domain rotor impedance $Z_2(p)$ (Figure 1c):

$$\frac{\omega_b}{pL_{1\delta}(p)} = \frac{\omega_b}{pL_\mu} + \frac{1}{Z_2(p)}. \tag{3}$$

The p-domain inductance $L_{1\delta}(p)$ can be derived from a solution of Maxwell's differential system of equations which are formulated, for instance, on the basis of an IM multi-layer model [17]. However, the p-domain inductance $L_{1\delta}(p)$ is not directly applicable in an analysis of IM transients due to the lack of possibility for inverse transformation of a Laplace transform including this inductance. The above-mentioned difficulty can be circumvent by the partial fraction decomposition of the inverse p-domain inductance $L_{1\delta}(p)$, which is an irrational function with an infinite number of negative real poles. This, in turn, leads to the rotor mathematical model in the form of a two-terminal network with constant $R_{2(n)}$, $L_{\sigma 2(n)}$ parameters [17]:

$$\frac{\omega_b}{pL_{1\delta}(p)} = \frac{\omega_b}{pL_\mu} + \sum_{n=1}^{\infty} \frac{1}{R_{2(n)} + \frac{1}{\omega_b} pL_{\sigma 2(n)}}. \tag{4}$$

An exact approximation of the reference p-domain inductance $L_{1\delta}(p)$ is obtained with an infinite number of poles of an approximative rational function (an infinite number of parallel connected two-terminals in the rotor mathematical model). For the sake of the desired simplicity of the IM mathematical model, the number of parallel connected two-terminals in the rotor equivalent circuit is limited to N two-terminals, and for achieving the required approximation accuracy of the irrational function $L_{1\delta}(p)$, the $(N+1)$th residual two-terminal with parameters $R_{2(0)}$, $L_{\sigma 2(0)}$ is included (Figure 2) [17]:

$$\frac{\omega_b}{pL_{1\delta}(p)} = \frac{\omega_b}{pL_\mu} + \sum_{n=1}^{N} \frac{1}{R_{2(n)} + \frac{1}{\omega_b} pL_{\sigma 2(n)}} + \frac{1}{R_{2(0)} + \frac{1}{\omega_b} pL_{\sigma 2(0)}}. \tag{5}$$

The methodology for determination of the residual two-terminal electromagnetic parameters $R_{2(0)}$, $L_{\sigma 2(0)}$ has been described previously [17]. The approximation accuracy of the reference p-domain inductance is evaluated by comparing its frequency characteristic with a characteristic $\underline{L}_{1\delta}(p = \mathrm{j}\omega_2)$ resulting from the IM mathematical model with rotor two-terminal network representation.

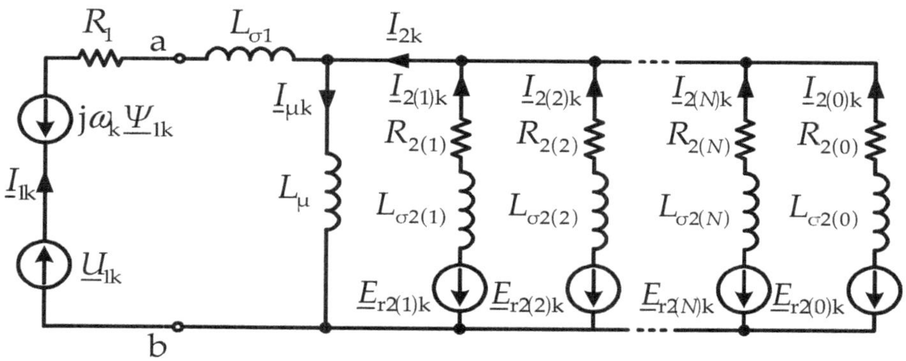

Figure 2. The equivalent circuit of an induction motor with rotor two-terminal network representation.

If the analytical formula describing the p-domain inductance $L_{1\delta}(p)$ is not known, which is the case, for instance, in experimental determination of the reference inductance frequency characteristic (IFCh) $\underline{L}_{1\delta}(\omega_2)$ [25,27], then the "synthetic" electromagnetic parameters of the IM equivalent circuit (Figure 2) can be identified as a result of an approximation of the reference IFCh $\underline{L}_{1\delta}(\omega_2)$ by the characteristic $\underline{L}_{1\delta}(p = \mathrm{j}\omega_2)$ derived from the adopted IM mathematical model. In such an approach, the residual two-terminal does not formally occur in Equation (5), and its participation in the approximation accuracy of the p-domain rotor impedance is smaller for larger the numbers N of parallel connected two-terminals in the rotor mathematical model:

$$\frac{1}{Z_2(p)} \cong \sum_{n=1}^{N} \frac{1}{R_{2(n)} + \frac{1}{\omega_b} p L_{\sigma 2(n)}}. \tag{6}$$

In this way, an IM of any rotor construction (e.g., squirrel-cage, double-cage, or solid rotors) can be represented with the use of the mathematical model in which the electromagnetic parameter variability resulting from the rotor deep-bar effect is approximated by the two-terminal network with constant parameters. The process of electromagnetic parameter identification for individual two-terminals in the rotor equivalent circuit, ensuring the required approximation accuracy of the reference IFCh $\underline{L}_{1\delta}(\omega_2)$, can be conveniently performed using selected optimization methods. In such an approach, the number N of the parallel connected two-terminals is determined empirically.

The IM mathematical model with rotor N-loop representation is described by the following system of equations:

$$\underline{U}_{1k} = R_1 \underline{I}_{1k} + T_N \frac{d}{dt} \underline{\Psi}_{1k} + j\omega_k \underline{\Psi}_{1k} \tag{7a}$$

$$\begin{cases} 0 = R_{2(1)} \underline{I}_{2(1)k} + T_N \frac{d}{dt} \underline{\Psi}_{2(1)k} + j(\omega_k - \omega_{sh}) \underline{\Psi}_{2(1)k} \\ \cdots \\ 0 = R_{2(N)} \underline{I}_{2(N)k} + T_N \frac{d}{dt} \underline{\Psi}_{2(N)k} + j(\omega_k - \omega_{sh}) \underline{\Psi}_{2(N)k} \end{cases} \tag{7b}$$

$$\underline{\Psi}_{1k} = L_1 \underline{I}_{1k} + L_\mu \underline{I}_{2k} \tag{7c}$$

$$\begin{cases} \underline{\Psi}_{2(1)k} = L_\mu \left(\underline{I}_{1k} + \underline{I}_{2k} \right) + L_{\sigma 2(1)} \underline{I}_{2(1)k} \\ \cdots \\ \underline{\Psi}_{2(N)k} = L_\mu \left(\underline{I}_{1k} + \underline{I}_{2k} \right) + L_{\sigma 2(N)} \underline{I}_{2(N)k} \end{cases} \tag{7d}$$

$$\underline{I}_{2k} = \sum_{n=1}^{N} \underline{I}_{2(n)k} \tag{7e}$$

$$\frac{d\omega_{sh}}{dt} = \frac{1}{T_M} (T_{em} - T_L) \tag{7f}$$

$$T_{em} = \mathrm{Im}\left(\underline{\Psi}_{1k}^* \underline{I}_{1k} \right) \tag{7g}$$

$$\underline{E}_{r1k} = j\omega_k \underline{\Psi}_{1k} \tag{7h}$$

$$\underline{E}_{r2(n)k} = j(\omega_k - \omega_{sh}) \underline{\Psi}_{2(n)k} \tag{7i}$$

where $\underline{U}_{1k}$ and $\underline{E}_{r1k}$ are the stator voltage and electromotive force space vectors, respectively, $\underline{I}_{2(n)k}$, $\underline{\Psi}_{2(n)k}$, and $\underline{E}_{r2(n)k}$ represent the rotor current, flux, and electromotive force, respectively, related to the nth two-terminal in the rotor equivalent circuit, T_{em} and T_L constitute the electromagnetic and load torque, respectively, $T_N = 1/\omega_b$, T_M is the motor mechanical time constant, $j^2 = -1$, * denotes the complex conjugate, and k indicates physical quantity space vectors expressed in an orthogonal coordinate system rotating at an arbitrary angular velocity ω_k.

3. The Novel Estimation Scheme of the Rotor Flux Space Vector

The variability of rotor electromagnetic parameters resulting from the rotor deep-bar effect can be represented in the rotor mathematical model by the two-terminal network with constant $R_{2(n)}$, $L_{\sigma 2(n)}$ parameters. Therefore, the application of such an IM mathematical model in an estimation scheme of the rotor flux space vector is justifiable, especially in the case of an IM characterized by the intense rotor deep-bar effect. However, the rotor flux estimation scheme, which would be based on the IM mathematical model of this type, has not been developed until now. This chapter presents the investigations leading to define the unique rotor flux space vector on the basis of the IM mathematical model with rotor two-terminal network representation.

The current space vector of the nth two-terminal in the rotor equivalent circuit (Figure 2) can be determined with the use of the transformed Equation (7d):

$$\underline{I}^{\bullet}_{2(n)\mathrm{k}} = \frac{1}{L^{\bullet}_{\sigma 2(n)}}\left(\underline{\Psi}^{\bullet}_{2(n)\mathrm{k}} - L_{\mu}\underline{I}_{\mu\mathrm{k}}\right) \tag{8a}$$

$$\underline{I}_{\mu\mathrm{k}} = \underline{I}_{1\mathrm{k}} + \underline{I}_{2\mathrm{k}} \tag{8b}$$

$$\underline{\Psi}_{\mu\mathrm{k}} = L_{\mu}\underline{I}_{\mu\mathrm{k}} \tag{8c}$$

where $\underline{I}_{\mu\mathrm{k}}$ and $\underline{\Psi}_{\mu\mathrm{k}}$ are the magnetizing current and flux space vectors, respectively.

Incorporation of the formulas describing the rotor two-terminal current space vectors (Equation (8a)) to the transformed voltage equation (Equation (7b)) associated with the nth two-terminal in the rotor multi-loop equivalent circuit, the model of the rotor flux space vector related to the nth two-terminal is obtained in the form of:

$$T_{2(n)}\frac{\mathrm{d}}{\mathrm{d}t}\underline{\Psi}_{2(n)\mathrm{k}} = L_{\mu}\underline{I}_{\mu\mathrm{k}} - \underline{\Psi}_{2(n)\mathrm{k}} + \mathrm{j}\omega_{\mathrm{b}}T_{2(n)}(\omega_{\mathrm{k}} - \omega_{\mathrm{m}})\underline{\Psi}_{2(n)\mathrm{k}} \tag{9a}$$

$$T_{2(n)} = \frac{L_{\sigma 2(n)}}{R_{2(n)}}T_{\mathrm{N}} \tag{9b}$$

where $T_{2(n)}$ constitutes the time constant of the nth two-terminal in the rotor equivalent circuit presented in Figure 2.

The magnetizing current space vector $\underline{I}_{\mu\mathrm{k}}$, which is required in Equation (9a), can be determined based on Equation (7c) where the stator flux space vector $\underline{\Psi}_{1\mathrm{k}}$ is obtained with the use of the stator voltage Equation (7a):

$$\underline{I}_{\mu\mathrm{k}} = \frac{1}{L_{\mu}}\left(\underline{\Psi}_{1\mathrm{k}} - L_{\sigma 1}\underline{I}_{1\mathrm{k}}\right). \tag{10}$$

In general, the rotor flux space vector can be expressed as the sum of the magnetizing flux (Equation (8c)) and rotor leakage flux space vectors. Concerning the IM mathematical model with rotor multi-loop representation, the equation takes the following form:

$$\underline{\Psi}_{2\mathrm{k}} = L_{\mu}\underline{I}_{1\mathrm{k}} + \left(L_{\mu} + L_{\sigma 2\mathrm{eq}}\right)\underline{I}_{2\mathrm{k}} \tag{11}$$

where $L_{\sigma 2\mathrm{eq}}$ is the equivalent rotor leakage inductance of the rotor two-terminal network (Figure 2).

The resultant rotor current space vector is the sum of the current space vectors of parallel connected two-terminals in the rotor equivalent circuit (Equation (7e)). Taking into account the formulas describing these current space vectors (Equation (8a)), Equation (11) is as follows:

$$\underline{\Psi}_{2\mathrm{k}} = L_{\mu}\underline{I}_{\mu\mathrm{k}} + L_{\sigma 2\mathrm{eq}}\sum_{n=1}^{N}\frac{\underline{\Psi}_{2(n)\mathrm{k}}}{L_{\sigma 2(n)}} - L_{\mu}\underline{I}_{\mu\mathrm{k}}L_{\sigma 2\mathrm{eq}}\sum_{n=1}^{N}\frac{1}{L_{\sigma 2(n)}}. \tag{12}$$

The magnetizing flux space vector $\underline{\Psi}_{\mu\mathrm{k}}$ has been included in the formulas representing the flux space vectors associated with the individual two-terminals in the rotor multi-loop equivalent circuit

(Equation (9a)), thus the magnetizing flux space vector is redundant in Equation (12) for the rotor flux space vector. The first and the third components of the sum in Equation (12), constituting and containing the magnetizing flux space vector, respectively, reduce each other in cases when the inverse equivalent rotor leakage inductance $L_{\sigma 2\text{eq}}$ equals the sum of the inverse leakage inductances of the individual rotor two-terminals:

$$\frac{1}{L_{\sigma 2\text{eq}}} = \sum_{n=1}^{N} \frac{1}{L_{\sigma 2(n)}}. \tag{13}$$

On the basis of the above reasoning, the derived formulas describe the equivalent rotor leakage inductance of the rotor equivalent circuit presented in Figure 2, which result from a parallel connection of leakage inductances of the individual rotor two-terminals:

$$L_{\sigma 2\text{eq}} = \lim_{p \to \infty} \frac{Z_2(p)}{p} \tag{14}$$

and the unique rotor flux space vector of the IM mathematical model with rotor multi-loop representation:

$$\underline{\Psi}_{2\text{k}} = L_{\sigma 2\text{eq}} \sum_{n=1}^{N} \frac{\underline{\Psi}_{2(n)\text{k}}}{L_{\sigma 2(n)}}. \tag{15}$$

According to the above, the voltage–current model of the rotor flux space vector can be formulated. When expressed in the orthogonal coordinate system (α–β) stationary with respect to the stator ($\omega_\text{k} = 0$, indicated by s), this model is represented by the following system of equations:

$$\underline{I}^{\text{e}}_{\mu\text{s}} = \frac{1}{L_\mu}\left(\frac{1}{T_\text{N}}\int_0^t \left(\underline{U}_{1\text{s}} - R_1\underline{I}_{1\text{s}}\right)\text{d}t - L_{\sigma 1}\underline{I}_{1\text{s}}\right) \tag{16a}$$

$$\begin{cases} T_{2(1)}\frac{\text{d}}{\text{d}t}\underline{\Psi}^{\text{e}}_{2(1)\text{s}} = L_\mu \underline{I}^{\text{e}}_{\mu\text{s}} - \underline{\Psi}^{\text{e}}_{2(1)\text{s}} + \text{j}\omega_\text{b} T_{2(1)}\omega_\text{m}\underline{\Psi}^{\text{e}}_{2(1)\text{s}} \\ \ldots \\ T_{2(N)}\frac{\text{d}}{\text{d}t}\underline{\Psi}^{\text{e}}_{2(N)\text{s}} = L_\mu \underline{I}^{\text{e}}_{\mu\text{s}} - \underline{\Psi}^{\text{e}}_{2(N)\text{s}} + \text{j}\omega_\text{b} T_{2(N)}\omega_\text{m}\underline{\Psi}^{\text{e}}_{2(N)\text{s}} \end{cases} \tag{16b}$$

$$\underline{\Psi}^{\text{e}}_{2\text{s}} = L_{\sigma 2\text{eq}} \sum_{n=1}^{N} \frac{\underline{\Psi}^{\text{e}}_{2(n)\text{s}}}{L_{\sigma 2(n)}} \tag{16c}$$

where "e" denotes the estimated rotor flux space vector.

Figure 3 presents a schematic diagram of the rotor flux estimation scheme, corresponding to Equations (16a)–(16c). The rotor angular velocity and the stator voltage and current space vector components constitute the input signals of the developed rotor flux estimation scheme. It should also be noted that when the rotor deep-bar effect is represented by the single two-terminal $N = 1$ in the rotor mathematical model, the voltage-current model of the rotor flux space vector remains valid and corresponds, in this case, to the IM classical mathematical model.

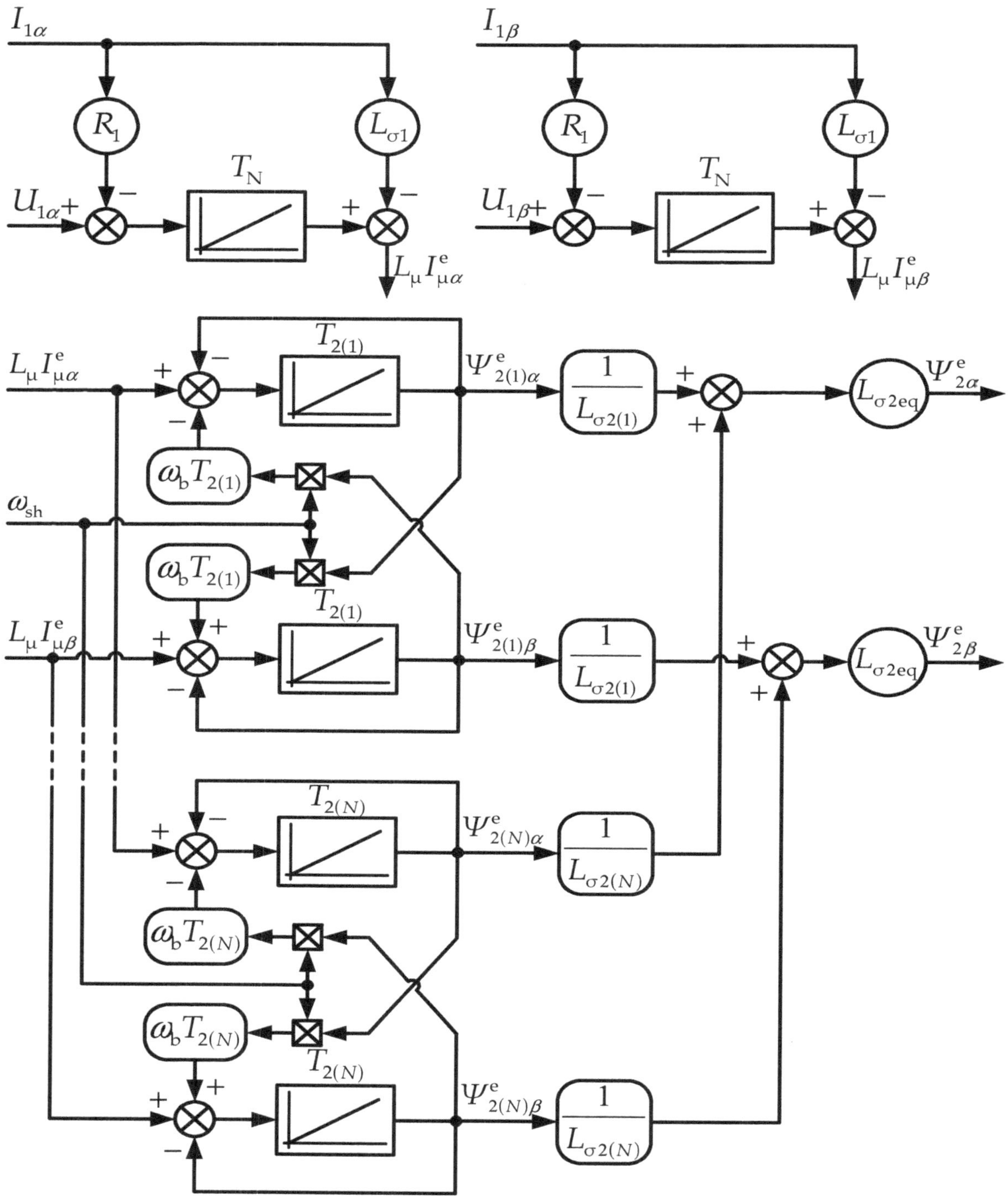

Figure 3. The schematic diagram of the voltage–current model of the rotor flux space vector, developed on the basis of the IM mathematical model with rotor two-terminal network representation, expressed in the orthogonal coordinate system (α–β) stationary with respect to the stator $\omega_k = 0$.

4. Laboratory Tests

4.1. Laboratory Test Bench

The verification of the novel rotor flux estimation scheme was performed with the laboratory test bench, of which a schematic diagram is shown in Figure 4. Figure 5 presents a picture of the laboratory test bench.

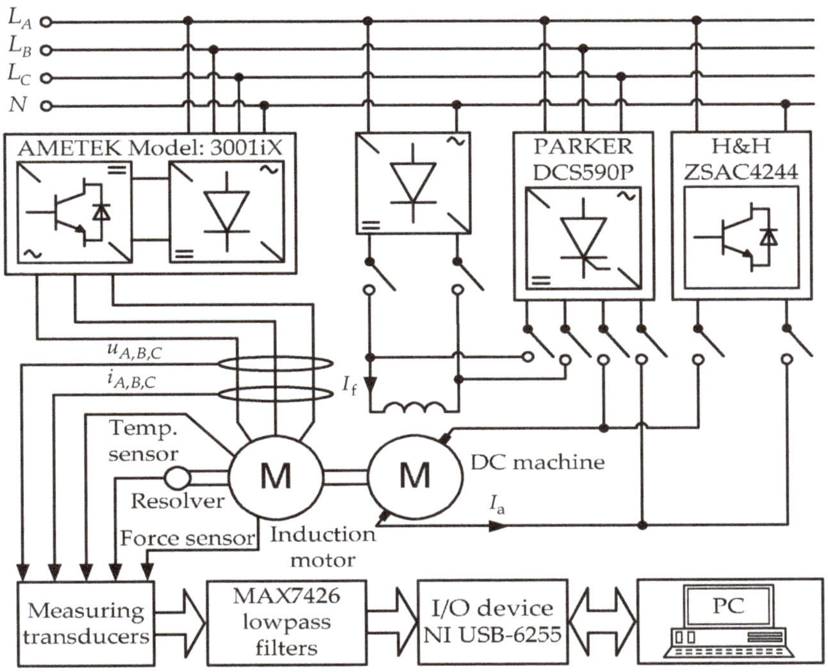

Figure 4. The schematic diagram of the laboratory test bench.

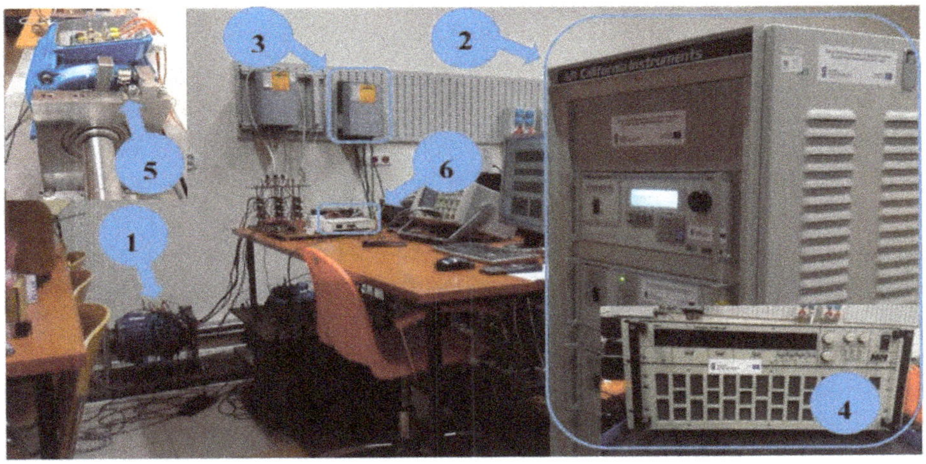

Figure 5. The laboratory test bench. The particular markers indicate: 1. tested induction motor, 2. programmable AC power source, AMETEK model: 3001iX, 3. 4Q thyristor converter, Parker DC590P, 4. electronic AC load ZSAC4244, Höcherl and Hackl GmbH, 5. tensometric force sensor, 6. multifunction I/O device NI USB-6255.

Investigations were conducted on the four-pole IMs of type Sg 132S-4 with a squirrel-cage rotor (CR-IM) and a solid rotor (SR-IM). The cross-section dimensions of the studied rotors are included in Appendix B.

The solid rotor was designed and manufactured only for the purpose of the presented study. This is because of the fact that such a rotor is characterized by the intense skin effect. This feature was conveniently used in the verification of the novel rotor flux estimation scheme. Such an approach allowed us to conduct experimental investigations with the use of low-power IMs, giving the background for the employment of the large squirrel-cage IMs in the next stage of the study. The tested SR-IM is marked by significant slip $s \approx 1$ corresponding to the breakdown torque at the stator supply voltage frequency of $f_1 = 50$ Hz. In order to reduce the breakdown slip of the SR-IM, the new operating points were adopted for both the CR-IM and SR-IM. The rated values of the tested IMs corresponding to the adopted machine operating points were set together in Appendix B.

The tested IMs (marker 1 in Figure 5) were powered by the programmable AC source AMETEK Model: 3001iX (marker 2 in Figure 5). The investigations were carried out at the three setpoints of stator voltage frequency for both considered IMs, maintaining a constant voltage/frequency ratio. Prior to the measurement tests, the CR-IM and SR-IM operated at the given load conditions for a period of time, allowing the stator winding temperature to stabilize at the assumed level. This aimed to reduce the influence of variability of rotor and stator resistances, resulting from winding temperature changes, on the identified electromagnetic parameters of the IM mathematical models and on the accuracy evaluation of the rotor flux space vector estimation.

A separately excited DC machine of type PCMb 54b served as a load for the investigated IMs in the presented study. During the no-load, blocked rotor, and load curve (LC) tests, conducted in order to identify the electromagnetic parameters of the considered IM mathematical models, the DC machine was powered by the 4Q thyristor converter Parker DC590P (marker 3 in Figure 5). Such a solution provided a wide adjustment range of load conditions for the tested IMs, enabling the measurement of demanded physical quantities in the generating, motoring, and ideal no-load modes of machine operation. In turn, the programmable electronic load ZSAC4244 – H&H GmbH (marker 4 in Figure 5) was used to control the armature current of the separately excited DC machine for the verification of the rotor flux estimation scheme. Such an approach enabled shaping of the desired dynamics of IM slip frequency changes in the assumed range, corresponding to the load adjustment range up to 1.30 of the stator rated current for both tested IMs. The slip frequency range considered during the verification of the rotor flux estimation scheme corresponded also to the slip frequency range when the LC tests were conducted. The power rating of the individual devices and DC machine which were used in the experimental investigations are included in Appendix B.

During the laboratory tests, the measurement of stator winding voltages, currents, and temperature, as well as shaft angular velocity, were carried out. Additionally, the shaft torque of the investigated IMs was determined on the basis of the force measurement realized by means of the force sensor (marker 5 in Figure 5). The accuracy class and measuring range of the used force sensor were 0.2 and 5 kN, respectively. The stator currents of the tested IMs were converted into voltage signals by means of non-inductive resistive voltage dividers with an accuracy class of 0.5 and a measuring range of 10 A. Similarly, the stator voltages were scaled by means of voltage dividers with a voltage ratio of 1000:1, composed of non-inductive resistors with an accuracy class of 0.2. The angular velocity measurement was carried with the use of a resolver. Data acquisition (DAQ) was performed by means of the National Instruments USB-6255 high-resolution, multifunction I/O device (marker 6 in Figure 5). The DAQ system was equipped with the MAX7426 5th-order, lowpass, elliptic, switched-capacitor filters. The configuration of the DAQ device and the acquisition of measurement data were carried out in the National Instruments LabView environment.

4.2. The Identification Procedure of Electormagnetic Parametersfor the IM Mathematical Model

The identification process for electromagnetic parameters of the IM mathematical model with rotor two-terminal network representation was conducted in conformity to a procedure described previously [25].The reference IFCh $\underline{L}_1(\omega_2)$ of the considered IMs were determined on the basis of the measurement data derived from the LC test [28] according to the following equations:

$$\frac{\omega_b}{j\omega_1}\left(\underline{Z}_1(\omega_2) - R_1\right) = \underline{L}_1(\omega_2) = L_{1\sigma} + \underline{L}_{1\delta}(\omega_2) \tag{17a}$$

$$\underline{Z}_1(\omega_2) = \frac{\left|\underline{U}_{1\text{ph}}\right|}{\left|\underline{I}_{1\text{ph}}(\omega_2)\right|}\left(\cos(\phi_1(\omega_2)) + \text{j}\sqrt{1 - [\cos(\phi_1(\omega_2))]^2}\right) \tag{17b}$$

$$\cos(\phi_1(\omega_2)) = \frac{P_1(\omega_2)}{3\left|\underline{U}_{1\text{ph}}\right|\left|\underline{I}_{1\text{ph}}(\omega_2)\right|} \tag{17c}$$

where $\underline{Z}_1(\omega_2)$ is the IM impedance expressed as a function of slip frequency, ω_1 represents the stator voltage angular frequency, $|\underline{U}_{1ph}|$ and $|\underline{I}_{1ph}(\omega_2)|$ constitute the measured root mean squared values of stator phase voltages and currents, respectively, and $P_1(\omega_2)$ and $\cos(\phi_1(\omega_2))$ are the measured stator power and calculated stator power factor, respectively.

Approximation of the reference IFCh $\underline{L}_{1\delta}(\omega_2)$ by means of the frequency-domain inductance $\underline{L}_{1\delta}(p = \mathrm{j}\omega_2)$ resulting from the adopted IM mathematical model, for instance, with N parallel connected two-terminals in the rotor equivalent circuit, in the form of the following equations:

$$\frac{1}{\underline{L}_{1\delta}(\omega_2)} \cong \frac{1}{\underline{L}_{1\delta}(p = \mathrm{j}\omega_2)} = \frac{1}{L_\mu} + \frac{\mathrm{j}\omega_2}{\omega_b}\frac{1}{\underline{Z}_2(p = \mathrm{j}\omega_2)} \tag{18a}$$

$$\frac{1}{\underline{Z}_2(p = \mathrm{j}\omega_2)} \cong \sum_{n=1}^{N} \frac{1}{R_{2(n)} + \frac{\mathrm{j}\omega_2}{\omega_b} L_{\sigma 2(n)}}. \tag{18b}$$

allows determination of the "synthetic" electromagnetic parameters of the IM mathematical model.

The stator phase winding resistance R_1 is identified through the DC line-to-line resistance measurement conducted according to the standards [28,29]. The parameters $L_{1\sigma}$, L_μ, $R_{2(n)}$, and $L_{\sigma 2(n)}$ are subject to the identification process which can be considered as a minimization issue of the evaluation function, defined as the sum of the mean squared errors of the reference characteristic approximation [25]:

$$\begin{aligned} F\left(\left|\underline{L}_1(\omega_2)\right|, \angle\underline{L}_1(\omega_2)\right) = \sum_{\omega_{2\min}}^{\omega_{2\max}} k_{\mathrm{mod}}\left(\frac{|\underline{L}_1(\omega_2)| - |\underline{L}_1(p=\mathrm{j}\omega_2)|}{|\underline{L}_1(\omega_2)|}\right)^2 + \\ + \sum_{\omega_{2\min}}^{\omega_{2\max}} k_{\mathrm{arg}}\left(\angle\underline{L}_1(\omega_2) - \angle\underline{L}_1(p = \mathrm{j}\omega_2)\right)^2 \end{aligned} \tag{19}$$

where $|\underline{L}_1(\omega_2)|$ and $\angle\underline{L}_1(\omega_2)$ are the modulus and argument of the IM reference IFCh, respectively, $|\underline{L}_1(p = \mathrm{j}\omega_2)|$ and $\angle\underline{L}_1(p = \mathrm{j}\omega_2)$ constitute the modulus and argument of the frequency-domain IM inductance, respectively, $\omega_{2\min}$ and $\omega_{2\max}$ represent the lower and upper limits of the considered slip frequency range, and k_{mod} and k_{arg} are the weighting factors of the individual components of the evaluation function.

Similarly to the studies presented previously [25], a minimization process of the adopted evaluation function was carried out with the use of the genetic algorithm by means of the Genetic Algorithms for Optimization Toolbox in the Matlab environment. The choice of the genetic algorithm was dictated by the effectiveness of this optimization tool, as indicated in numerous scientific publications concerned the identification of electromagnetic parameters for IM mathematical models [30,31].

The criterion adopted in the identification process of electromagnetic parameters for the CR-IM and SR-IM mathematical models with rotor multi-loop representation, assumed the approximation of the reference IFCh modulus with an error not exceeding 2% in the considered range of slip frequency, while maintaining a possible minimum approximation error of the reference IFCh argument and a minimum number N of two-terminals in a rotor equivalent circuit. The criterion was met with the use of the IM mathematical models with two $N = 2$ and three $N = 3$ parallel connected two-terminals in the cage and solid rotor network representations, respectively. The electromagnetic parameters of the IM mathematical models with multi-loop representation of the cage and solid rotors are denoted as CR-RML and SR-RML, respectively. These parameters are listed in Tables 1 and 2. Tables 1 and 2 also include the electromagnetic parameters of the IM classical mathematical model (T-type equivalent circuit parameters), which were identified based on selected procedures described in the standard 1 [28] (the designations are CR-T Std 1, SR-T Std 1) and standard 2 [29] (the designations are CR-T Std 2, SR-T Std 2) in the vicinity of the adopted operating points of the tested IMs (Appendix B). The stator phase resistance, determined according to the guidelines included in standard 1 [28], after correction to the reference winding temperature of 25 °C, equalled $R_1 = 2.9597\ \Omega$.

Table 1. Electromagnetic parameters (p.u.) of the considered squirrel-cage rotor induction motor (CR-IM) mathematical models. The individual resistances were corrected to the reference winding temperature of 25 °C.

Electromagnetic Parameters	$L_{1\sigma}$	L_{μ}	$R_{2(1)}$	$L_{\sigma 2(1)}$	$R_{2(2)}$	$L_{\sigma 2(2)}$
CR-T Std 1	0.0949	3.0978	0.0309	0.1428	–	–
CR-T Std 2	0.0910	3.1235	0.0335	0.1358	–	–
CR-RML	0.1090	3.0203	0.0395	0.0888	0.1326	1.3294

Table 2. Electromagnetic parameters (p.u.) of the considered solid rotor induction motor (SR-IM) mathematical models. The individual resistances correspond to the average temperature of stator winding of 55 °C registered under the load curve(LC) test.

Electromagnetic Parameters	$L_{1\sigma}$	L_{μ}	$R_{2(1)}$	$L_{\sigma 2(1)}$	$R_{2(2)}$	$L_{\sigma 2(2)}$	$R_{2(3)}$	$L_{\sigma 2(3)}$
SR-T Std 1	0.2597	2.8323	0.1329	0.4550	–	–	–	–
SR-T Std 2	0.1645	3.1736	0.2271	0.2456	–	–	–	–
SR-RML	0.1977	3.4401	0.2852	0.2313	0.4695	2.1032	0.1518	7.4735

As an example, in Figure 6, the SR-IM reference IFCh $\underline{L}_1(\omega_2)$ are set together with the approximative characteristics, which were determined on the basis of the IM mathematical model with single and three parallel connected two-terminals in the solid rotor equivalent circuit. Figure 7 presents the modulus relative errors and the argument absolute errors between the reference and approximative characteristics [25].

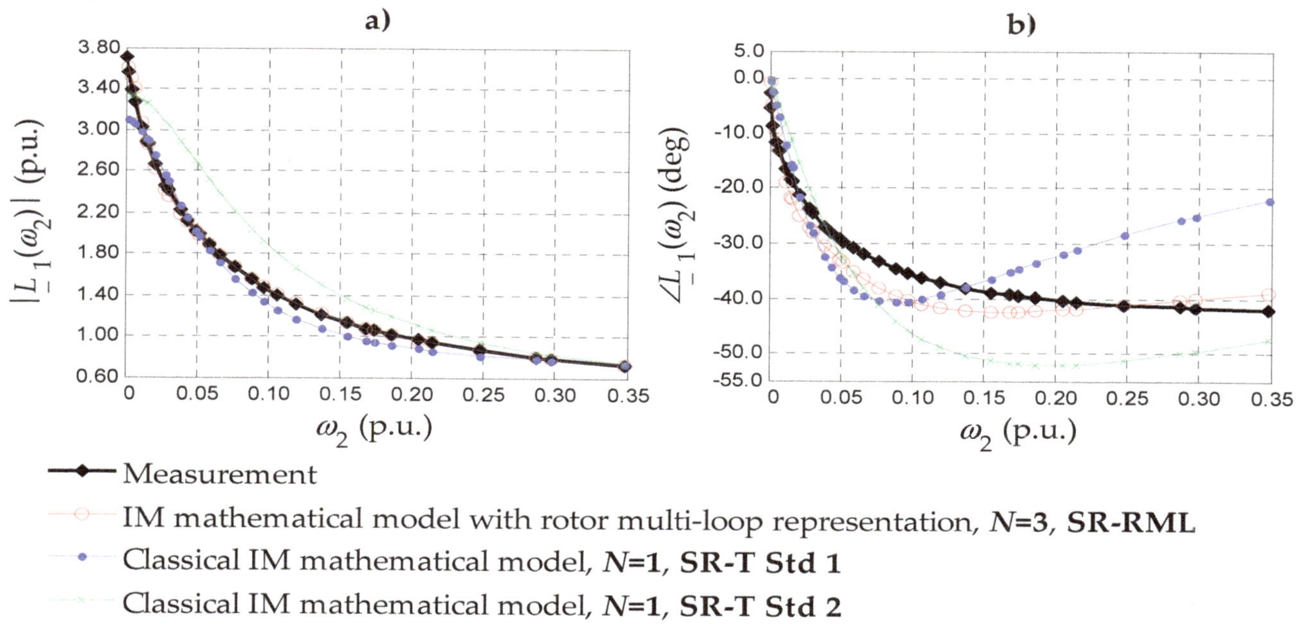

Figure 6. The SR-IM reference inductance frequency characteristic (IFCh) and its approximation by means of the IM mathematical models with single and three two-terminals in the solid rotor equivalent circuit: (**a**) Modulus and (**b**) argument.

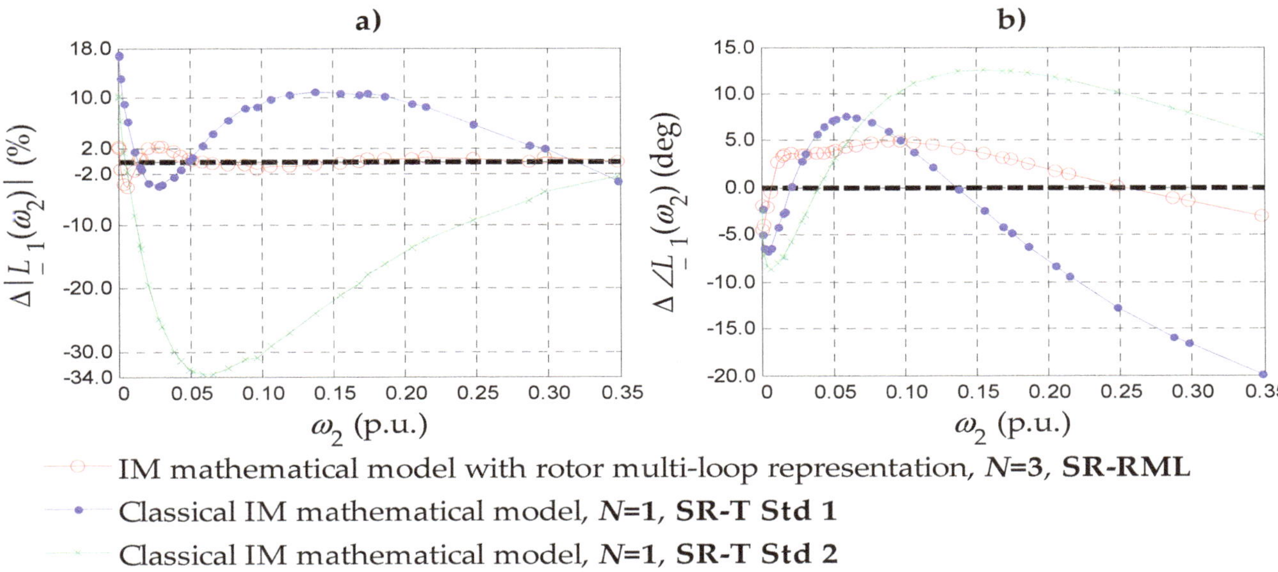

Figure 7. Approximation errors of the SR-IM reference IFCh: (**a**) Modulus relative errors and (**b**) argument absolute errors.

It is worth nothing that the use of the IM mathematical model with three parallel connected two-terminals in the solid rotor equivalent circuit allowed the most accurate approximation of the SR-IM reference IFCh to be achieved, in comparison to the approximation accuracy obtained by the IM classical mathematical model. The required approximation accuracy of the SR-IM reference IFCh modulus was met with argument approximation error not exceeding five degrees in the considered range of slip frequency (Figure 7b). Theoretically, approximation accuracy of the reference IFCh could be further improved by incorporating additional two-terminals in the rotor mathematical model, but the adopted criterion also assumed their minimal number. This resulted from the desirable simplicity of the rotor flux estimation scheme, which is intended for the IM rotor-field-oriented control implementation. Ultimately, the approximation of the SR-IM reference IFCh realized by the IM mathematical model with three parallel connected two-terminals in the solid rotor equivalent circuit was considered to be sufficiently accurate for the purpose of the rotor flux estimation.

4.3. Estimation of the Rotor Flux Space Vector

In the presented study, the estimation accuracy of the rotor flux space vector was evaluated in the slip frequency range corresponding to the IM load adjustment range up to 1.30 of the stator rated current. The range of slip frequency exceeding the typical operating range of slip frequency for the field-oriented controlled IM, was adopted in this study for the robustness verification of the novel rotor flux estimation scheme against electromagnetic parameter variability resulting from the rotor deep-bar effect. This is the reason why the developed rotor flux estimation scheme was not employed in the rotor-flux-oriented control at this stage of the study. Since the direct measurement of rotor flux is not realized in practice, the verification of the rotor flux estimation accuracy was conducted indirectly, based on the reference quantities registered with the laboratory test bench.

During the study, the every endeavor was made to conduct the measurements of the reference quantities in a manner to assure the minimal possible measurement uncertainty. Due to the high resolution of the DAQ device and the precision of the sensors and measuring transducers, the registered shaft torque T_{sh} and the determined power losses ΔP were considered as the reference quantities in the verification of the developed rotor flux estimation scheme. The verification investigations used the fact that the shaft torque can be determined as the quotient of the motor shaft power and angular velocity:

$$T^{e}_{sh} = \frac{P^{e}_{sh}}{\omega_{sh}} \tag{20a}$$

$$P_{sh}^{e} = T_{em}^{e}\omega_{sh} - \Delta P \tag{20b}$$

where T_{sh}^{e} and T_{me}^{e} represent the estimated shaft and electromagnetic torque, respectively and P_{sh}^{e} is the estimated shaft power.

The power losses ΔP occurring in Equation (20b) were determined according to Equations (21a)–(21c) based on the measurement data derived from the LC tests. In order to eliminate the necessity to split up individual components of the power losses ΔP at the considered load conditions of the tested IMs, core losses were not erased from the IM input power whilst calculating the airgap power P_δ (Equation (21c)). Such an approach is in line with the adopted IM mathematical model described by Equations (7a)–(7i), in which the resistance associated with core losses is not included. In the presented study, the power losses ΔP were considered as any power losses determining the difference between the power transferred to the shaft $(1-s)P_\delta$ and the shaft power P_{sh}:

$$\Delta P(s) = (1-s)P_\delta(s) - P_{sh}(s) \tag{21a}$$

$$P_{sh}(s) = \frac{\omega_1}{\omega_b}(1-s)T_{sh}(s) \tag{21b}$$

$$P_\delta(s) = \frac{3}{2}\mathrm{Re}\left(\underline{U}_1\underline{I}_1^*(s)\right) - \frac{3}{2}R_1\left|\underline{I}_1(s)\right|^2 \tag{21c}$$

where $P_\delta(s)$constitutes the airgap power and s represent the motor slip.

Figure 8 presents the variability of the power losses, the shaft power, and the power transferred to the shaft expressed as a function of angular velocity of the tested CR-IM (Figure 8a) and SR-IM (Figure 8b), at the stator voltage frequency of $\omega_1 = 1.0$ (p.u.). Due to slight changes of the CR-IM power losses in the considered slip frequency range (Figure 8a), a constant value of these losses $\Delta P = 0.0863$ (p.u.) was assumed in the verification studies of the novel rotor flux estimation scheme. In relation to the SR-IM, on account of significant changes of the power losses in the considered slip frequency range (Figure 8b), the power loss variability was approximated by means of the second order polynomial:

$$\Delta P(\omega_{sh}) = -0.6551\omega_{sh}^2 + 0.8076\omega_{sh} - 0.0232. \tag{22}$$

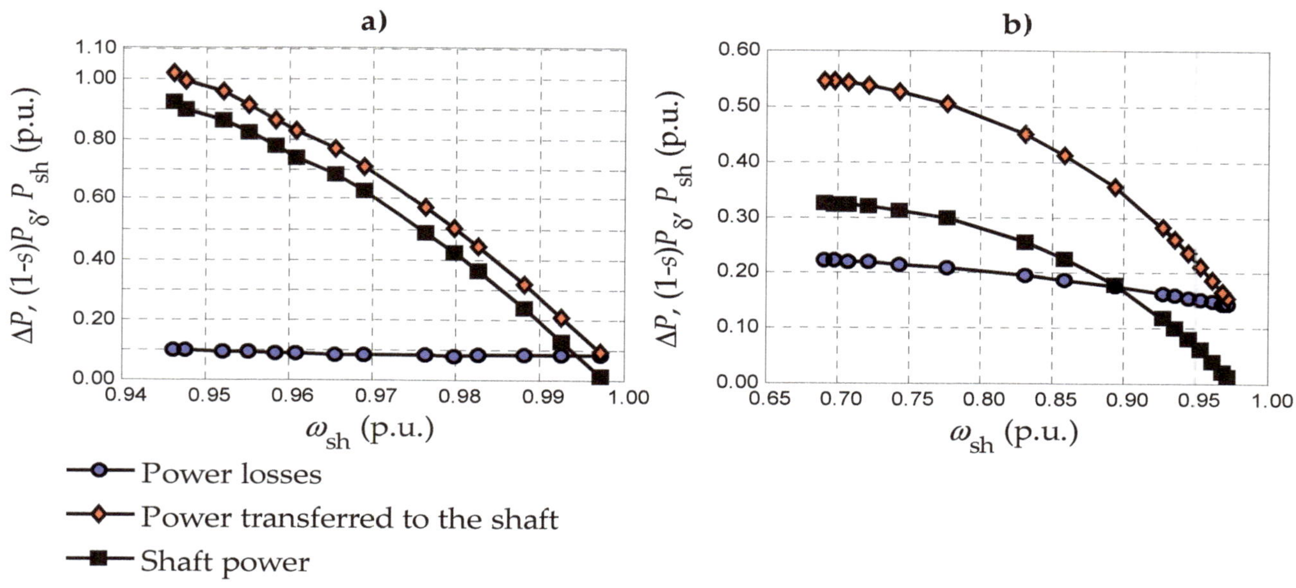

Figure 8. The variability of the power losses, the power transferred to the shaft and the shaft power expressed as a function of angular velocity of the investigated: (**a**) CR-IM and (**b**) SR-IM.

The electromagnetic torque required in Equation (20b) was determined with the use of the estimated rotor flux space vector in conformity with the following equations:

$$T_{\text{em}}^{\text{e}} = \frac{L_{\mu}}{L_{2\text{eq}}} \text{Im}\left(\underline{I}_{1\text{k}} \left(L_{\sigma 2\text{eq}} \sum_{n=1}^{N} \frac{\underline{\Psi}_{2(n)\text{k}}}{L_{\sigma 2(n)}} \right)^{*} \right) \tag{23a}$$

$$L_{2\text{eq}} = L_{\mu} + L_{\sigma 2\text{eq}}. \tag{23b}$$

The estimation precision of the registered shaft torque was verified based on the absolute estimation errors:

$$\Delta T_{\text{sh}}^{\text{e}} = T_{\text{sh}} - T_{\text{sh}}^{\text{e}}. \tag{24}$$

Additionally, Tables 3 and 4 present the maximum and mean absolute errors of the registered shaft torque estimation which were determined in accordance with the following equations:

$$\max\left|\Delta T_{\text{sh}}^{\text{e}}\right| = \max\left|T_{\text{sh}} - T_{\text{sh}}^{\text{e}}\right| \tag{25a}$$

$$\text{M}\left|\Delta T_{\text{sh}}^{\text{e}}\right| = \frac{1}{n}\sum_{i=1}^{n}\left|T_{\text{sh},i} - T_{\text{sh},i}^{\text{e}}\right| \tag{25b}$$

where $T_{\text{sh},i}$ and $T_{\text{sh},i}^{\text{e}}$ represent the ith samples of the registered and estimated shaft torque, respectively, and n is the number of samples.

For the sake of comparison, the shaft torque estimated through the use of the so called full order open-loop flux observer [32] was also considered in the presented verification. Research results presented previously [32] indicate that this rotor flux estimation scheme is characterized by limited sensitivity to erroneous identification or variability of the rotor electromagnetic time constant in a wide range of slip frequency changes, in comparison to the commonly known current model of the rotor flux space vector. The full order open-loop flux observer was formulated on the basis of the IM classical mathematical model, and is represented by the following system of equations:

$$T_{\text{N}} \frac{\text{d}}{\text{d}t} \underline{I}_{1\text{s}}^{\text{e}} = \frac{1}{\sigma L_1} \left[\underline{U}_{1\text{s}} - \left(R_1 + \left(\frac{L_{\mu}}{L_2} \right)^2 R_2 \right) \underline{I}_{1\text{s}}^{\text{e}} + \frac{L_{\mu}}{L_2} \left(\frac{R_2}{L_2} - \text{j}\omega_{\text{m}} \right) \underline{\Psi}_{2\text{s}}^{\text{e}} \right] \tag{26a}$$

$$T_2 \frac{\text{d}}{\text{d}t} \underline{\Psi}_{2\text{s}}^{\text{e}} = L_{\mu} \underline{I}_{1\text{s}}^{\text{e}} - \underline{\Psi}_{2\text{s}}^{\text{e}} + \text{j}\omega_{\text{b}} T_2 \omega_{\text{m}} \underline{\Psi}_{2\text{s}}^{\text{e}} \tag{26b}$$

$$T_2 = \frac{L_2}{R_2} T_{\text{N}} \tag{26c}$$

where T_2 is the rotor electromagnetic time constant.

The verification of the rotor flux estimation schemes additionally includes the shaft torque, which was estimated with the help of the elaborated voltage–current model (Equations (16a)–(16c)) and formulated on the basis of the IM classical mathematical model.

Figure 9 presents a block diagram of the algorithm used in the accuracy evaluation of the rotor flux space vector estimation. The algorithm was implemented in the Matlab environment.

In the presented study, the investigated estimation schemes of the rotor flux space vector were fed by the registered angular velocity and stator voltages and currents. This case corresponded to the operation of the tested estimation schemes in an IM drive with angular velocity measurements. Due to the similar accuracy of the shaft torque estimation, obtained through the considered estimation schemes of the rotor flux space vector at each setpoint of stator voltage frequency for both tested IMs, this paper presents the study results conducted at the nominal stator voltage frequency $\omega_1 = 1.0$ (p.u.).

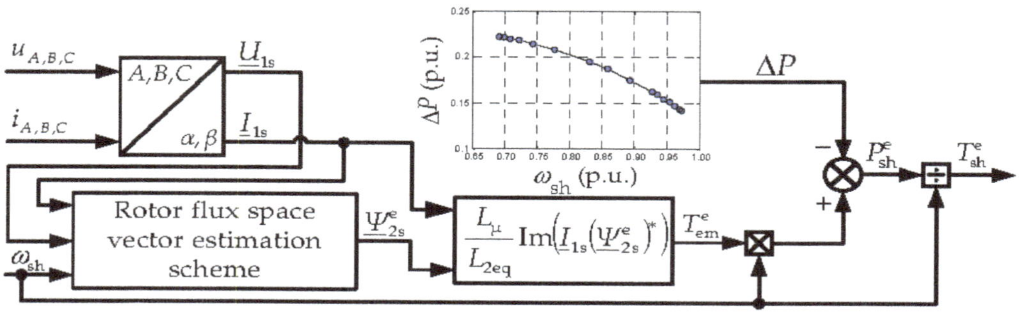

Figure 9. The block diagram of the algorithm used in the accuracy evaluation of the rotor flux space vector estimation.

Figure 10a presents the registered SR-IM shaft torque together with the shaft torque generated with the use of the tested rotor flux estimation schemes which were formulated on the basis of the IM mathematical model with single and three parallel connected two-terminals in the rotor equivalent circuit. The absolute errors of the registered shaft torque estimation are shown in Figure 10b, whereas Table 3 lists the maximal and mean absolute estimation errors. Figure 10 includes the shaft torque obtained with the use of the rotor flux estimation schemes based on the IM classical mathematical model in the structure of which the electromagnetic parameters SR-Std 2 were applied. The use of these parameters allowed for a more accurate estimation of the registered SR-IM shaft torque in relation to the estimation accuracy acquired by means of the employed schemes with the electromagnetic parameters SR-Std 1 (Table 3).

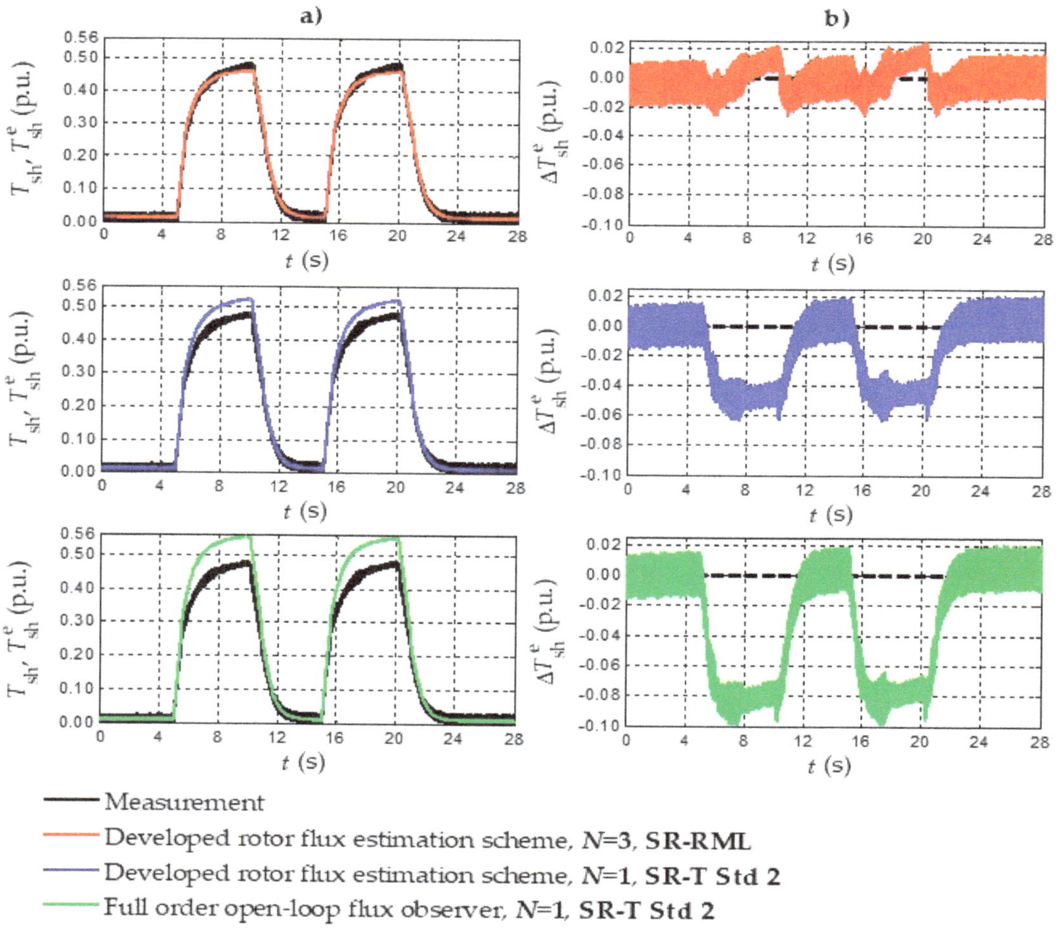

Figure 10. The SR-IM shaft torque estimation: (**a**) The registered and estimated shaft torque and (**b**) the absolute errors of the registered shaft torque estimation.

Table 3. Maximum and mean absolute errors (p.u.) of the registered shaft torque estimation of the tested SR-IM.

Rotor Flux Estimation Schemes	Full Order Open-Loop Flux Observer (26a)–(26c)		Voltage-Current Model (16a)–(16c) $N = 1$		Voltage-Current Model (16a)–(16c) $N = 3$
Electromagnetic parameters	SR-T Std 1	SR-T Std 2	SR-T Std 1	SR-T Std 2	SR-RML
max$\|\Delta T^{e}_{sh}\|$	0.3191	0.0986	0.2877	0.0631	0.0262
M$\|\Delta T^{e}_{sh}\|$	0.0845	0.0346	0.0712	0.0209	0.0075

Figure 11a presents the registered and estimated shaft torque of the tested CR-IM, whereas Figure 11b shows the absolute estimation errors. The maximal and mean absolute estimation errors are listed in Table 4.

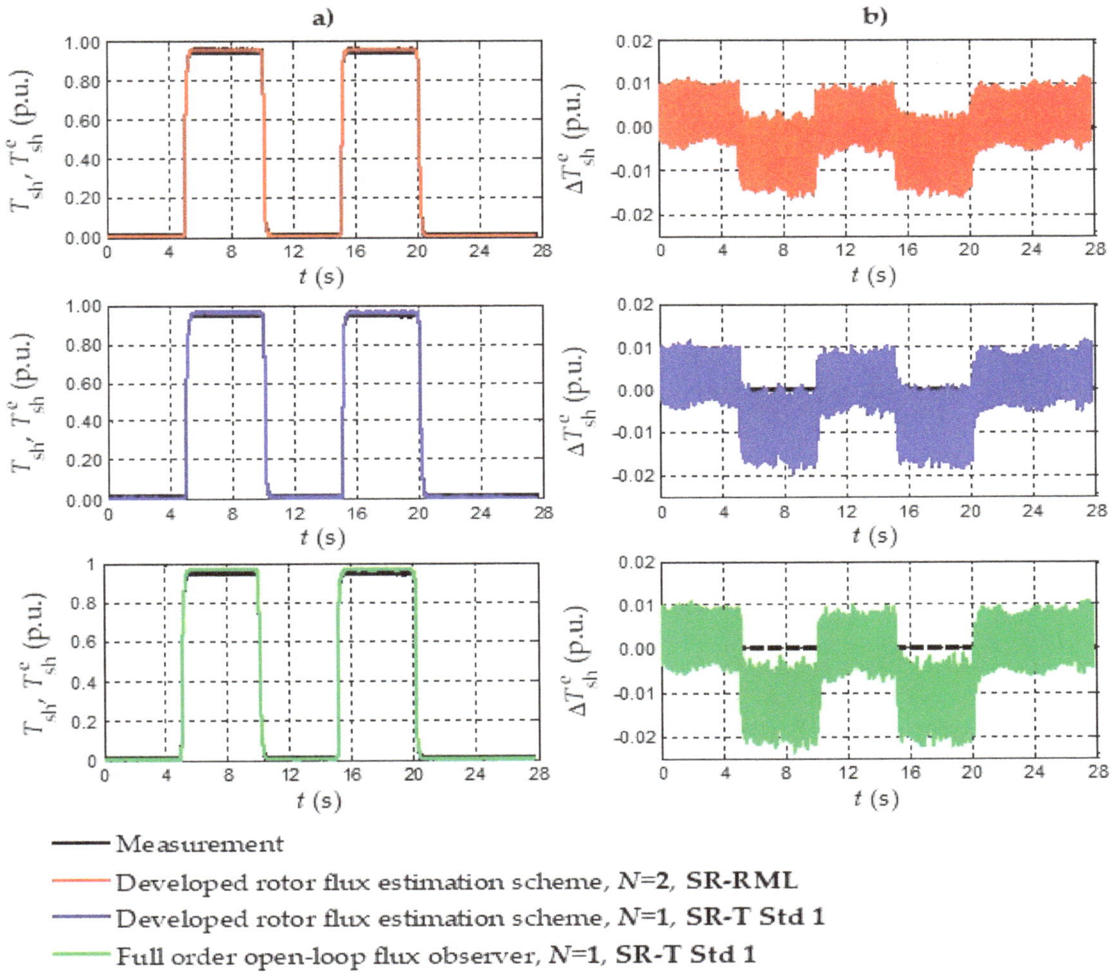

Figure 11. The CR-IM shaft torque estimation: (**a**) The registered and estimated shaft torque and (**b**) the absolute errors of the registered shaft torque estimation.

Table 4. Maximum and mean absolute errors (p.u.) of the registered shaft torque estimation of the tested CR-IM.

Rotor Flux Estimation Schemes	Full Order Open-Loop Flux Observer (26a)–(26c)		Voltage-Current Model (16a)–(16c) $N = 1$		Voltage-Current Model (16a)–(16c) $N = 2$
Electromagnetic parameters	CR-T Std 1	CR-T Std 2	CR-T Std 1	CR-T Std 2	CR-RML
max$\|\Delta T^{e}_{sh}\|$	0.0239	0.0345	0.0196	0.0211	0.0164
M$\|\Delta T^{e}_{sh}\|$	0.0065	0.0105	0.0056	0.0063	0.0047

As regards the CR-IM, a more accurate estimation of the registered shaft torque was achieved using the electromagnetic parameters CR-T Std 1 in the considered estimation schemes based on the IM classical mathematical model, in comparison to when the electromagnetic parameters CR-T Std 2 were applied (Table 4). For this reason, the shaft torque generated through the rotor flux estimation schemes described by Equations (26a)–(26c) and (16a)–(16c) (single two-terminal rotor representation $N = 1$) with the electromagnetic parameters CR-T Std 1 are included in Figure 11.

5. Conclusions

This paper presents a novel estimation scheme of the rotor flux space vector which has been developed on the basis of the IM mathematical model with rotor multi-loop representation.

In regards to the tested SR-IM, the use of the rotor flux estimation scheme, in which the rotor skin effect was modeled by three parallel connected two-terminals in the rotor equivalent circuit, enabled a multiple reduction of the registered shaft torque estimation errors in relation to the estimation errors obtained through the considered estimation schemes based on the IM classical mathematical model (Table 3). The absolute estimation errors of the registered SR-IM shaft torque achieved by using the elaborated voltage–current model ($N = 3$) did not exceed the level of ± 0.0262 (p.u.) (Figure 10b, Table 3).

The results of the presented study indicate considerable improvement in the accuracy of the rotor flux space vector estimation of the tested SR-IM, which was obtained by the estimation scheme elaborated on the IM mathematical model with rotor two-terminal network representation, in comparison with the estimation precision acquired by the schemes formulated on the IM classical mathematical model.

It should also be noted that even for the tested CR-IM, which does not show a substantial deep-bar effect, the superiority of the novel rotor flux estimation scheme (with two parallel connected two-terminals $N = 2$ in the rotor equivalent circuit) over the estimation schemes based on the IM classical mathematical model can be observed (Figure 11, Table 4).

The results of the conducted study indicate that the developed voltage–current model enables accurate estimation of the rotor flux of IMs characterized by intense deep-bar effect, in the operating range of the slip frequency. Considering the above, the novel rotor flux estimation scheme can be applied for the rotor-flux-oriented control of the IMs with any rotor construction, including squirrel-cage, double-cage, and solid rotors. Moreover, the elaborated voltage–current model of the rotor flux space vector can be employed as the adjustable model of the Model Reference Adaptive System based estimator for speed-sensorless IM drive applications.

Future work will include experimental studies of the IM rotor-flux-oriented control with the novel rotor flux estimation scheme.

Author Contributions: Conceptualization, G.U. and J.R.; Formal analysis, G.U.; Investigation, J.R.; Methodology, G.U. and J.R.; Supervision, A.K.; Validation, G.U. and J.R.; Visualization, G.U.; Writing—original draft, G.U., J.R. and A.K.

Appendix A

In the presented study, the total apparent electrical power was adopted as base apparent power (input voltampere base).The base values are defined in accordance to the contents of Table A1:

Table A1. The per-unit system base values.

Base Quantity	Symbol	Unit	Formula
Apparent power	S_b	voltampere (V·A)	$3 \cdot U_{1ph} \cdot I_{1ph}$
Frequency	ω_b	radian per second (rad/s)	$2 \cdot \pi \cdot f_{1N}$
Angular velocity	ω_{mb}	radian per second (rad/s)	$2 \cdot \pi \cdot f_{1N} \cdot (p_p)^{-1}$
Magnetic flux	Ψ_b	volt second per radian (V·s/rad)	$U_{1ph} \cdot (2 \cdot \pi \cdot f_{1N})^{-1}$
Impedance	Z_b	ohm (Ω)	$U_{1ph} \cdot (I_{1ph})^{-1}$
Inductance	L_b	henry per radian (H/rad)	$U_{1ph} \cdot (I_{1ph} \cdot 2 \cdot \pi \cdot f_{1N})^{-1}$
Torque	T_b	newton meter per radian (N·m/rad)	$3 \cdot U_{1ph} \cdot I_{1ph} \cdot p_p \cdot (2 \cdot \pi \cdot f_{1N})^{-1}$

Where U_{1ph} and I_{1ph} are the nominal stator phase voltage and current, respectively, f_{1N} stands for the nominal frequency of stator voltages, and p_p is the number of pole pairs.

Appendix B

Figure A1 presents the cross-section dimensions (millimeters) of the tested squirrel-cage (Figure A1a) and solid (Figure A1b) rotors.

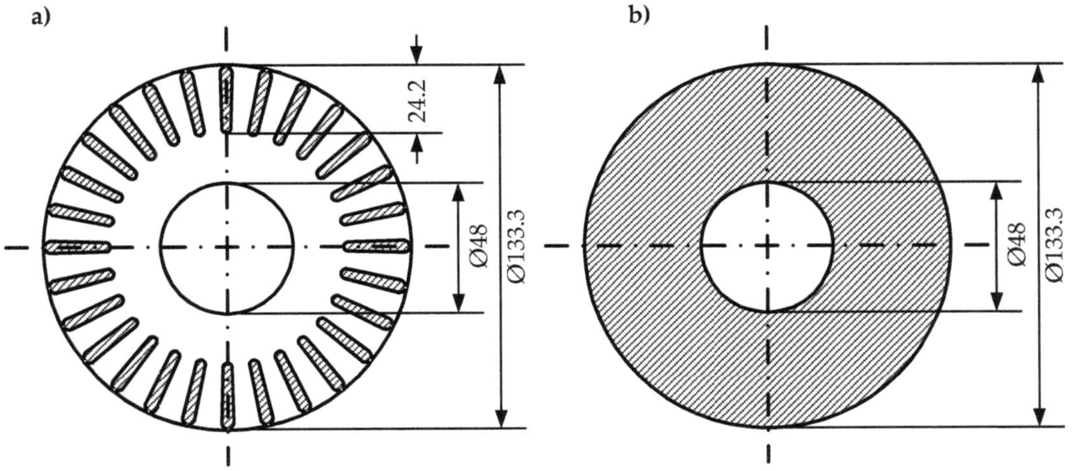

Figure A1. The cross-section dimensions of: (**a**) CR-IM and (**b**) SR-IM.

The rated values of the tested IMs corresponding to the adopted machine operating points are included in Table A2. These rated values were determined for the purpose of the presented study and aimed to reduce the breakdown slip of the SR-IM. The rated values were settled so as to maintain approximately equal stator flux amplitudes of the CR-IM and SR-IM, bearing in mind the limitations resulting from rated values of the programmable AC source (AMETEK Model: 3001iX) powering the investigated IMs.

Table A2. The rated values of the CR-IM and SR-IM corresponding to the adopted operating points.

Rating	Unit	CR-IM	SR-IM
Output power	kilowatt (kW)	2.358	1.992
Stator voltage	volt (V)	400 (wye)	391 (delta)
Stator frequency	hertz (Hz)	50	85
Stator current	ampere (A)	4.536	7.785
Torque	newton meter (N·m)	15.53	9.39
Rotational speed	revolution per minute (r/m)	1450	2030
Power factor	(-)	0.8819	0.6698
Efficiency	(-)	0.8525	0.5641
Stator flux	weber (Wb)	0.973	0.995

Table A3. The power rating of the individual devices and DC machine employed within the laboratory test bench.

Name	Description	Unit	Rated Power
AMETEK Model: 3001iX	programmable AC source	kilowatt (kW)	9.0
DC590P	4Q thyristor converter	kilowatt (kW)	7.5
ZSAC4244 – H&H GmbH	programmable electronic load	kilowatt (kW)	4.2
PCMb 54b	separately excited DC machine	kilowatt (kW)	6.5

References

1. Blaschke, F. The principle of fields-orientation as applied to the transvector closed loop control system for rotating-field machines. *Siemens Rev.* **1972**, *34*, 217–220.
2. Hasse, K. Drehzahlregelverfahren fur schnelle Umkehrantriebe mit stromrichtergespeisten Asynchron-Kurzschlusslaufermotoren. *Regelungstechnik* **1972**, *20*, 60–66. [CrossRef]
3. Takahashi, I.; Noguchi, T. A new quick-response and high efficiency control strategy of an induction motor. *IEEE Trans. Ind. Appl.* **1986**, *5*, 820–827. [CrossRef]
4. Garces, L.J. Parameter adaption for the speed-controlled static ac drive with a squirrel-Cage induction motor. *IEEE Trans. Ind. Appl.* **1980**, *2*, 173–178. [CrossRef]
5. Matsuo, T.; Lipo, T.A. A rotor parameter identification scheme for vector-controlled induction motor drives. *IEEE Trans. Ind. Appl.* **1985**, *3*, 624–632. [CrossRef]
6. Krishnan, R.; Doran, F.C. Study of parameter sensitivity in high-performance inverter-fed induction motor drive systems. *IEEE Trans. Ind. Appl.* **1987**, *4*, 623–635. [CrossRef]
7. Orlowska-Kowalska, T. Application of extended Luenberger observer for flux and rotor time-constant estimation in induction motor drives. *IEE Proc. (Control Theory Appl.)* **1989**, *6*, 324–330. [CrossRef]
8. Garcia Soto, G.; Mendes, E.; Razek, A. Reduced-order observers for rotor flux, rotor resistance and speed estimation for vector controlled induction motor drives using the extended Kalman filter technique. *IEE Proc. (Electr. Power Appl.)* **1999**, *3*, 282–288. [CrossRef]
9. Marčetić, D.P.; Vukosavić, S.N. Speed-sensorless AC drives with the rotor time constant parameter update. *IEEE Trans. Ind. Electron.* **2007**, *5*, 2618–2625. [CrossRef]
10. Proca, A.B.; Keyhani, A. Sliding-mode flux observer with online rotor parameter estimation for induction motors. *IEEE Trans. Ind. Electron.* **2007**, *2*, 716–723. [CrossRef]
11. Salmasi, F.R.; Najafabadi, T.A.; Maralani, P.J. An adaptive flux observer with online estimation of DC-link voltage and rotor resistance for VSI-based induction motors. *IEEE Trans. Power Electron.* **2010**, *5*, 1310–1319. [CrossRef]
12. Wang, S.; Dinavahi, V.; Xiao, J. Multi-rate real-time model-based parameter estimation and state identification for induction motors. *IET Electr. Power Appl.* **2013**, *1*, 77–86. [CrossRef]
13. Zaky, M.S.; Metwaly, M.K. Sensorless torque/speed control of induction motor drives at zero and low frequencies with stator and rotor resistance estimations. *IEEE J. Emerg. Sel. Top. Power Electron.* **2016**, *4*, 1416–1429. [CrossRef]
14. Kim, J.K.; Ko, J.S.; JangHyeon, L.; Lee, Y.K. Rotor flux and rotor resistance estimation using Extended Luenberger-Sliding Mode Observer (ELSMO) for three phase induction motor control. *Can. J. Electr. Comput. Eng.* **2017**, *3*, 181–188. [CrossRef]
15. Yang, S.; Ding, D.; Li, X.; Xie, Z.; Zhang, X.; Chang, L. A novel online parameter estimation method for indirect fieldoriented induction motor drives. *IEEE Trans. Energy Convers.* **2017**, *4*, 1562–1573. [CrossRef]
16. Babb, D.S.; Williams, J.E. Network analysis of A-C machine conductors. *Trans. Am. Inst. Electr. Eng.* **1951**, *2*, 2001–2005. [CrossRef]
17. Paszek, W.; Kaplon, A. Induction machine with anisotropic multilayer rotor modelling the electromagnetic and the electrodynamic states of a symmetrical machine with deep bar cage in solid iron rotor core. *Electromagn. Fields Electr. Eng.* **1988**, 205–210. [CrossRef]
18. Levy, W.; Landy, C.H.; McCulloch, M.D. Improved models for the simulation of deep bar induction motors. *IEEE Trans. Energy Convers.* **1990**, *2*, 393–400. [CrossRef]

19. De Doncker, R.W. Field-oriented controllers with rotor deep bar compensation circuits. *IEEE Trans. Ind. Appl.* **1992**, *5*, 1062–1071. [CrossRef]
20. Healey, R.C.; Williamson, S.; Smith, A.C. Improved cage rotor models for vector controlled induction motors. *IEEE Trans. Ind. Appl.* **1995**, *4*, 812–822. [CrossRef]
21. Smith, A.C.; Healey, R.C.; Williamson, S. A transient induction motor model including saturation and deep bar effect. *IEEE Trans. Energy Convers.* **1996**, *1*, 8–15. [CrossRef]
22. Seok, J.K.; Sul, S.K. Pseudorotor-flux-oriented control of an induction machine for deep-bar-effect compensation. *IEEE Trans. Ind. Appl.* **1998**, *3*, 429–434. [CrossRef]
23. Monjo, L.; Kojooyan-Jafari, H.; Córcoles, F.; Pedra, J. Squirrel-cage induction motor parameter estimation using a variable frequency test. *IEEE Trans. Energy Convers.* **2015**, *2*, 550–557. [CrossRef]
24. Benzaquen, J.; Rengifo, J.; Albánez, E.; Aller, J.M. Parameter estimation for deep-bar induction machines using instantaneous stator measurements from a direct startup. *IEEE Trans. Energy Convers.* **2017**, *2*, 516–524. [CrossRef]
25. Rolek, J.; Utrata, G. An identification procedure of electromagnetic parameters for an induction motor equivalent circuit including rotor deep bar effect. *Arch. Electr. Eng.* **2018**, *2*, 279–291. [CrossRef]
26. McKinnon, D.J.; Seyoum, D.; Grantham, C. Investigation of parameter characteristics for induction machine analysis and control. In Proceedings of the Second International Conference on Power Electronics, Machines and Drives, Edinburgh, UK, 31 March–2 April 2004. [CrossRef]
27. Willis, J.R.; Brock, G.J.; Edmonds, J.S. Derivation of induced motor models from standstill frequency response tests. *IEEE Trans. Energy Convers.* **1989**, *4*, 608–615. [CrossRef]
28. *2013 Rotating Electrical Machines—Part 28: Test Methods for Determining Quantities of Equivalent Circuit Diagrams for Three-Phase Low Voltage Cage Induction Motors*; Standard PN-EN 60034-28; Polski Komitet Normalizacyjny: Warszawa, Poland, 2013.
29. *Standard Test Procedure for Polyphase Induction Motors and Generators*; IEEE Standard 112™-2004; IEEE Power Engineering Society: New York, NY, USA, 4 November 2004. [CrossRef]
30. Alonge, F.; D'ippolito, F.; Ferrante, G.; Raimondi, F.M. Parameter identification of induction motor model using genetic algorithms. *IEE Proc. Control Theory Appl.* **1998**, *6*, 587–593. [CrossRef]
31. Kumar, P.; Dalal, A.; Singh, A.K. Identification of three phase induction machines equivalent circuits parameters using multi-objective genetic algorithms. In Proceedings of the 2014 International Conference on Electrical Machines (ICEM), Berlin, Germany, 2–5 September 2014. [CrossRef]
32. Jansen, P.L.; Lorenz, R.D. A physically insightful approach to the design and accuracy assessment of flux observers for field oriented induction machine drives. *IEEE Trans. Ind. Appl.* **1994**, *1*, 101–110. [CrossRef]

12

Design and Optimization of a Magnetically Levitated Inductive Reaction Sphere for Spacecraft Attitude Control

Liming Yuan [1,2,†], Jie Zhang [1,2,†], Si-Lu Chen [1,*], Yusheng Liang [1], Jinhua Chen [1], Chi Zhang [1] and Guilin Yang [1]

1 Zhejiang Key Laboratory of Robotics and Intelligent Manufacturing Equipment Technology, Ningbo Institute of Materials Technology and Engineering, Chinese Academy of Sciences, Ningbo 315201, China; yuanliming@nimte.ac.cn (L.Y.); zhangjie@nimte.ac.cn (J.Z.); liangyusheng@nimte.ac.cn (Y.L.); chenjinhua@nimte.ac.cn (J.C.); zhangchi@nimte.ac.cn (C.Z.); glyang@nimte.ac.cn (G.Y.)
2 University of Chinese Academy of Sciences, Beijing 100049, China
* Correspondence: chensilu@nimte.ac.cn.
† These authors contributed equally to this work.

Abstract: The inductive reaction sphere (RS) brings the benefit of simple, economical, and miniaturized design, and it is capable of multi-DOF torque generation. Thus, it is a suitable choice for the angular momentum exchange actuator in attitude control of micro-spacecrafts. To synthesize symmetric distribution of eddy currents and improve the speed and stability of rotation, a novel 4-pole winding design is proposed. However, the developed simplified analytical model shows that reduced pole number degrades the torque generation. To enhance the output torque of 4-pole RS, its curved cores and electromagnets are redesigned to enable the side teeth to be functional. As the analytical torque model for the RS with the slotted cores is not available, a constrained optimization problem is formulated, and the optimized parameters are calculated based on the prediction model from supported vector machine and finite element analysis. The lab prototypes are developed to validate the proposed design and test the speed performance. The experimental results show that the 4-pole RS prototype obtains a stable rotation over 700 rpm about X, Y and Z axis respectively with the angular momentum of 0.08 kg·m^2/s, being superior to the 6-pole counterpart.

Keywords: reaction sphere; spherical motor; structural design; torque density optimization; support vector machines; finite element method

1. Introduction

Besides the orbit control, the attitude control is essential to prevent rollover of the spacecraft and to ensure its antennae consistently directs to a fixed point on the Earth surface [1]. Since spacecrafts are free-floating in space, one way to adjust their attitudes is to transfer the momentum from a rotating actuator to the spacecraft body back and forth. When the actuator changes its rotational speed, the spacecraft will counter-rotate to remain the conservation of angular momentum.

The concept of using a spherical rotor as a momentum exchanger for spacecrafts attitude control starts in the 1960s [2–4]. This is where the name "reaction sphere (RS)" comes from. Compared with other momentum exchange actuators such as the flywheel [5] and the control moment gyroscope [6], the RS can generate 3-axis output torque within one single actuator, thus giving a compact design. In addition, the torques generated in the RS around different axes are naturally decoupled, which simplify the controller design. However, due to the difficulties in designing a working prototype, the RS had not gained much attention for decades. With the rapid development of micro-spacecrafts from the early of this century, the demand for minimization design of the attitude control system has helped the reaction spheres get noticed again [7,8].

A RS can be driven by several single axis motors or a single spherical motor. The former scheme drives the reaction sphere with multiple wheels rotating at desired directions [9–12]. These wheels are independently driven by their corresponding single-axis motors. The later scheme considers the reaction sphere as the rotor of a spherical motor, where the stator of the motor is designed in a spherically symmetric way. No mechanical output shaft exists in these spherical motors. The driving principle can be inductive [13–19], permanent magnet (PM) [7,8,20,21], hysteresis [22,23], variable reluctance [24] and ultrasonic [25,26].

Among all types of reaction spheres, the inductive RSs have the advantages of simple design and high reliability, making them be more competitive in commercial applications. Inductive RSs follow the same driving principle as the traditional single-DOF induction motors. The stator of an induction reaction sphere is excited with alternating currents. Then the generated rotating magnetic fields cut the rotor conductor to produce the induced currents. The interaction between the induced currents on the rotor and the magnetic fields in the air gap produces Lorentz force, which drives the rotor to rotate in the direction of the rotating magnetic fields. Compared with conventional induction motors, inductive RSs feature the capacity of generating multi-DOF rotating magnetic fields.

One class of inductive reaction spheres obtains this capacity via distributed electromagnets [16–18]. A 3D reaction sphere design is proposed in [17], where 3 pairs of electromagnets are put orthogonally in space. A one-DOF inductive RS prototype is developed, with 4 electromagnets for rotation and 1 electromagnet for levitation. With the similar design, a well-developed one-DOF inductive RS prototype is presented in [18]. It reaches the maximum speed of 13,500 rpm and the maximum torque of 0.7 Nm in the performance test experiments. However, this prototype has the disadvantage of large stator. The stator occupies much more space than the rotor. An inductive RS design with 20 distributed electromagnets is proposed in [16]. Subsequently, a one-DOF prototype is developed, but no performance data are reported. The control strategy can be very complicated if the suspension and rotation are driven by the same electromagnets.

The other class of inductive RSs gains this capacity via curved inductors. This type of inductive RSs tends to have a more compact design over the ones with distributed electromagnets, for the coils of curved inductors are extended along the rotor surface but coils of electromagnets are extended radially. The curved inductors in inductive RSs can be arranged to be vortex-like [13,19] and orthogonal [13–15]. The vortex-like arrangement can get the spherical rotor half exposed. This feature makes it be suitable for applications like mobile platforms and robots and the case with no requirement of full angle rotation. The design and control aspects of a 3D spherical induction motor for mobile robots is presented in [19], featuring the vortex-like arranged four curved inductors and the half exposed spherical rotor. The angular velocity and orientation control studies enable a large prototype to rotate along arbitrary axes under the speed of 300 rpm, which validate the feasibility of the vortex-like arrangement. In orthogonal arrangement, the curved inductors go all around the rotor and form the closed structure. It is well fit for the devices working through momentum exchange. Patent [14] proposes a magnetic bearing inductive RS design with three pairs of curved inductors orthogonally arranged. In this design, the AC windings for rotation and the DC windings for suspension are placed on the same teeth of curved cores. No relevant feasibility research is found of this design.

In this paper, a compact design of inductive RS is proposed, featuring 12 curved inductors in orthogonal arrangement for rotating and 6 electromagnets in pairs for magnetic bearing. The overall design enables the RS to generate three-DOF rotation under magnetic levitation. To synthesize symmetric distribution of eddy currents and improve the stability during high-speed rotation, the 4-pole winding is proposed instead of the intuitive 6-pole design. However, the analytical torque model with simplified slotless assumption indicates that reduced pole number degrades the torque generation. To enhance the capability of the torque synthesis in the 4-pole winding design, the slotted iron-core is imposed in design of curve stators. In addition, the electromagnets in the RS are redesigned to be the cross-shape, so that two more slots near the notches of the iron-core curve stator become functional. Such design improves the uniformity of the eddy current and magnetic field induced along the stator circle. As the analytical torque model for the RS with slotted curve stator is not available, a constrained optimization problem is formulated for torque maximization in the RS. From here, the data-based regression algorithms are applied to find the optimal design parameters. Speed test experiments on the developed prototype are conducted to show the superior performance of the optimized 4-pole inductive RS design over the conventional 6-pole counterpart.

2. Structure Design

2.1. Rotor

The attitude adjustment of the spacecraft is realized through the exchange of angular momentum between the rotating sphere and spacecraft body. The rotor of the spherical motor is the carrier of angular momentum. To store more angular momentum, large moment of inertia (MOI) is preferred with a fixed amount of mass. In rotor structural design, MOI of a hollow and uniform sphere is calculated by

$$J = \frac{2}{5}M\frac{R^5 - R_0^5}{R^3 - R_0^3}, \quad 0 \le R_0 \le R \tag{1}$$

where J refers to the MOI, M is the mass of the rotor, R and R_0 are the outer and inner radius of the sphere.

Consider the case given the fixed mass and the fixed outer radius. As plotted in Figure 1, it can be seen that the thinner spherical shell has higher MOI. For example, the MOI of a hollow sphere with $0.1R$ thickness is 1.365 times higher than the one with $0.5R$ thickness. In the extreme case, the inertia of a solid sphere is 60% of the spherical shell with its thickness arbitrarily approaching to zero. The brief mathematical proof sees Appendix A. When R_0 increases, the density is expected to be raised given a certain mass. Therefore, materials with high density get priority in machining the rotor. Furthermore, consider the case given the same mass and the same material, it can be concluded that the outer mass gives more MOI than the inner mass through calculation. For example, the inertial of a hollow sphere with radius of R and thickness of $0.1R$ is 3.6 times higher than the solid sphere with radius of $0.647R$. Therefore, hollow sphere is preferred when designing the reaction sphere.

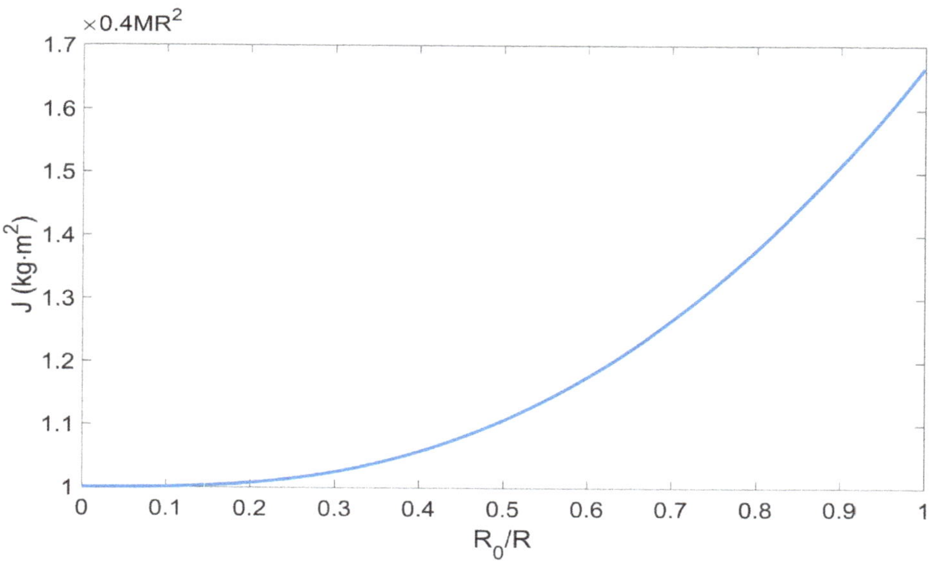

Figure 1. MOI of a hollow sphere with a fixed amount of mass.

Eventually, the rotor is designed as a hollow sphere, consisting of three layers of metals, as shown in Figure 2. The inner layer is made of steel, which is used to form the magnetic circuits with the stator cores through the air gap. The surface of the steel is electroplated with a copper film, forming the middle layer, which is used to raise the conductivity and enlarge the electromagnet torque induced by the eddy currents on the surface of the rotor. Additionally, to prevent the copper from oxidation and leading to conductivity reduction, a thin nickel layer is deposited on the copper by electroplating, forming the outer layer.

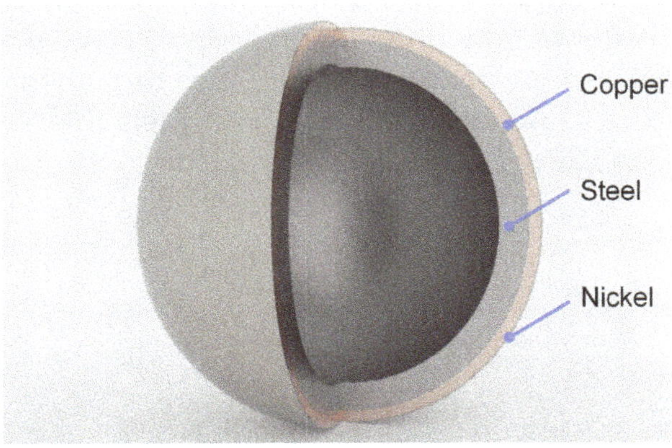

Figure 2. Multilayer spherical shell.

2.2. Stator

The stator of the reaction sphere is fixed with the spacecraft body, which consists of the curved stators and the electromagnets. They are responsible for rotation motion and magnetic suspension respectively. The curved cores and the electromagnet cores are used to conduct the magnetic flux generated by the stator coils. To meet the requirement of generating tri-axial torques, the stator must be designed as a spherically symmetric structure. As shown in Figure 3, the stator of the reaction is formed by three circles that are orthogonal to each other. Each circle is composed of 4 curved cores and enables the rotor to rotate about one certain axis. The output torque about any axis can be realized by combining the torques generated by the three circles. On the other hand, in order to eliminate the mechanical friction and extend its working life, six electromagnets are set in the position of the six intersection of three circles to get the spherical rotor magnetically levitated in the center of the stator.

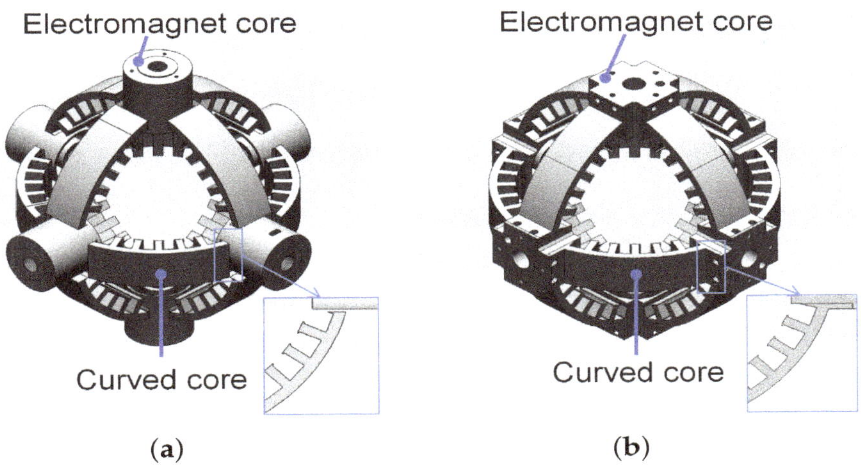

Figure 3. Structural design of the stator cores: (**a**) original design and (**b**) improved design.

In the original design with round-shape electromagnet cores, as shown in Figure 3a, only five complete teeth on the curve iron-core are functional, while the two teeth near the side of the curve stator are not. The notches are formed by connecting three pieces of iron-core curve stator to form one complete stator circle. It causes non-uniformity eddy current and magnetic field induced along the stator circle. To reduce such effect and get the stator magnetic fields utilized in a more sufficient way, the structure of the electromagnet cores are revised to be the cross-shape. In this way, the curved iron-core and electromagnet core can be assembled together and so that two extra teeth on the side are functional, as shown in Figure 3b.

2.3. Winding

For the convenience of speed adjustment and the feature of self-starting, the stator coils are rolled as three-phase windings. In order to generate the equivalent tri-axial torques, the stator should be symmetric in both mechanical and electrical designs. Since the uniaxial plane is divided into four equal parts, as shown in Figure 4, 12 coils are missing in each circle due to the occupation of the electromagnets. All in all, 1/3 of the driving coils are replaced by electromagnet coils.

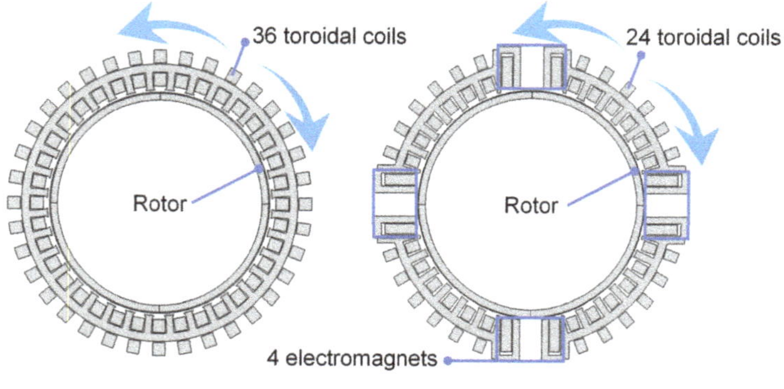

(**a**) (**b**)

Figure 4. Section diagram of the reaction sphere (one-DOF): (**a**) 36 toroidal coils. (**b**) 24 toroidal coils and 4 electromagnet coils.

As can be seen in Figure 5a,c, the three-phase windings come to two types, 6-pole and 4-pole. 6-pole windings can be regarded as removing 12 coils directly from the complete set of three-phase windings, which does not make any adjustment of the rest windings. As for 4-pole windings, the phase of next coil starts right from the end of the last one so that the windings on the four curved cores can achieve consistency.

With finite element method (FEM) modeling in COMSOL Multiphysics 5.3, the fields and currents distribution on the surface of the rotor can be visually analyzed. As shown in Figure 5b,d, the gray lines stand for the magnetic line of force, the red arrows refer to the induced currents and the surface colors refer to the radius magnetic flux density. In this simulation, copper thickness, steel thickness, ampere turns, tooth thickness and tooth width are set as 0.3 mm, 3 mm, 2.4 kA, 3 mm and 25 mm respectively. In this case, the driving torques of the 6-pole arrangement and 4-pole arrangement are 33.8 mNm and 27.0 mNm, where the former is 1.25 times larger than the latter. However, it can be seen that, in 6-pole scheme, two poles are much larger than the rest four and the field distortions in the regions that face the electromagnets are serious. Also, it induces an asymmetric distribution of eddy currents by the adjacent curved stators. These features increase the difficulty of multi-DOF motion control of this 6-pole reaction sphere. Furthermore, from the radius magnetic flux density in Figure 5b,d, it can be seen that, in 6-pole arrangement it is the electromagnet cores that mainly contribute to the driving torque, while in 4-pole arrangement it is the curved cores that mainly contribute to the driving torque. The 4-pole arrangement makes the curved cores better exploited. Additionally, the case without electromagnets cores is simulated, where the torque of 4-pole scheme declines from 27 mNm to 18.4 mNm, and the torque of 6-pole scheme declines from 33.8 mNm to 9.89 mNm. This phenomenon reflects the generated torque is stronger dependent on the electromagnet cores in 6-pole scheme. The flux generated by this 6-pole arrangement mainly passes through the inner core of the electromagnets, which can interfere the magnetic suspension and increase the risk of magnetic saturation.

In summary, while reducing the output torque (20% reduction rate in case of the Figure 5), 4-pole arrangement brings the benefits in three aspects compared with 6-pole arrangement. Firstly, it generates a more uniform distribution of magnetic field and eddy currents on the surface of the rotor. Secondly, it is beneficial for larger angular momentum storage due to its higher synchronous speed with the same power frequency. Thirdly, it reduces the risk of magnetic saturation when the curved inductors and electromagnets are working at the same time. Thus, the 4-pole scheme is adopted for the reaction sphere windings.

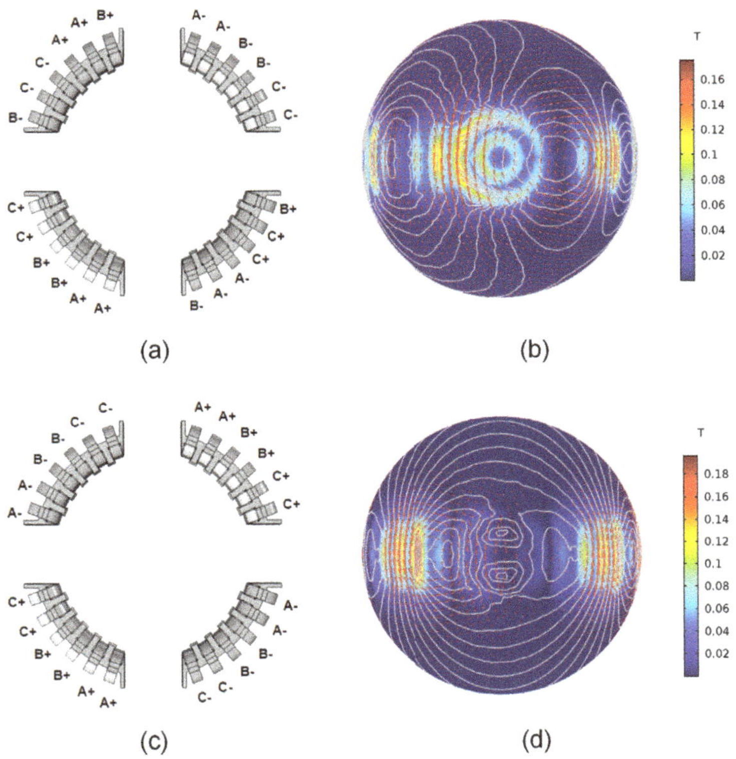

Figure 5. Winding mode and simulation of magnetic field and induced currents on the surface of the rotor: (**a**) 6-pole scheme. (**b**) Simulated distribution of 6-pole scheme. (**c**) 4-pole scheme. (**d**) Simulated distribution of 4-pole scheme.

3. Analytical Model

This section analyzes the magnetic fields on the surface of the rotor, which leads to the solution to electromagnetic driving torque. This also allows us to analyze the generated torque with respect to some of the key design parameters, such as number of poles. Several simplifications are made here. (1) The magnetic permeability of the rotor core is infinite. Thus, no magnetic saturation is encountered. (2) The length of air-gap is evenly distributed. (3) Displacement currents are neglected. Thus, the current law are simplified as Ampere's law.

Generally speaking, the magnetic flux density can be calculated by computing the curl of magnetic vector potential. In spherical coordinates (r, θ, φ), the formula can be expressed as

$$\overrightarrow{B} = \frac{1}{r^2 \sin\theta} \begin{vmatrix} \overrightarrow{e_r} & r\overrightarrow{e_\theta} & r\sin\theta\overrightarrow{e_\varphi} \\ \frac{\partial}{\partial r} & \frac{\partial}{\partial \theta} & \frac{\partial}{\partial \varphi} \\ A_r & rA_\theta & r\sin\theta A_\varphi \end{vmatrix},$$

where $\overrightarrow{e_r}$, $\overrightarrow{e_\theta}$ and $\overrightarrow{e_\varphi}$ refer to the unit vectors, A_r, A_θ and A_φ refer to components of the vector potential in spherical coordinate system.

The general torque $\overrightarrow{\tau}$ of the reaction sphere is the vector summation of 3 torques $\overrightarrow{\tau_x}$, $\overrightarrow{\tau_y}$, and $\overrightarrow{\tau_z}$ generated by 3 identical circles of the stator.

$$\overrightarrow{\tau} = \overrightarrow{\tau_x} + \overrightarrow{\tau_y} + \overrightarrow{\tau_z}.$$

The following analysis will take torque about z-axis $\overrightarrow{\tau_z}$ for example. It is the longitudinal magnetomotive force (mmf) that induces the z-axis rotating magnetic field, in which case $A_\varphi = A_r = 0$. Therefore, the magnetic flux density can be simplified as

$$\overrightarrow{B} = -\frac{\overrightarrow{e_r}}{r\sin\theta}\frac{\partial A_\theta}{\partial\varphi} + \frac{\overrightarrow{e_\varphi}}{r}\frac{\partial rA_\theta}{\partial r} \equiv B_r\overrightarrow{e_r} + B_\varphi\overrightarrow{e_\varphi}. \tag{2}$$

In spherical coordinate system, by Ampere's law $\overrightarrow{J} = \Delta \times \overrightarrow{H}$,

$$\overrightarrow{J} = \frac{1}{r^2\sin\theta}\begin{vmatrix} \overrightarrow{e_r} & r\overrightarrow{e_\theta} & r\sin\theta\overrightarrow{e_\varphi} \\ \frac{\partial}{\partial r} & \frac{\partial}{\partial\theta} & \frac{\partial}{\partial\varphi} \\ H_r & rH_\theta & r\sin\theta H_\varphi \end{vmatrix}.$$

In this case, only those currents flowing in θ-direction induce torque about z-axis,

$$J_\theta = -\frac{1}{r\sin\theta}\left(\frac{\partial r\sin\theta H_\varphi}{\partial r} - \frac{\partial H_r}{\partial\varphi}\right) = -\frac{1}{\mu_0 r\sin\theta}\left(\frac{\partial r\sin\theta B_\varphi}{\partial r} - \frac{\partial B_r}{\partial\varphi}\right). \tag{3}$$

Combine (3) and (2),

$$J_\theta = -\frac{1}{\mu_0 r}\frac{\partial^2 rA_\theta}{\partial r^2} - \frac{1}{\mu_0 r^2\sin^2\theta}\frac{\partial^2 A_\theta}{\partial\varphi^2}. \tag{4}$$

Express the magnetic potential A_θ written in complex form $e^{j\omega t}$ [27], where ω is the angular frequency of input currents, j is the imaginary unit. Then,

$$J_\theta = \sigma E = -\sigma\frac{\partial A}{\partial t} = -j\sigma\omega A_\theta. \tag{5}$$

Combine (4) and (5), the differential equation for magnetic field distribution on the surface of the rotor is obtained as

$$\sigma j\omega A_\theta = \frac{1}{\mu_0 r}\frac{\partial^2 rA_\theta}{\partial r^2} + \frac{1}{\mu_0 r^2\sin^2\theta}\frac{\partial^2 A_\theta}{\partial\varphi^2}. \tag{6}$$

To make (6) solvable, apply the the method of separation of variables as

$$A_\theta(r,\theta,\varphi) = X(r)Y(\theta,\varphi). \tag{7}$$

Substitute (7) into (6),

$$\sigma j\omega = \frac{1}{\mu_0 rX}\frac{\partial^2 rX}{\partial r^2} + \frac{1}{\mu_0 r^2\sin^2\theta Y}\frac{\partial^2 Y}{\partial\varphi^2}. \tag{8}$$

With high-order time and space harmonics neglected, it is assumed that $e^{j\omega t}$ only has the first mmf space harmonic. In this way,

$$\frac{1}{Y\sin^2\theta}\frac{\partial^2\theta Y}{\partial\varphi^2} = -p^2, \tag{9}$$

where p is the number of pair pole. The analytical solution to (9) is given as

$$Y = C_1 e^{jp\varphi\sin\theta} + C_2 e^{-jp\varphi\sin\theta}. \tag{10}$$

From (8) and (9), it yields

$$r^2\frac{\partial^2 X}{\partial r^2} + 2r\frac{\partial X}{\partial r} - X(r^2\mu\sigma j\omega + p^2) = 0. \tag{11}$$

Set $X(r,\theta) = \frac{1}{\sqrt{r}}X_1(r,\theta)$. Then, (11) are transformed to modified Bessel's Equation (12) [28],

$$r_1{}^2\frac{\partial^2 X_1}{\partial r_1{}^2} + r_1\frac{\partial X_1}{\partial r_1} - X_1(r_1{}^2 + p^2 + \frac{1}{4}) = 0, \tag{12}$$

where $r_1 = r\sqrt{\mu\sigma j\omega}$. The analytical solution to (12) is as follows:

$$X(r) = \frac{1}{\sqrt{r}}X_1(r,\theta) = \frac{1}{\sqrt{r}}C_3 I_{\sqrt{p^2+\frac{1}{4}}}(r\sqrt{\mu\sigma j\omega}) + \frac{1}{\sqrt{r}}C_4 K_{\sqrt{p^2+\frac{1}{4}}}(r\sqrt{\mu\sigma j\omega}), \tag{13}$$

where $I_p(r_1)$ and $K_p(r_1)$ are the two linearly independent solutions to the modified Bessel's Equation [28]:

$$\begin{aligned} I_\alpha(x) &= \sum_{m=0}^{\infty}\frac{1}{m!\Gamma(m+\alpha+1)}(\frac{x}{2})^{\alpha+2m}, \\ K_\alpha(x) &= \frac{\pi}{2}\frac{I_{-\alpha}(x) - I_\alpha(x)}{\sin\alpha\pi}. \end{aligned} \tag{14}$$

Subsequently, the magnetic flux density can be calculated through (15), with boundary conditions of $H_\varphi|_{r=R_s} = 0$ and $H_{\varphi\text{copper}}|_{r=R_c} = H_{\varphi\text{air}}|_{r=R_c}$, $B_{r\text{copper}}|_{r=R_c} = B_{r\text{air}}|_{r=R_c}$ for continuity, where R_c is the radius of rotor outer surface and R_s is the radius of rotor inner surface. The nickel layer is neglected because it is much thinner than copper layer and steel layer.

$$\begin{aligned} B_r &= -\frac{X(r)}{r\sin\theta}\frac{\partial Y(\theta,\varphi)}{\partial\varphi}, \\ B_\varphi &= \frac{X(r)Y(\theta,\varphi)}{r} + Y(\theta,\varphi)\frac{\partial X(r)}{\partial r}. \end{aligned} \tag{15}$$

Based on the flux density distribution, the driving torque can be evaluated by Lorentz forces induced by eddy currents on the copper layer of the rotor. The electromagnetic torque arising from the rotor steel is negligible [27].

$$T_{eL} = -\int_0^{2\pi}\int_{R_s}^{R_c}\int_0^{\pi} J_\theta B_r r^3 \sin^2\theta\, d\theta dr d\varphi. \tag{16}$$

In case of proposed design, 16 of the 36 teeth in circle are occupied by 4 electromagnets (see Figure 4a). After structural optimization (see Figure 3b), each side tooth of these curved cores can be approximated as a half tooth concerning its combination with the electromagnet core. This helps form 4 equivalent teeth. So the output torque generated by the four curved inductors is 2/3 of the complete electromagnetic torque. Then, substitute (5), (7) and (10) into (16),

$$T_{eLRS} = \frac{2}{3}T_{eL} \tag{17}$$
$$= \frac{2}{3}\int_0^{2\pi}\int_{R_s}^{R_c}\int_0^{\pi} J_\theta \frac{\partial A_\theta}{\partial \varphi} r^2 \sin\theta \, d\theta dr d\varphi \tag{18}$$
$$= \frac{2}{3}\sigma p\omega \int_{R_s}^{R_c} X^2(r) r^2 dr \int_0^{2\pi}\int_0^{\pi} Y^2(\theta,\varphi)\sin^2\theta d\theta d\varphi \tag{19}$$
$$= \frac{\pi^2}{3}\sigma p\omega \int_{R_s}^{R_c} X^2(r) r^2 dr. \tag{20}$$

From (20), it can be seen that higher electrical conductivity and pole numbers can enhance the output torque. So it is reasonable to adopt high conductivity materials like copper as the outer layer of the rotor. Also, the torque is expected to be reduced when choosing the 4-pole winding instead of 6-pole winding. To raise the output torque, the design parameters like tooth thickness and tooth width are required to be optimized in developing the prototypes.

However, the analytical torque model (20) does not include these parameters. To authors' best knowledge, all the proposed analytical fields and torque models of inductive reaction spheres are developed based on the assumptions that the stator is slotless [27,29–31]. Otherwise, the differential equation will be too complex to find an analytical solution. Due to the nonlinear dependency of design parameters and the 3-D fields distribution, it is too complicated to derive a purely analytical torque model if the stator contains slots. Hence, data-based modeling technique is used to predict the output torque with design parameters considered to be optimized in the following section.

4. Torque Density Optimization

4.1. Data-Based Torque Model

In this optimization, by dimension constraints, the air gap length is fixed as 1 mm, the outside diameter of the rotor is fixed as 99 mm and the outside diameter of the stator is fixed as 123 mm. The goal is to raise the output torque under the constraints of mass and MOI by searching the optimal tooth thickness (S), tooth width (B), ampere turns (NI), copper thickness (T_{co}) and steel thickness (T_{st}) as shown in Figure 6. Due to the time consuming of FEM calculation, the comprehensive design is not practical. Thus, Taguchi design of five factors five levels is adopted to select the representative samples data necessary for building learning models as shown in Table 1 [32]. Consider the overall size and the process feature, the ranges of T_{co}, T_{st}, NI, S and B are set as 0.1–0.5 mm, 1–5 mm, 1.5–2.7 kA, 2–4 mm and 20–30 mm respectively.

Eight groups of the Taguchi design form the 200 trials with no repeated experiments. Each group consists of 25 samples, which selects the representative samples from the whole parameter space by following the two rules. (1) The value in column has the same number of occurrences. (2) Any pair of values between two arbitrary columns have the same number of occurrences. Subsequently, of all the eight groups, seven groups are used to train the learning model and the rest one is used to test the prediction performance of the model. The output torque is obtained by FEM calculation. To reveal the uneven distribution of the air gap length and raise the accuracy of the calculation, 3D FEM is adopted instead of 2D one. The whole mass is estimated by the CAD model volume and its density. The materials used in FEM simulation are listed in Table 2. The rated current is set to 1.4 A. The power source is 50 Hz.

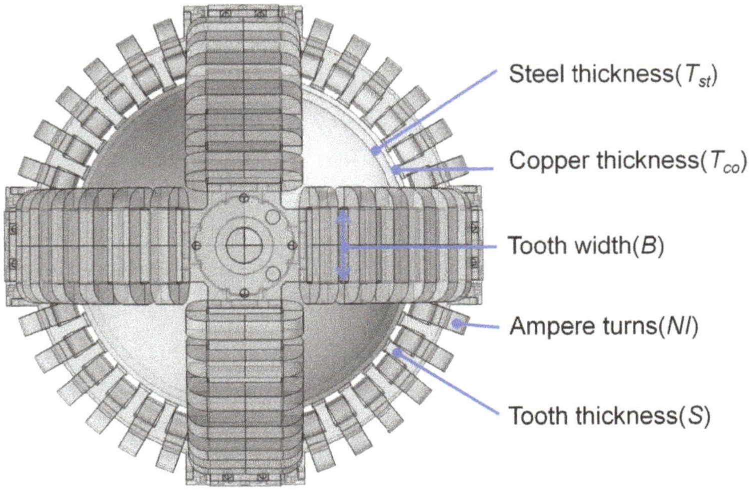

Figure 6. Design parameters to be optimized.

Table 1. The parameters of sample space.

Element	Level 1	Level 2	Level 3	Level 4	Level 5
T_{co} (mm)	0.1	0.2	0.3	0.4	0.5
T_{st} (mm)	1	2	3	4	5
NI (kA)	1.5	1.8	2.1	2.4	2.7
S (mm)	2	2.5	3	3.5	4
B (mm)	20	22.5	25	27.5	30

Table 2. Material components of FEM and lab prototype.

Component	Material
Rotor outer layer	Copper
Rotor inner layer	Low Carbon Steel 1010
Curved core	Silicon Steel NGO M-22
Electromagnet core	Iron
Winding	Copper

The various parameters of the inductive RS have high nonlinear relationships due to the 3-D field distribution and electromagnetic conversion. Fitting neural network (FNN) has the strength of realizing the complex non-linear mapping between the input and the output. So it is suitable for solving the problem of inductive RS torque prediction with complicated internal mechanism. Due to the large calculation and time consumption of 3D finite element simulation, obtaining a large number of training samples are not practical. Support vector machines (SVM) fit well for the small or medium size of training samples. It determines the final results with only a few support vectors. The major redundant samples are eliminated after training. Random forest (RF) can handle data with many features and have high training speed with independent decision trees. This fits the efficient multi-variable optimization of the inductive RS. Also, it can reduce the risk of over-fitting by increasing tree numbers. It is hard to give strict conclusion of application domains for the three methods. In practice, we choose the appropriate method according to the specific situation.

To enhance the prediction performance, the tree numbers of RF, the hidden layer size of the FNN and parameters of the SVM are optimized. As shown in Figure 7, increasing the number of trees generally improves the accuracy of RF and prediction tends to be stable. It reaches the highest average accuracy of 0.0022 Nm with 44 trees. As for fitting neural networks, it takes a row vector of N hidden layer sizes, and a BP training function, and returns a feed-forward neural network with N + 1 layers. Theoretically, multi-layer fitting neural networks can fit any finite input-output relationship well given enough hidden neurons. However, the risk of over fitting increases with the increase of hidden neurons, where it may fit well for the training data but badly for the test data. Thus, the hidden layers should be chosen properly. As shown in the second figure of Figure 7, it reaches the highest average accuracy of 0.0011 Nm with 10 hidden layers. As for SVM, the radial basis function is set as the kernel function considering the nonlinearity of the torque model. The SVM parameter cost (c), radial basis function parameter (g) and epsilon in loss function (p) are chosen to be optimized. This optimization is carried out by genetic algorithms. To avoid over fitting, c, g and p are constrained within 100, 100 and 1 respectively. It shows that the SVM prediction model reaches the highest average accuracy of 0.00093 Nm when c p and g are 70.716, 0.019 and 0.066 respectively.

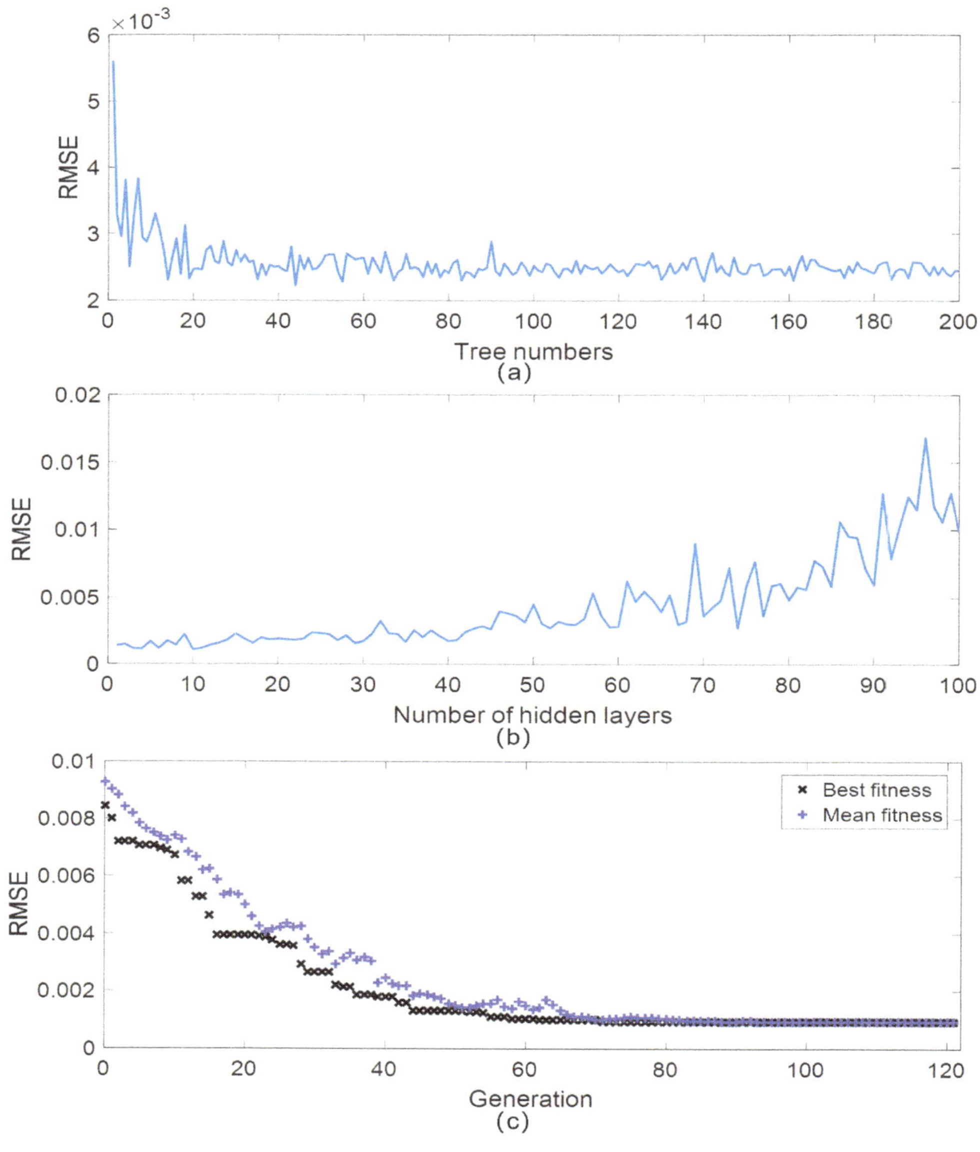

Figure 7. Data-based models parameters optimization: (**a**) RF regression with increasing trees, best root-mean-square error (RMSE) of 0.0022 Nm with 44 trees. (**b**) NN regression with increasing hidden layers, best RMSE of 0.0011 Nm with 10 hidden layers. (**c**) SVM regression with genetic algorithm optimizing parameters, best RMSE of 0.0009 Nm with c (70.716), p (0.019) and g (0.066).

All in all, the regression results of the test set show that the SVM model achieves the least RMSE over the other two in the case of small-sample database. The optimal predictions by RF, FNN and SVM are shown in Figure 8a, with these test samples arranged by Taguchi design of five factors five levels. Figure 8b shows the prediction errors of RF, FNN and SVM for each sample. It can be seen that the

SVM torque model has a stable prediction, with all errors restricted in 2 mNm. Thus, the SVM model is chosen to be the objective function of the optimization in next subsection.

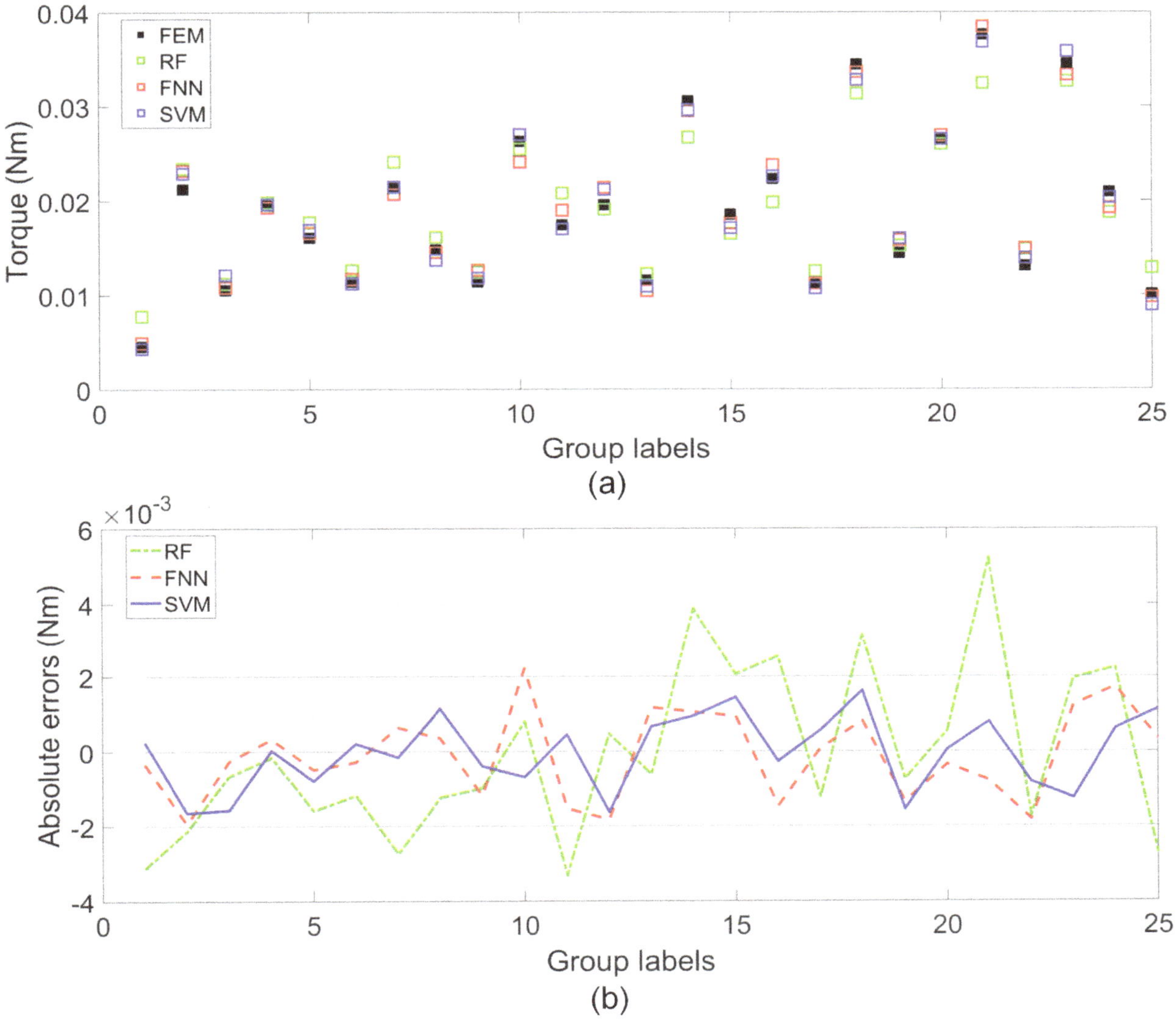

Figure 8. RF, FNN and SVM models prediction performance with test data: (**a**) torque predictions and (**b**) absolute errors.

4.2. Design Parameters Optimization

The reaction sphere design parameters optimization is a multi-parameter optimization problem with constraints. The objective function is the SVM torque prediction model established above. The constraints include the box constraints, linear inequality constraints and nonlinear inequality constraints. The box constraints (22)–(26) are set to be consistent with the range of the sample spaces shown in Table 1. The linear inequality constraints are considered between coil turns, tooth thickness and tooth width: The coil turns and tooth thickness are limited by the tooth pitch (27); The coil turns and tooth width are limited by the fixed stator arc length (28). The parameters of the linear inequality constraints are validated by the lab possessed parts shown in Figure 9. The nonlinear inequality constraints consider the application in micro spacecrafts: the MOI of the rotor is larger than 0.001 kg·m^2 (29) and the whole mass of the actuator is no more than 3 kg (30).

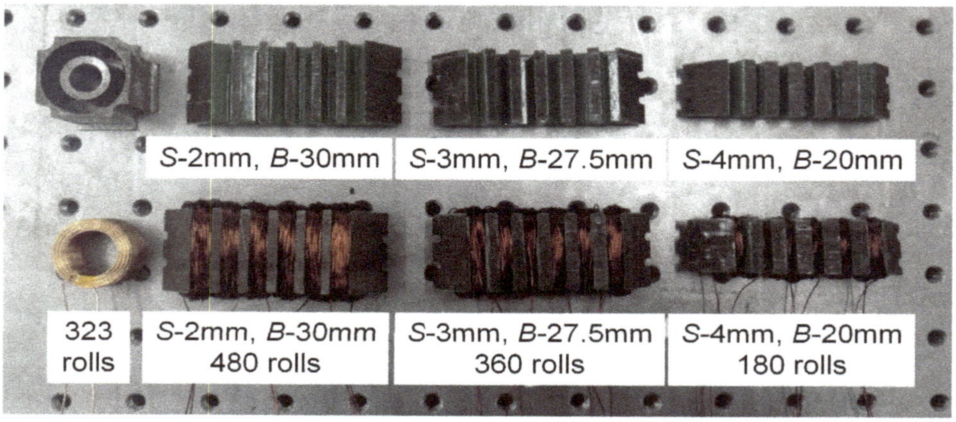

Figure 9. Parts of the reaction sphere prototype for constraints and masses validation.

The mass function $M(T_{co}, T_{st}, NI, S, B)$ evaluates mass of the whole actuator, consisting of rotor, stator core, toroidal winding and electromagnets. The mass of the rotor and stator core is estimated by their measured volume of the digital model and their corresponding densities. The mass of the toroidal winding is calculated by the product of turn number and mass of each turn. The mass of the electromagnets is a constant, which is weighed with prototypes. Several sample components are selected to validate its accuracy as shown in Table 3. Due to the insulating coating, the mass test results of the curved cores are slightly larger than the estimated ones.

Table 3. Mass estimation and experimental validation.

Samples	Estimated Mass (kg)	Tested Mass (kg)
Rotor(T_{co}0.3 mm, T_{st}3 mm)	0.7396	0.7400
Curved core(S2 mm,B30 mm)	0.0795	0.0805
Curved core(S3 mm,B27.5 mm)	0.0813	0.0820
Curved core(S4 mm,B20 mm)	0.0652	0.0660
Curved core(S2 mm,B30 mm), 480 rolls	0.1482	0.1370
Curved core(S3 mm,B27.5 mm), 360 rolls	0.1296	0.1225
Curved core(S4 mm,B20 mm), 180 rolls	0.0759	0.0750
Electromagnets core	\	0.0625
Electromagnets coil	\	0.027

Based on the analysis above, the mathematical description of the optimization problem can be expressed as:

$$\max\ T = \text{svmpredict}(T_{co}, T_{st}, NI, S, B) \tag{21}$$
$$s.t.\quad 0.1\ \text{mm} \leq T_{co} \leq 0.5\ \text{mm}, \tag{22}$$
$$1\ \text{mm} \leq T_{st} \leq 5\ \text{mm}, \tag{23}$$
$$1.5\ \text{kA} \leq NI \leq 2.7\ \text{kA}, \tag{24}$$
$$2\ \text{mm} \leq S \leq 4\ \text{mm}, \tag{25}$$
$$20\ \text{mm} \leq B \leq 30\ \text{mm}, \tag{26}$$
$$NI/33.6 \leq 140 - 20S, \tag{27}$$
$$NI/33.6 \leq 340 - 10S - 8B, \tag{28}$$
$$I_{co} + I_{st} \geq 0.001\ \text{kg}\cdot\text{m}^2, \tag{29}$$
$$M(T_{co}, T_{st}, NI, S, B) \leq 3\text{kg}. \tag{30}$$

This constrained optimization is carried out by interior point methods on MATLAB2016a optimization toolbox, with parameter setting as follows: function tolerance and constraints tolerance:

10^{-6}, max iterations: 1000. Inner max iterations: 100, relative tolerance: 10^{-2}. The optimization results are listed in Table 4. The iteration process is shown in Figure 10. The final T_{co}, T_{st}, NI, S, B for developed 4-pole prototype are rounded as 0.5 mm, 2.4 mm, 2.688 kA, 3 mm and 27.5 mm respectively. The driving torque is 0.0371 Nm (FEM) with the whole mass of 3 kg and MOI of 0.001 kg·m^2.

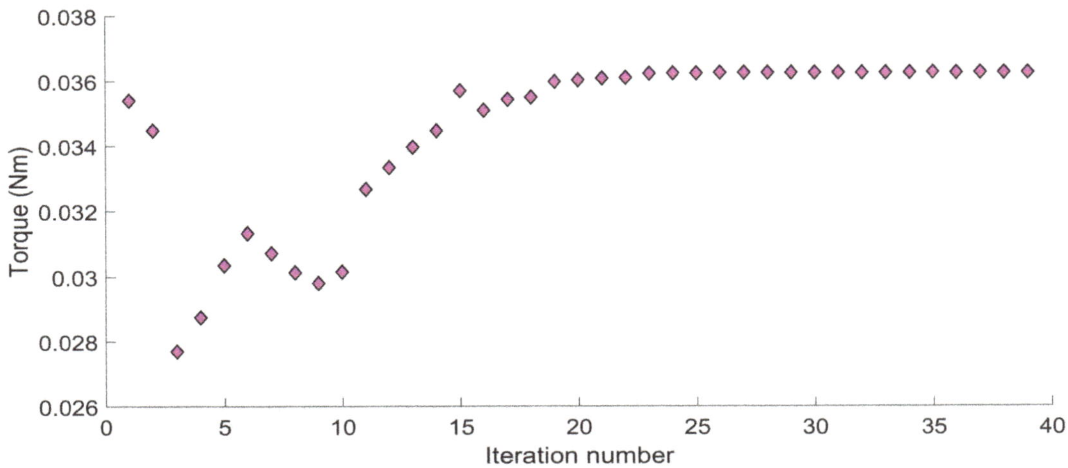

Figure 10. Design parameters optimization by interior point method for better output torque.

Table 4. Constrained optimization results by interior point methods.

T_{co}	T_{st}	NI	S	B	Torque	Density	Moment of Inertia
0.5 mm	2.395 mm	2.7 kA	2.982 mm	27.671 mm	0.0363 Nm	0.0121 Nm/kg	0.001 kg· m^2

5. Feasibility Test with Lab Prototypes

To validate the feasibility of the proposed design, the experimental setup is developed to test the speed performance of the lab prototypes, which is composed of converters, DC power supply, reaction sphere actuators, controllers and amplifiers as shown in Figure 11. The three converters are responsible for driving the reaction sphere actuators in the three orthogonal axis respectively. The DC power supply provides direct currents for amplifiers and controllers for magnetic levitation control.

In this magnetic levitation system, to get the feedback position signal in the closed-loop control, an eddy current displacement sensor made of tiny coils is placed at the center of each electromagnet. These sensors provide the signal of air gap distance to the controller. From here, a negative feedback control loop is set up by two serial-link PD controller to regulate the levitation distance. Specifically, the controller uses STM32F051K6T6 (Suzhou Joint-Tech Mechanical and Electrical Technology Co., LTD, Suzhou, China), which collects the signal of eddy current displacement sensors and generates PWM signal to drive the power amplifier. Each driver amplifies the control signal and drives the two electromagnetic coils on the same axis. What should be mentioned is that the electromagnets about Z-axis need to overcome the gravity of the rotor. They require higher currents compared with the counterparts about the other two axes. A 4-pole reaction sphere prototype and a 6-pole reaction sphere prototype are developed to test the speed performance with the same controllers and drivers. A binary painting strategy for speed measurement is presented as shown in Figure 12. The four silvery white circles on the surface of the rotor are located on the four vertexes of a tetrahedron. In case of measuring the X, Y and Z rotational speed, the laser focus will come across the high reflective areas, which is the nickel protection layer shown in Figure 2, 3 times when the rotor turn around once. Therefore, the output speed can be estimated as the one-third of counts per minute. The frequency of AC is set to 60 Hz in X, Y and Z axis respectively of the 4-pole reaction sphere prototype and in X and Y axis respectively of the 6-pole reaction sphere prototype. The Z axis of the 6-pole reaction sphere is set to 54 Hz, for higher frequency cannot generate a stable rotation with its significantly increasing vibration.

This results from the parts manufacturing and assembling error, as well as the dynamical stability of the sphere about Z-axis rotation under gravity force for 6-pole RS, probably due to its unequal pole. The speed test results are listed in Table 5. It is shown that the 4-pole reaction sphere prototype has the maximum of 775 rpm, 791 rpm and 795 rpm in X, Y and Z axis respectively, which obtains a better speed performance over the 6-pole one as predicted. As for speed uncertainties, the 4-pole prototype shows a relatively smaller fluctuations, which may benefit from the uniform fields distributions.

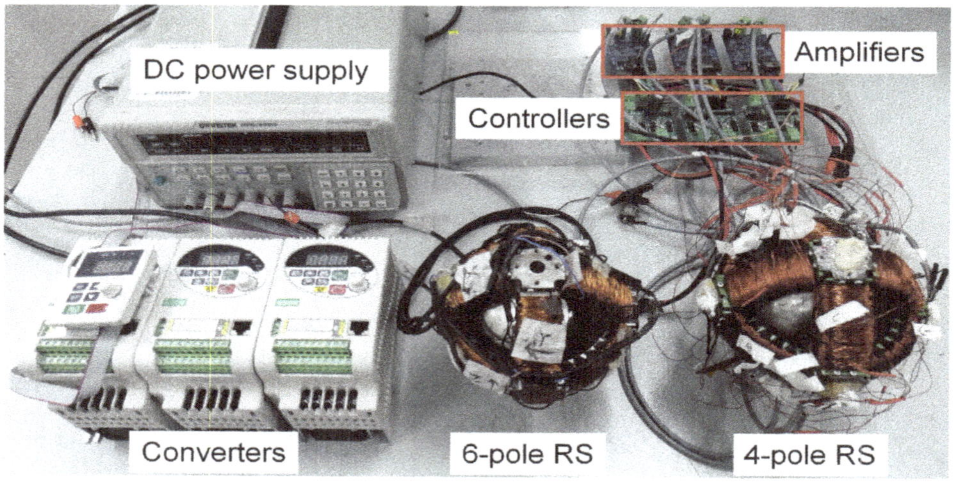

Figure 11. Experimental setup for driving the 4-pole reaction sphere and 6-pole reaction sphere under magnetic levitation.

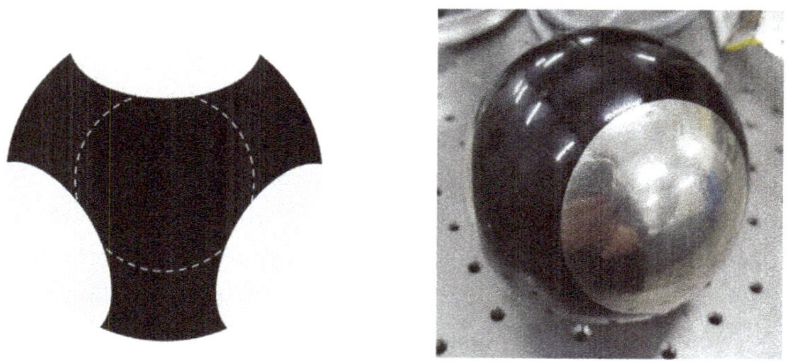

Figure 12. Spherical rotor with black painting for speed counts: (**a**) diagram and (**b**) prototype.

Table 5. Maximum speed of the four-pole reaction sphere prototype and the six-pole reaction sphere prototype.

Axis	Four-Pole Reaction Sphere			Six-Pole Reaction Sphere		
	X (60 Hz)	Y (60 Hz)	Z (60 Hz)	X (60 Hz)	Y (60 Hz)	Z (54 Hz)
Tachometer counts1	2278	2290	2349	1667	1447	963
Tachometer counts2	2351	2413	2365	1712	1322	851
Tachometer counts3	2344	2412	2439	1597	1362	810
Average counts	2324	2372	2384	1659	1377	875
Average speed (rpm)	775	791	795	553	459	292
Speed uncertainty (rpm)	24.3	41.0	30.0	38.3	42.7	51.0
Angular momentum (kg·m^2/s)	0.081	0.083	0.083	0.058	0.048	0.031

The speed control experiment is not so successful. The tested output speed is much less than the synchronization speed (1800 rpm @ 60 Hz). When the frequency is over 60 Hz, the increasing vibration leads to the mechanical collision between the rotor and the stator. One reason lies in the quality of the prototypes. The magnetic levitation demands for the high precision of parts processing and assembling, which is not well guaranteed in the developed prototypes. This results in the non-unique performance in different axes, which is particularly evident between Z axis and the other two axes in the 6-pole prototype.

A more essential reason lies in a drawback of the proposed design: When producing the driving torque, the rotation windings on the curved inductor will generate an extra thrust in the meantime. This extra thrust will push the electromagnets to generate the flux and attract the rotor to be back in the center of the stator. Then, additional damping torques will be induced by the interaction between the rotating rotor and the electromagnets flux, which leads to speed reduction.

6. Conclusions

This paper presents the new design, modeling and parameter optimization methods of a magnetically levitated inductive reaction sphere. It provides a simple, low-cost and miniaturized solution to spacecraft attitude control via momentum exchange. The compact design of electromagnets and curved inductors has the dimensional ratio of 1:1.4 between the rotor and the stator assembly. The simplified, analytical driving torque model is derived based on Ampere's law and separation of variables, showing that the conductivity of the rotor outer layer and pole number of the stator fields contribute to the driving torque. To raise the agility of attitude adjustment of satellites, it is crucial to improve the torque density of this 4-pole inductive RS besides introducing slots in the iron-core curve stator. Here, its curved inductors and electromagnets are redesigned as the cross-shape, allowing two more slots near the notches of the curve stator to be functional. This also improves the uniformity of the induced eddy current and magnetic field along the stator circle. As the full analytical torque model is not available for inductive RS with slotted curve stators, a constrained optimization problem is formulated to maximize the output torque. In addition, the design parameters are efficiently solved by SVM with only small volume of training data. The lab prototype is developed for feasibility test. It shows that the optimized RS can rotate over 700 rpm in X, Y and Z axis with the angular momentum of 0.08 $kg \cdot m^2/s$.

Small volume and small mass are two essential issues in reaction spheres design. The proposed compact design enables the developed prototype to have the rotor mass of 0.66 kg and the whole assembly of 3 kg under the dimension constraints of 136 mm $\times$ 136 mm $\times$ 136 mm. The necessary angular momentum of the actuator is expected to be within 0.1–1 $kg \cdot m^2/s$ for actual micro-spacecraft (10–100 kg) attitude control [33,34], and the practical design target is set as 0.4 $kg \cdot m^2/s$. If we maintain the design parameters in this paper, the rotor has to spin about 3800 rpm to achieve this, demanding a high level of dynamic balance. Due to the current restraint of motor drivers and the error during part fabrication and assembly, it is not practical to realize such speed with our prototypes. A more practical way is to increase both the output speed and the mass of the steel layer of the rotor appropriately. For example, we can achieve the angular momentum of 0.4 $kg \cdot m^2/s$ by increasing the speed to 1500 rpm and the steel thickness of the hollow sphere to 9.5 mm respectively. Remarkably, the design and optimization methods proposed in this paper are still applicable in the practical reaction sphere development.

Author Contributions: L.Y. derived the electromagnetic modeling, optimization method and prepared the manuscript; J.Z. and J.C. proposed the original inductive RS design; S.-L.C. advised the theoretical study, designed the experiments and refined the manuscript; Y.L. and L.Y. assembled the prototypes and carried out experimental test; C.Z. and G.Y. supervised the research throughout.

Appendix A. Proof: (A1) is a Monotonically Increasing Function

The MOI of a hollow and uniform sphere is

$$J = J(R_0) = \frac{2}{5}M\frac{R^5 - R_0^5}{R^3 - R_0^3}, \quad 0 \le R_0 \le R, \tag{A1}$$

where R_0 is a variable, R and M are constants, $R_0, R, M \in \mathbb{R}^+$. Take logarithms on both sides of (A1),

$$\ln(J) = \ln\left(\frac{2}{5}M\right) + \ln(R^5 - R_0^5) + \ln(R^3 - R_0^3). \tag{A2}$$

Take the derivative of (A2) with respect to R_0, it yields

$$\frac{J'}{J} = \frac{R_0^2(-5R_0^2R^3 + 2R_0^5 + 3R^5)}{(R^5 - R_0^5)(R^3 - R_0^3)}. \tag{A3}$$

Set $H(R_0) = -5R_0^2R^3 + 2R_0^5 + 3R^5$, so $H(R_0)|_{R_0=R} = 0$. In addition, $\frac{dH}{dR_0} = 10R_0(R_0^3 - R^3) < 0$. In this way, for $R_0 < R$, $H(R_0) > 0$. This yields $J' = JH(R_0)/[(R^5 - R_0^5)(R^3 - R_0^3)] > 0$.

This concludes $J(R_0)$ is monotonically increasing when $R_0 \in [0, R)$. Additionally, $J_{\min} = J(0) = \frac{2}{5}MR_0^2$, $J < J(R) = \lim\limits_{R_0 \to R} \frac{2}{5}M\frac{R^5 - R_0^5}{R^3 - R_0^3} = \frac{2}{3}MR_0^2$, $J(0) = 60\%J(R)$.

References

1. Maini, A.K.; Agrawal, V. *Satellite Technology: Principles and Applications*; John Wiley & Sons: New York, NY, USA, 2011.
2. Ormsby, R.D. Capabilities and limitations of reaction spheres for attitude control. *ARS J.* **1961**, *31*, 808–812. [CrossRef]
3. Haeussermann, W. *The Spherical Control Motor for Three Axis Attitude Control of Space Vehicles*; NASA TM X-50071; NASA: Greenbelt, MD, USA, 1959.
4. Haeussermann, W. Space Vehicle Attitude Control Mechanism. U.S. Patent 3,017,777, 23 June 1962.
5. Bitterly, J.G. Flywheel technology: Past, present, and 21st century projections. *IEEE Aerosp. Electron. Syst. Mag.* **1998**, *13*, 13–16. [CrossRef]
6. Kurokawa, H. *A Geometric Study of Single Gimbal Control Moment Gyros*; Report of Mechanical Engineering Laboratory, no. 175; Agency of Industrial Technology and Science: Tokyo, Japan, 1998.
7. Stagmer, E. Reaction Sphere for Stabilization and Control in Three Axes. U.S. Patent 9,475,592, 25 October 2016.
8. Rossini, L. Electromagnetic mOdeling and Control Aspects of a Reaction Sphere for Satellite Attitude Control. Ph.D. Thesis, EPFL, Lausanne, Switzerland, 2014.
9. Craveiro, A.A.; Sequeira, J.S. Reaction sphere actuator. *IFAC-PapersOnLine* **2016**, *49*, 212–217. [CrossRef]
10. Zhang, M.; Zhu, Y.; Chen, A.L.; Yang, K.M.; Cheng, R. A Maglev Reaction Sphere Driven by Magnetic Wheels. CN Patent 105207430 A, 30 December 2015.
11. Keshtkar, S.; Moreno, J.A.; Kojima, H.; Hernández, E. Design concept and development of a new spherical attitude stabilizer for small satellites. *IEEE Access* **2018**, *6*, 57353–57365. [CrossRef]
12. Takehana, R.; Paku, H.; Uchiyama, K. Attitude control of satellite with a spherical rotor using two-degree-of-freedom controller. In Proceedings of the 2016 7th IEEE International Conference on Mechanical and Aerospace Engineering (ICMAE), London, UK, 18–20 July 2016; pp. 352–357.
13. Hollis, R.L., Jr.; Kumagai, M. Spherical Induction Motor. U.S. Patent 9,853,528, 26 December 2017.
14. Fan, D.; Chun, C.S.; He, Y.; Song, J.; Zhang, N. Magnetization Suspension Inductive Reaction Sphere. CN Patent 105,775,169, 20 July 2016.
15. Isely, W.H. Magnetically Supported and Torqued Momentum Reaction Sphere. U.S. Patent 4,611,863, 16 September 1986.

16. Wampler-Doty, M.P.; Doty, J. A reaction sphere for high performance attitude control. In Proceedings of the 8th Annual CubeSat Developers' Workshop, San Luis Obispo, CA, USA, 20–22 April 2011.
17. Iwakura, A.; Tsuda, S.-I.; Tsuda, Y. Feasibility study on three dimensional reaction wheel. *Proc. Sch. Sci. Tokai Univ. Ser.* **2008**, *33*, 51–57.
18. Kim, D.-K.; Yoon, H.; Kang, W.-Y.; Kim, Y.-B.; Choi, H.-T. Development of a spherical reaction wheel actuator using electromagnetic induction. *Aerosp. Sci. Technol.* **2014**, *39*, 86–94. [CrossRef]
19. Kumagai, M.; Hollis, R.L. Development and control of a three dof spherical induction motor. In Proceedings of the 2013 IEEE International Conference on Robotics and Automation, Karlsruhe, Germany, 6–10 May 2013; pp. 1528–1533.
20. Chételat, O. Torquer Apparatus. U.S. Patent 8,164,294, 24 April 2012.
21. Chabot, J.; Schaub, H. Spherical magnetic dipole actuator for spacecraft attitude control. *J. Guid. Control. Dyn.* **2016**, *39*, 911–915. [CrossRef]
22. Zhou, L.; Nejad, M.I.; Trumper, D.L. *Hysteresis Motor Driven One Axis Magnetically Suspended Reaction Sphere*; American Society for Precision Engineering: Raleigh, NC, USA, 2014.
23. Zhou, L.; Nejad, M.I.; Trumper, D.L. One-axis hysteresis motor driven magnetically suspended reaction sphere. *Mechatronics* **2017**, *42*, 69–80. [CrossRef]
24. Liu, J. Structural Design and Three-Dimensional Magnetic Field Analysis of Magnetic Levitation Switched Reluctance Spherical Motor. Master's Thesis, Yangzhou University, Yangzhou, China, 2011.
25. Paku, H.; Uchiyama, K. Satellite attitude control system using a spherical reaction wheel. In *Applied Mechanics and Materials*; Trans Tech Publ: Zurich, Switzerland, 2015; Volume 798, pp. 256–260.
26. Mashimo, T.; Awaga, K.; Toyama, S. Development of a spherical ultrasonic motor with an attitude sensing system using optical fibers. In Proceedings of the 2007 IEEE International Conference on Robotics and Automation, Roma, Italy, 10–14 April 2007; pp. 4466–4471.
27. Zhu, L.; Guo, J.; Gill, E. Analytical field and torque analysis of a reaction sphere. *IEEE Trans. Magn.* **2018**, *54*, 1–11. [CrossRef]
28. Stegun, A. Handbook of Mathematical Functions. US Government Printing Office: October, 1973. Available online: http://people.math.sfu.ca/~cbm/aands/page_374.htm (accessed on 14 January 2019).
29. Spałek, D. Spherical Induction Motor with Anisotropic Rotor-Analytical Solutions for Electromagnetic Field Distribution, Electromagnetic Torques and Power Losses. 2007. Available online: http://www.compumag.org/jsite/images/stories/TEAM/problem34.pdf (accessed on 15 November 2018).
30. Davey, K.; Vachtsevanos, G.; Powers, R. The analysis of fields and torques in spherical induction motors. *IEEE Trans. Magn.* **1987**, *23*, 273–282. [CrossRef]
31. Davey, K.; Vachtsevanos, G.; Powers, R. Analysis of a spherical induction motor. *Electr. Mach. Power Syst.* **1987**, *12*, 206–223. [CrossRef]
32. Mark A. Tschopp, DOE with MATLAB. Available online: https://icme.hpc.msstate.edu/mediawiki/index.php/DOE_with_MATLAB_3 (accessed on 8 October 2018).
33. Scharfe, M.; Roschke, T.; Bindl, E.; Blonski, D. Design and development of a compact magnetic bearing momentum wheel for micro and small satellites. In Proceedings of the 15th Annual/USU Conference on Small Satellites, Logan, UT, USA, 13–16 August 2001.
34. Grillmayer, G.; Falke, A.; Roeser, H.-P. Technology demonstration with the micro-satellite flying la'pt op. In *Small Satellites for Earth Observation: Selected, Proceedings of the 5th International Symposium of the International Academy of Astronautics, Berlin, Germany, 4–8 April 2005*; Walter de Gruyter: Berlin, Germany 2011; p. 419.

13

Comparative Analysis of High Frequency Signal Injection Based Torque Estimation Methods for SPMSM, IPMSM and SynRM

Maria Martinez *, David Reigosa, Daniel Fernandez and Fernando Briz

Electrical, Electronic, Computers and Systems Engineering, University of Oviedo, 33204 Oviedo, Spain; diazdavid@uniovi.es (D.R.); fernandezalodaniel@uniovi.es (D.F.); fbriz@uniovi.es (F.B.)
* Correspondence: martinezgmaria@uniovi.es.

Abstract: Torque estimation in permanent magnet synchronous machines and synchronous reluctance machines is required in many applications. Torque produced by a permanent magnet synchronous machine depends on the permanent magnets' flux and *dq*-axes inductances, whereas torque in synchronous reluctance machines depends on the *dq*-axes inductances. Consequently, precise knowledge of these parameters is required for proper torque estimation. The use of high frequency signal both for permanent magnets' flux and *dq*-axes inductances estimation has been recently shown to be a viable option. This paper reviews the physical principles, implementation and performance of high-frequency signal injection based torque estimation for permanent magnet synchronous machines and synchronous reluctance machines.

Keywords: torque estimation; online parameters estimation; permanent magnet synchronous machines; synchronous reluctance machines; high frequency signal injection

---

1. Introduction

The use of Permanent Magnet Synchronous Machines (PMSMs) has substantially increased in industrial applications, electric traction, renewables energies, etc. due to their superior performance compared with other types of electrical machines, such as Induction Machines (IM), in terms of power density, torque density, dynamic response, wide speed range, simplicity of the control and efficiency. However, the high and often unpredictable price of rare-earth materials and the risk of demagnetisation due to excessive operating currents/temperatures are a concern for this type of machines. Synchronous Reluctance Machines (SynRM) have gained popularity over the last years as a viable alternative thanks to their lower cost and high tolerance to overcurrents and/or overheating and overall increased robustness, resulting from the absence of magnets in the rotor [1–7].

Torque production capability of PMSMs depends on the magnetisation state of the permanent magnets (PMs) and the saliency ratio, i.e., the difference between *d*- and *q*-axes inductances. Magnet strength and inductances can change during normal operation of the machine due to changes in the fundamental current and/or PM temperature [8–10]. An increase of the PM temperature reduces the PM remanent flux (i.e., magnetisation state), and consequently the machine torque for a given stator current. In addition, PM remanent flux variation changes the *d*-axis saturation level (*d*-axis assumed to be aligned with PM flux), and therefore the *d*-axis inductance [11]. Also injection of fundamental current will change the saturation level, resulting therefore into inductance variations [11,12]. Torque in SynRMs is function of *d* and *q*-axes inductances exclusively, which can vary significantly with the current level due to saturation of the core material.

Precise knowledge of the torque produced by the machine is required in many applications. Torque measurement is problematic. Torque transducers based on strain gauges are likely the preferred option [13–17]. However, this type of sensor can introduce resonances into the system, are highly sensitive to electromagnetic interference, requires precise mounting and calibration to ensure accuracy and their cost could account for a significant portion of the drive cost [18]. Less popular options are torque measurement systems based on torsional displacement [19] or magnetoelastic effect [20]. Torque measurement systems based on torsional displacement are less sensitive to electromagnetic noise but use optical probes, which are expensive and require accurate calibration [19]. Torque measurement based on magneto-elastic sensors is simpler and uses non-contact technology without the need of calibration, but it requires special shaft materials and can incur in torque measurement errors due to shaft thermal expansion. Regardless of the method being used, precise torque measurement is expensive, and requires extra elements (sensor, cables, connectors, ...), what can reduce the reliability of the drive. These concerns have boosted the interest in torque estimation methods.

Torque estimation methods can be roughly separated into (a) methods based on the torque equation [21,22] and (b) indirect estimation methods [23–31]. The first type can use the General Torque Equation (GTE) assuming constant motor parameters [21], Flux estimation with Compensation Scheme (FCS) where *dq*-flux linkage is estimated by the measurement of stator voltage, currents and rotor position [21] or look-up-tables, which are used to adjust the machine parameters according to machine operating conditions [22]. A large number on indirect torque estimation methods have been proposed, which go from a simple power balance with known electric power and rotor speed [23], observer based methods [24–29] (e.g., sliding mode observers [24], model reference adaptive systems [25,26], model reference observers, reduced order observers [27], recursive least square parameters estimation [28] or affine projection algorithms parameters estimation [29]), methods requiring additional sensors, e.g., giant magnetoresistance effect (GMR) based methods [30], or neural networks based methods [31]. All these methods [21–31] require previous knowledge of certain machine parameters (e.g., resistances or inductances) and/or its operating condition (e.g., temperature).

Injection of a high-frequency (HF) signal has been recently proposed for on-line estimation of the machine parameters used by the torque equation. Consequently, they belong to category (a) discussed above [12,32,33]. The HF signal is injected via inverter on top of the fundamental voltage responsible of torque production, meaning that it does not interfere with the normal operation of the machine. Furthermore, no additional sensors or cabling are needed. The HF signal can be either a current or a voltage signal and have different shapes: pulsating, square-wave, sinusoidal, etc. Although the physical principles are the same in all the cases, this choice can result in significant differences in the implementation. Injection of a voltage signal is easier in principle as the inverter is a voltage source. On the contrary, current injection will require the use of current controllers, but will be shown to improve the reliability of the method. Note that instead of periodic HF signal, PWM pulses could also be used, either transient currents or neutral voltage following each inverter pulse being measured and processed in this case. However, its implementation implies changes in the hardware (e.g., additional sensors, access to the neutral of the machine, etc.), increasing the implementation cost and difficulty.

This paper reviews the use of HF signal injection based parameter estimation methods aimed to improve the accuracy of torque estimation for Interior PMSMs (IPMSMs), Surface PMSMs (SPMSMs) and SynRMs. Pros and cons of each method will be discussed in terms of implementation requirements, accuracy and suitability depending on the machine design. A key aspect for the proposed method will be modelling the relationship between the incremental inductances obtained from the injected HF signal and the apparent inductance used in conventional torque equation. The paper is organised as follows. The fundamental model of the synchronous machine (SM) is presented in Section 2, while the HF model is presented in Section 3. Parameter identification and torque estimation using different form of HF signal injections is discussed in Section 4. Finally, experimental results are provided in Section 5.

2. Fundamental Model of a Synchronous Motor

The fundamental model of a synchronous machine in a reference frame synchronous with the rotor is given by (1) [8], where R_d, R_q, L_d and L_q are the d and q-axes resistances and inductances, respectively; ω_r is the mechanical rotational speed; λ_{pm} is PM flux linkage; and p is the differential operator. In the nomenclature used in this paper, superscripts "s" and "r" indicate the stationary and rotor synchronous reference frames, respectively; subscripts "s" and "r" indicate stator and rotor variables, respectively; and subscripts "dq" and "HF" indicate fundamental and HF components respectively. Finally, "$*$" is used to indicate commanded values, and "^" indicates estimated values.

$$\begin{bmatrix} v_{sd}^r \\ v_{sq}^r \end{bmatrix} = \begin{bmatrix} R_d + pL_d & -\omega_r L_q \\ \omega_r L_d & R_q + pL_q \end{bmatrix} \begin{bmatrix} i_{sd}^r \\ i_{sq}^r \end{bmatrix} + \begin{bmatrix} 0 \\ \lambda_{pm}\omega_r \end{bmatrix} \tag{1}$$

The output torque in the steady state can be expressed by (2), where P stands for the number of poles [19]. The first term on the left-hand side of (2), T_{syn} is the electromagnetic/synchronous torque due to the PM flux linking magnet and stator coils, whereas the second term on the right-hand side of (2), T_{rel}, is the reluctance torque due to saliency of the machine, i.e., the difference between d- and q-axis inductances.

$$T_{out} = \frac{3}{2}\frac{P}{2}\left[\lambda_{pm} i_{sq}^r + \left(L_d - L_q\right) i_{sd}^r i_{sq}^r\right] = T_{syn} + T_{rel} \tag{2}$$

It is observed from (2) that estimation of the synchronous torque T_{syn} requires knowledge of λ_{pm}, whereas estimation of the reluctance torque T_{rel} requires knowledge of the differential inductance $(L_d - L_q)$.

As already mentioned, this paper will address torque estimation for three machine designs: IPMSM, SPMSM and SynRM. Their schematic cross sections and main characteristics being shown in Figure 1 and Table 1, respectively. From the combined analysis of (2) and Table 1, the mechanisms for torque production of each machine design becomes evident.

Torque produced by IPMSMs and SPMSMs is determined both by the magnetic torque due to the PMs (synchronous torque) and the reluctance torque due to different L_d and L_q inductances, i.e., saliency ratio. Note that for the case of SPMSM, the saliency ratio is often negligible. Torque produced by SynRM will be determined only by the saliency ratio. Proper torque estimation will, therefore, require precise estimation of λ_{pm} and/or L_d and L_q.

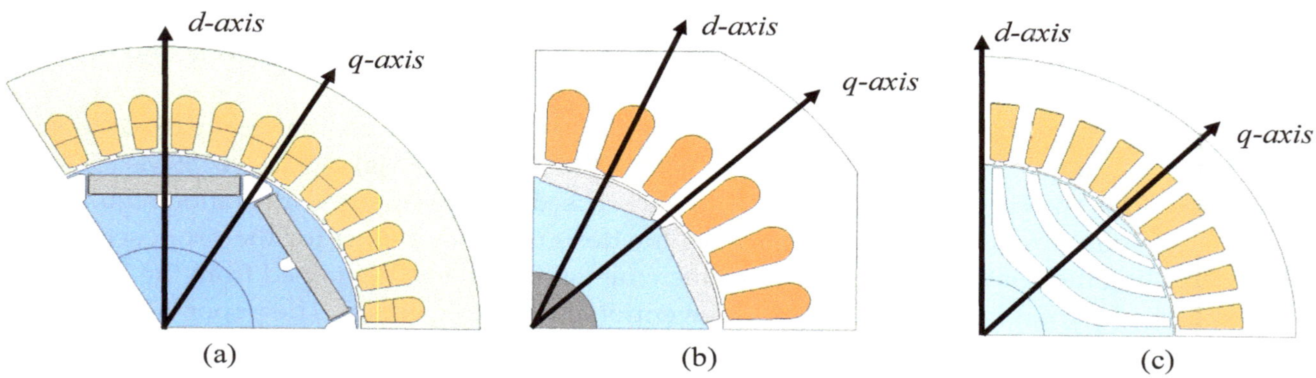

Figure 1. Partial cross-sectional view of the motors being analysed: (**a**) IPMSM, (**b**) SPMSM and (**c**) SynRM.

Table 1. Comparison of motors parameters.

	IPMSM	SPMSM	SynRM
Magnetic flux linkage	$\lambda_{pm} > 0$	$\lambda_{pm} > 0$	$\lambda_{pm} = 0$
d and q inductances	$L_d < L_q$	$L_d \approx L_q$	$L_d >> L_q$

Parameters Variation with Operating Conditions of the Machine

Torque equation is function of inductances and magnet strength, which can change with the operating point of the machine. q- and d-axes inductances can be modelled as (3) and (4).

$$L_{q(I_{sd},I_{sq})} = L_{q0}\left(1 + \alpha_{I_{sd}} I^r_{sd} + \alpha_{I_{sq}} I^r_{sq}\right) \tag{3}$$

$$L_{d(I_{sd},I_{sq},T_r)} = L_{d0}\left(1 + \alpha_{I_{sd}} I^r_{sd} + \alpha_{I_{sq}} I^r_{sq} + \alpha_{T_r}\left(T_r - T_{r0}\right)\right) \tag{4}$$

From Equation (3), q-axis inductance is seen to be function of the fundamental component of the d-axis current due to cross-coupling, and of the q-axes current due to saturation [11]. Similarly, d-axis inductance (4) is function of the fundamental d- and q-axes currents, but also of the PMs remanent flux, which varies with temperature T_r [8,11,12]. Variation of d-axis inductance with PM flux λ_{pm} has been shown to be almost linear, and can therefore be modelled as (5) [12], where λ_{pm0} and L_{d0} are the base value of PM flux and d-axis inductance, i.e., at room temperature (T_{r0}) and when there is no fundamental current, and $L_{d(I_{sd},I_{sq},T_r)}$ is the d-axis inductance for a given magnet temperature T_r and current I_{sdq}.

$$\lambda_{pm} = \lambda_{pm0}\frac{L_{d0}}{L_{d(I_{sd},I_{sq},T_r)}} \tag{5}$$

Substituting (3)–(5) into (2), the general torque equation can be written as (6), which evidences the dependence of torque with the operating condition of the machine, and highlights the incorrectness of assuming constant inductances. Parameters estimation using HF signal injection is discussed following.

$$T_{out} = \frac{3}{2}\frac{P}{2}\left[\left(\lambda_{pm0}\frac{L_{d0}}{L_{d(I_{sd},I_{sq},T_r)}}\right) i^r_{sq} + \left(L_{d(I_{sd},I_{sq},T_r)} - L_{q(I_{sd},I_{sq})}\right) i^r_{sd} i^r_{sq}\right] \tag{6}$$

3. HF Model of the Synchronous Machine

The model representing the behaviour of a synchronous machine when the stator is fed at a frequency sufficiently higher than the rotor frequency (7) can be deduced from (1) by neglecting magnet flux, as it does not contain any HF component [12].

$$\begin{bmatrix} v^r_{sdHF} \\ v^r_{sqHF} \end{bmatrix} = \begin{bmatrix} R_{dHF} + pL_{dHF} & -\omega_r L_{qHF} \\ \omega_r L_{dHF} & R_{qHF} + pL_{qHF} \end{bmatrix} \begin{bmatrix} i^r_{sdHF} \\ i^r_{sqHF} \end{bmatrix} \tag{7}$$

As the HF signal being applied is a voltage, Equation (7) can be solved for the high frequency current as

$$\begin{bmatrix} i^r_{sdHF} \\ i^r_{sqHF} \end{bmatrix} = \frac{1}{(R_{dHF}+pL_{dHF})(R_{qHF}+pL_{qHF})+\omega_r^2 L_{dHF}L_{qHF}} \begin{bmatrix} R_{qHF} + pL_{qHF} & \omega_r L_{qHF} \\ -\omega_r L_{dHF} & R_{dHF} + pL_{dHF} \end{bmatrix} \begin{bmatrix} v^r_{sdHF} \\ v^r_{sqHF} \end{bmatrix} \tag{8}$$

Torque in Equation (6) is a function of absolute (apparent) inductances. However, the inductances of the HF model (7) and (8) are incremental inductances [34], i.e., inductances estimated by means of HF signal injection will be incremental. It is needed therefore the establish the relationship between absolute and incremental inductances.

The inductance of a stator winding is defined as the relationship between flux linkage λ divided by the stator current I producing that flux [35], where N is the number of turns of the stator winding.

$$L = \frac{\lambda}{I} = \frac{N\phi}{I} \quad (9)$$

Flux ϕ in the core which can be defined as (10), R being the reluctance of the magnetic circuit.

$$\phi = \frac{NI}{R} \quad (10)$$

Combining (9) and (10), the inductance can be written as (11), where μ is the permeability of the material ($\mu = \mu_0\mu_r$); A is the mean cross-sectional area of the magnetic circuit and l is the mean length of the magnetic circuit.

$$L = \frac{N^2}{R} = \frac{N^2\mu A}{l} \quad (11)$$

It is seen from (11) that the inductance is proportional to the permeability [35].

$$L(H) = k\mu(H) \quad (12)$$

Static and dynamic permeabilities can be used to analyse the B-H curve of ferromagnetic materials (see Figure 2). The static permeability is defined as the slope of a straight line from the origin to the actual operating point A, i.e., the ratio of flux density (B) vs. field intensity (H) at every operating point (13). The static permeability gives the absolute inductance of the machine.

$$\mu_s = \frac{1}{\mu_0}\frac{B}{H} \quad (13)$$

On the other hand, the slope (AB) of the B–H curve (14) is denoted as relative, differential or dynamic permeability.

$$\mu_d = \frac{1}{\mu_0}\frac{\partial B}{\partial H} \quad (14)$$

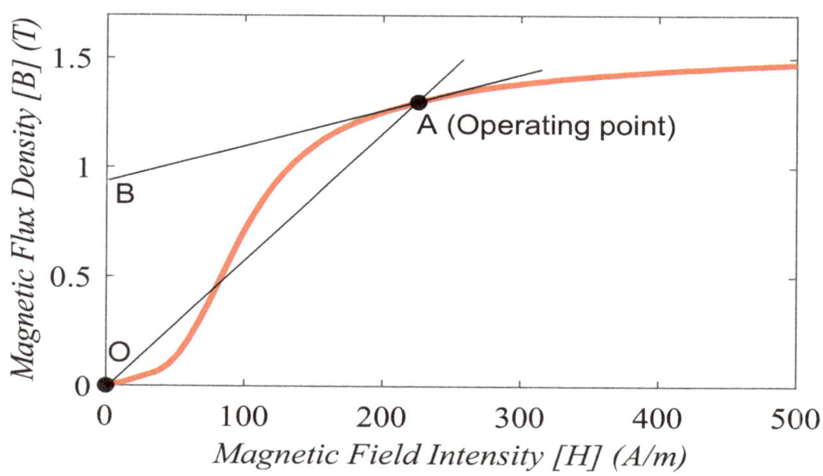

Figure 2. BH curve for NSSMC 50H470 material, and static and dynamic permeability definition.

As an example, Figure 3 shows the BH curve and corresponding static and dynamic permeabilities μ_s and μ_d for the core material NSSMC 50H470. It is observed that for very low excitation levels static and dynamic permeabilities have a ratio close to one, whereas for high current levels the ratio is almost constant (~15 in the figure, note the logarithmic scale). It was deduced from (12) that absolute and incremental inductances will mirror the behaviour of static and dynamic permeabilities, i.e., the relationship between absolute and incremental inductances can be modeled as (15), with k_μ being defined as (16).

$$L_{dq(I_{sd},I_{sq},T_r)} = k_\mu L_{dqHF(I_{sd},I_{sq},T_r)} \tag{15}$$

$$k_\mu = \frac{\mu_s}{\mu_d} \tag{16}$$

It is apparent from Figure 4 that at high excitation levels, the relationship between both inductances is almost constant. However, to model k_μ in the whole operating range, some type of polynomial function would be needed.

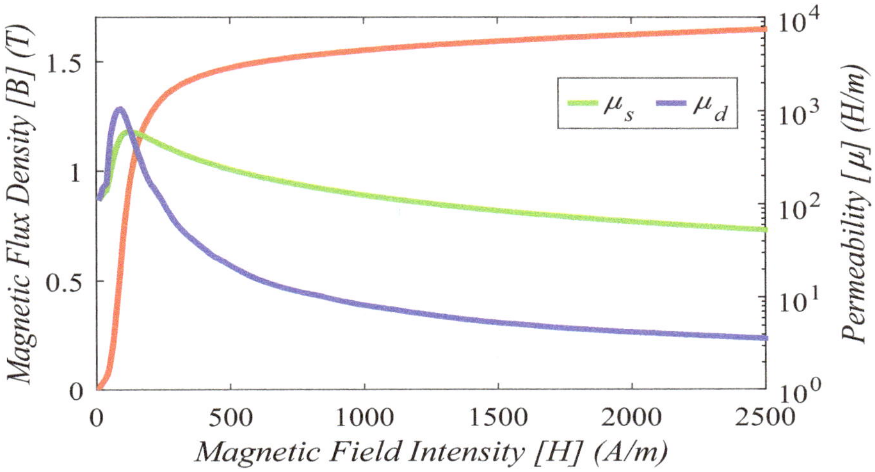

Figure 3. BH curve of ferromagnetic material and corresponding static and dynamic permeabilities (see Figure 2).

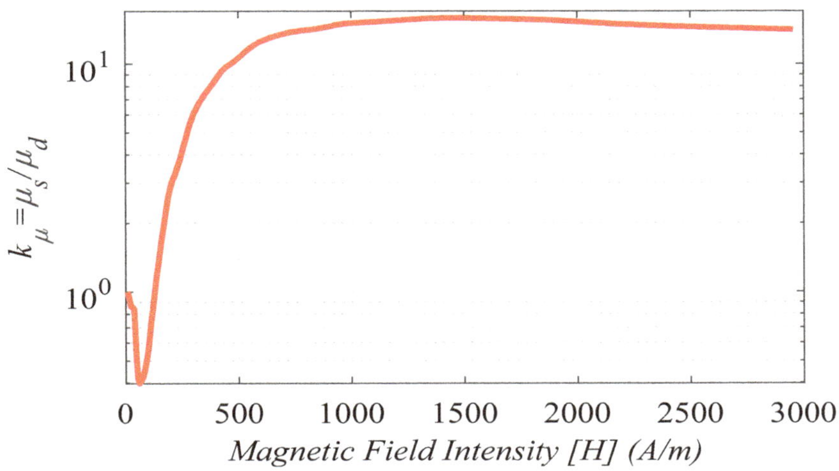

Figure 4. Ratio between static and dynamic permeabilities of the NSSMC 50H470 core material in Figure 3.

λ_{pm} variation shown in (5) can be therefore rewritten as (17), the output torque can be finally estimated combining (15) and (17) as (18).

$$\lambda_{pm} = \lambda_{pm0} \frac{L_{dHF_0}}{L_{dHF(I_{sd},I_{sq},T_r)}} \tag{17}$$

$$\hat{T}_{out} = \frac{3}{2}\frac{P}{2}\left[\left(\lambda_{pm_0}\frac{L_{dHF_0}}{L_{dHF(I_{sd},I_{sq},T_r)}}\right)i^r_{sq} + \left(k_\mu L_{dHF(I_{sd},I_{sq},T_r)} - k_\mu L_{qHF(I_{sd},I_{sq})}\right)i^r_{sd}i^r_{sq}\right] \quad (18)$$

4. HF Inductances Estimation

Different forms of HF excitation can be used to estimate the HF inductances. Generally speaking, voltage injection results in easier implementations as the inverter is a voltage source, but will be sensitive to speed, also being affected by the non-purely inductive machine behaviour. Those drawbacks can be overcome by injecting a HF current, at the price of an increase in the complexity of implementation. Rotating Voltage Injection, Pulsating Voltage Injection and Pulsating dq-axes Current Injection are discussed in the next subsections, followed by a comparative analysis. It has to be noted that other types of periodic high frequency signal injection (e.g., square wave) could be used. These options have not been discussed in this paper due to room constrains.

4.1. Pulsating Voltage Injection

HF inductances L_{dHF} and L_{qHF} can be estimated by injecting a voltage HF signal in a predefined direction referred to the d-axes. This form of excitation is commonly referred as pulsating voltage.

The HF model in the synchronous reference frame (7) can be written as function of the mean ΣL (20) and differential ΔL (21) inductances (19). Note that the resistive component has been neglected, as at high frequency it is significantly smaller than the inductive terms and the p operator in (7) has been replaced by $j\omega_{HF}$. Furthermore, rotor speed dependent terms in (7) have been neglected assumed that the HF voltage signal has a frequency much higher than the fundamental rotating frequency (i.e., $\omega_{HF} >> \omega_r$).

$$\begin{bmatrix} v^r_{sdHF} \\ v^r_{sqHF} \end{bmatrix} = \begin{bmatrix} j\omega_{HF}(\Sigma L - \Delta L) & 0 \\ 0 & j\omega_{HF}(\Sigma L + \Delta L) \end{bmatrix} \begin{bmatrix} i^r_{sdHF} \\ i^r_{sqHF} \end{bmatrix} \quad (19)$$

$$\Sigma L = \frac{L_{dHF} + L_{qHF}}{2} \quad (20)$$

$$\Delta L = \frac{L_{dHF} - L_{qHF}}{2} \quad (21)$$

The currents induced in the stator terminals expressed in the stator reference frame can be derived from the inverse matrix as (22)

$$\begin{bmatrix} i^s_{sdHF} \\ i^s_{sqHF} \end{bmatrix} = \frac{1}{j\omega_{HF}\left(\Sigma L^2 - \Delta L^2\right)} \begin{bmatrix} \Sigma L + \Delta L\cos(2\theta_r) & \Delta L\sin(2\theta_r) \\ \Delta L\sin(2\theta_r) & \Sigma L - \Delta L\cos(2\theta_r) \end{bmatrix} \begin{bmatrix} v^s_{sdHF} \\ v^s_{sqHF} \end{bmatrix} \quad (22)$$

If a pulsating HF voltage is injected in the stator terminals of the machine (23), the resulting HF current can be derived substituting in (23) into (22) as (26),

$$\begin{bmatrix} v^{s*}_{sdHF} \\ v^{s*}_{sqHF} \end{bmatrix} = V^*_{HF}\sin(\theta_{HF}) \begin{bmatrix} \cos(\hat{\theta}_r) \\ \sin(\hat{\theta}_r) \end{bmatrix} \quad (23)$$

$$\theta_{HF} = \omega_{HF}t \quad (24)$$

$$\hat{\theta}_r = \theta_r + \varphi \quad (25)$$

$$\begin{bmatrix} i^s_{sdHF} \\ i^s_{sqHF} \end{bmatrix} = \frac{V^*_{HF}\sin(\theta_{HF})}{j\omega_{HF}\left(\Sigma L^2 - \Delta L^2\right)} \begin{bmatrix} \Sigma L\cos(\hat{\theta}_r) + \Delta L\cos(2\theta_r - \hat{\theta}_r) \\ \Sigma L\sin(\hat{\theta}_r) + \Delta L\sin(2\theta_r - \hat{\theta}_r) \end{bmatrix} \quad (26)$$

where V^*_{HF} is the magnitude of the injected HF signal, θ_{HF} is the phase of the HF signal (24), ω_{HF} is the frequency of the HF signal, $\hat{\theta}_r$ is the injection angle of the pulsating HF voltage (25) and φ is an arbitrary angle; e.g., if $\varphi = 0$, the pulsating HF voltage will be injected in the d-axis of the machine, whereas if $\varphi = \pi/2$, it will be injected in the q-axis.

By synchronization with the injection reference frame "$\hat{\theta}_r$", the stator currents (26) are transformed into (27)

$$\begin{bmatrix} i_{sdHF}^{\hat{\theta}_r} \\ i_{sqHF}^{\hat{\theta}_r} \end{bmatrix} = \frac{V_{HF}^{*}\sin(\theta_{HF})}{j\omega_{HF}\left(\Sigma L^2 - \Delta L^2\right)} \begin{bmatrix} \Sigma L + \Delta L\cos(2(\theta_r - \hat{\theta}_r)) \\ \Delta L\sin(2(\theta_r - \hat{\theta}_r)) \end{bmatrix} \tag{27}$$

Finally, if the HF is injected between d and q-axes, i.e., $\hat{\theta}_r = \theta_r + \pi/4$, (27) can be simplified into (28) [32]. From (28) the HF inductances (29) and (30) are readily obtained.

$$\begin{aligned} \begin{bmatrix} i_{sdHF}^{\hat{\theta}_r} \\ i_{sqHF}^{\hat{\theta}_r} \end{bmatrix} &= \frac{V_{HF}^{*}\sin(\theta_{HF})}{j\omega_{HF}\left(\Sigma L^2 - \Delta L^2\right)} \begin{bmatrix} \Sigma L + \Delta L\cos(-\pi/2) \\ \Delta L\sin(-\pi/2) \end{bmatrix} \\ &= \frac{V_{HF}^{*}\sin(\theta_{HF})}{j\omega_{HF}\left(\Sigma L^2 - \Delta L^2\right)} \begin{bmatrix} \Sigma L \\ -\Delta L \end{bmatrix} \end{aligned} \tag{28}$$

$$L_{dHF} = \frac{V_{HF}^{*}\sin(\theta_{HF})}{j\omega_{HF}\left(i_{sdHF}^{\hat{\theta}_r} - i_{sqHF}^{\hat{\theta}_r}\right)} \tag{29}$$

$$L_{qHF} = \frac{V_{HF}^{*}\sin(\theta_{HF})}{j\omega_{HF}\left(i_{sdHF}^{\hat{\theta}_r} + i_{sqHF}^{\hat{\theta}_r}\right)} \tag{30}$$

Figure 5 shows the inverter control block diagram and the signal processing needed for torque estimation when pulsating HF voltage injection at an arbitrary angle of injection, $\hat{\theta}_r$, is used for dq-axes HF inductance estimation. The HF voltage v_{sdqHF1}^{r*}, is injected in open-loop. A High-Pass Filter (HPF) is needed to isolate the HF components of the overall stator currents. Inputs to the signal processing block are the commanded HF voltage v_{sdqHF1}^{r*}, the induced HF currents i_{sdqHF1}^{r} and the fundamental commanded current I_{sdq}^{r*}. A Band Pass Filter (BPF) is used to isolate the positive sequence component of the HF induced current. The d and q-axis HF inductances are estimated using (29) and (30), the PM flux is estimated using (17) and the output torque, $\hat{T_{out}}$, is finally estimated using (18). As the estimated dq-axes inductances already reflect the effects of temperature on PM magnetisation state as well as the effects of fundamental current, knowledge of PM temperature is not needed.

Appealing properties of HF pulsating voltage injection are its simplicity and the fact that a single frequency allows the estimation of d- and q-axes HF inductances.

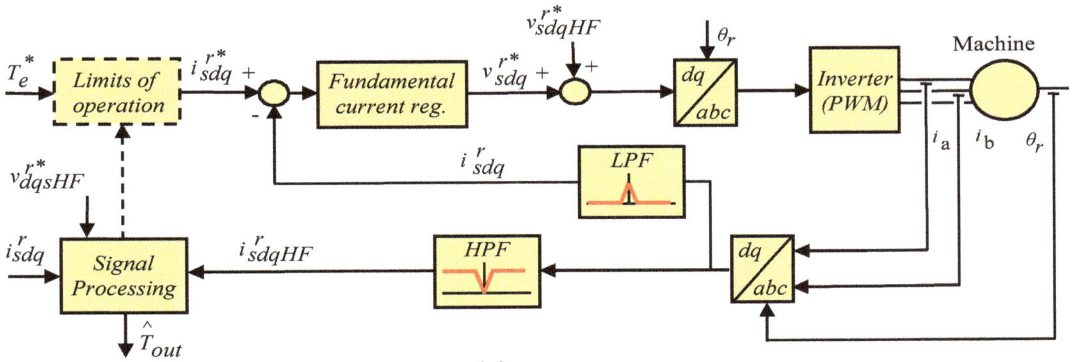

(a)

Figure 5. *Cont.*

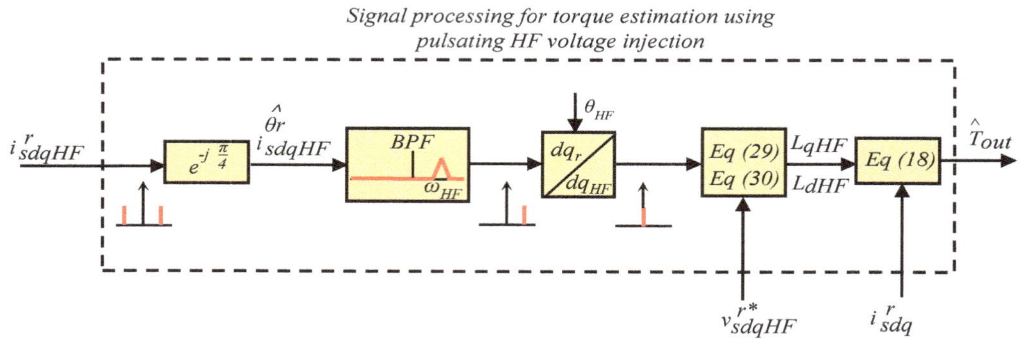

Figure 5. Implementation using pulsating HF voltage injection at an arbitrary angle of injection, $\hat{\theta}_r$. (**a**) Power converter control. (**b**) Signal processing for torque estimation.

4.2. Rotating Voltage Injection

While the assumption of pure inductive behaviour at high frequencies can be acceptable for most PMSMs designs, it can be arguable for SynRMs. In this case, the use of a HF pulsating voltage at 45° could incur in large errors in the estimated inductances, and consequently in the estimated torque. The use of a rotating HF voltage can be advantageous in this case.

When a rotating HF voltage (31) is injected in a synchronous machine, the HF currents induced in stator windings can be obtained by inserting (31) into (8).

$$v^{r*}_{sdqHF} = \begin{bmatrix} \bar{V}^{r*}_{sdHF} \\ \bar{V}^{r*}_{sqHF} \end{bmatrix} = \begin{bmatrix} V^*_{HF} \cos(\omega_{HF} t) \\ V^*_{HF} \sin(\omega_{HF} t) \end{bmatrix} = V^*_{HF} e^{j(\omega_{HF} t)} = v^{r*}_{sdqHFpc} \tag{31}$$

By solving (8), the stator HF currents in the rotor synchronous reference frame are obtained as

$$i^r_{sdHF} = \frac{R_{qHF} + j\omega_{HF} L_{qHF}}{(R_{dHF} + j\omega_{HF} L_{dHF})(R_{qHF} + j\omega_{HF} L_{qHF}) + \omega_r^2 L_{dHF} L_{qHF}} \left[v^r_{sdHF} + \frac{\omega_r L_{qHF} v^r_{sqHF}}{R_{qHF} + j\omega_{HF} L_{qHF}} \right] \tag{32}$$

$$i^r_{sqHF} = \frac{R_{dHF} + j\omega_{HF} L_{dHF}}{(R_{dHF} + j\omega_{HF} L_{dHF})(R_{qHF} + j\omega_{HF} L_{qHF}) + \omega_r^2 L_{dHF} L_{qHF}} \left[v^r_{sqHF} - \frac{\omega_r L_{dHF} v^r_{sdHF}}{R_{dHF} + j\omega_{HF} L_{dHF}} \right] \tag{33}$$

It can be seen from (32) and (33) that obtaining L_{dHF} and L_{qHF} is not straightforward due to cross-coupling between dq-axes. However, if the frequency of the injected HF signal is sufficiently higher than the rotor frequency, the rotor speed dependent terms can be safely neglected, the HF currents induced in the stator simplifying to (34) and (35). An orientative value for this assumption can be $\omega_{HF} > \omega_r + 2 \cdot \pi \cdot 500$ rad/s [12].

$$i^r_{sdHF} = \frac{v^r_{sdHF}}{(R_{dHF} + j\omega_{HF} L_{dHF})} = \frac{V^*_{HF} \cos(\omega_{HF} t)}{(R_{dHF} + j\omega_{HF} L_{dHF})} \tag{34}$$

$$i^r_{sqHF} = \frac{v^r_{sqHF}}{(R_{qHF} + j\omega_{HF} L_{qHF})} = \frac{V^*_{HF} \sin(\omega_{HF} t)}{(R_{dHF} + j\omega_{HF} L_{dHF})} \tag{35}$$

Estimation of L_{dHF} and L_{qHF} can be obtained from the imaginary part of the dq-axes impedance (36) and (37).

$$Z_{dHF} = R_{dHF} + j\omega_{HF} L_{dHF} = \frac{V^*_{HF} \cos(\omega_{HF} t)}{i^r_{sdHF}} \tag{36}$$

$$Z_{qHF} = R_{qHF} + j\omega_{HF}L_{qHF} = \frac{V_{HF}^{*}\sin(\omega_{HF}t)}{i_{sqHF}^{r}} \tag{37}$$

Figure 6 shows the inverter control and signal processing for torque estimation using rotating HF voltage injection. As for the implementation shown in Figure 5, the commanded HF voltage, v_{sdqHF}^{r*} is injected in open-loop, with no additional controllers required. A High-Pass Filter (HPF) is also needed to isolate the HF components of the overall stator currents. Inputs of the signal processing block are the commanded HF voltage v_{sdqHF}^{r*}, the induced HF currents i_{sdqHF}^{r}, and the fundamental commanded current I_{sdq}^{r*}. The d and q-axis HF inductances are estimated using (36) and (37), the PM flux is estimated using (17) and the output torque, $\hat{T_{out}}$, is finally estimated using (18).

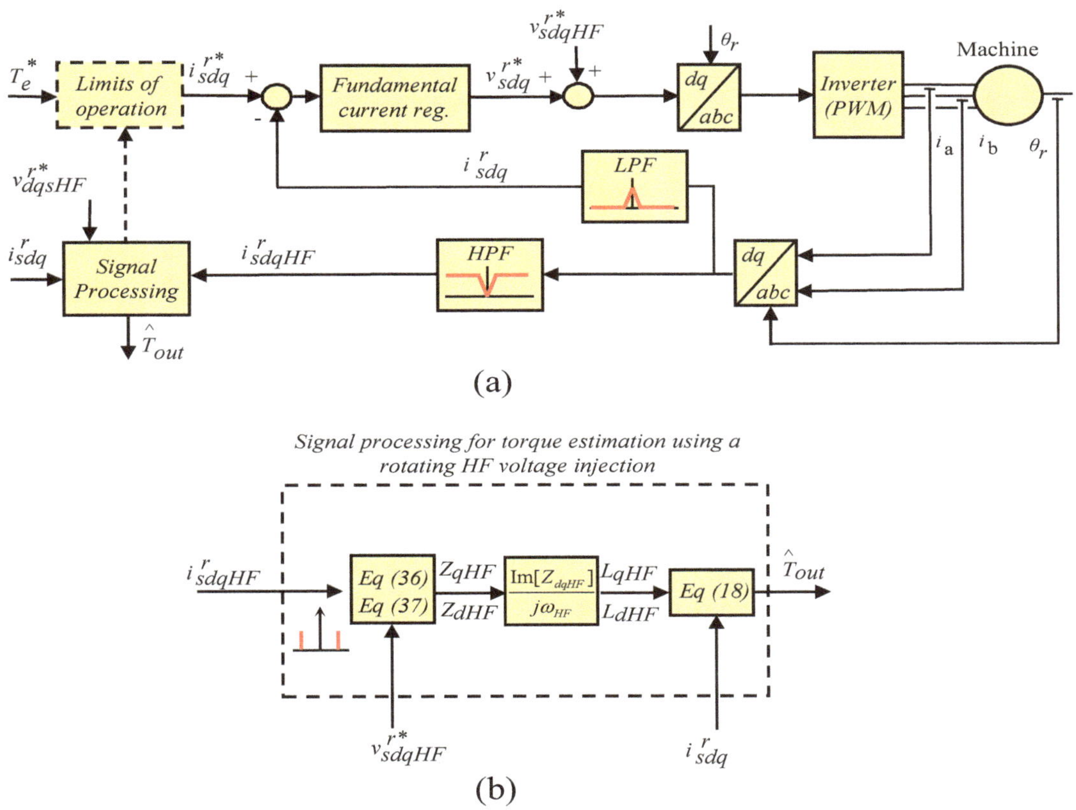

Figure 6. Implementation using rotating HF voltage injection. (**a**) Power converter control. (**b**) Signal processing for torque estimation.

The frequency of the HF signal should be sufficiently higher than the rotor frequency to safely neglect the rotor speed dependent (i.e., cross-coupling) terms. This can be problematic in the case of IPMSM designed to operate at very high-speeds. Furthermore, as the HF voltage is injected in a reference frame synchronous with the rotor, the effective frequency of the HF frequency signal will be $\omega_r + \omega_{HF}$. It must be guaranteed that it is smaller than half of the switching frequency (Nyquist frequency).

4.3. Pulsating Dq-Axes Current Injection

Injection of a pulsating HF current has been proposed as a viable mean to avoid issues due to the assumption of a purely inductive behaviour of the windings, as well as the problems with high speed machines discussed in the preceding subsections. HF inductances can be estimated by injecting two pulsating HF currents at different frequencies into the d- and q-axis. HF current controllers are needed for this purpose [12]. The process to estimate L_{dHF} is described following, and identical procedure is used to obtain L_{qHF}.

If a HF current is injected into the d-axis of the machine and the q-axis current is force to be equal to zero (38), the HF voltages commanded by the HF current controller will be of the form shown in (39). A fictitious HF voltage vector can be defined consisting only of the d-axis component $v^{r'}_{sdqHF1}$ (40). Both HF currents and voltages in (38) and (40) can be separated into positive sequence ($i^{r*}_{sdqHF1pc}$ and $v^{r'}_{sdqHF1pc}$) and negative sequence ($i^{r*}_{sdqHF1nc}$ and $v^{r'}_{sdqHF1nc}$) components, (41) and (42), of magnitude equal to half of the original signal. The d-axis HF impedance (43) can be obtained from the positive or negative sequence indistinctly. The d-axis HF inductance is finally obtained as the imaginary part of (43) and (44).

$$i^{r*}_{sdqHF1} = \begin{bmatrix} \bar{I}^{r*}_{sdHF1} \\ \bar{I}^{r*}_{sqHF1} \end{bmatrix} = \begin{bmatrix} I^{*}_{HF} \cos\left(\omega_{dHF} t\right) \\ 0 \end{bmatrix} \tag{38}$$

$$v^{r*}_{sdqHF1} = \begin{bmatrix} \bar{V}^{r*}_{sdHF1} \\ \bar{V}^{r*}_{sqHF1} \end{bmatrix} = \begin{bmatrix} \left(R_{dHF} + j\omega_{dHF} L_{dHF}\right) \bar{I}^{r}_{sdHF1} \\ \omega_r L_{dHF} \bar{I}^{r}_{sdHF1} \end{bmatrix} \tag{39}$$

$$v^{r'}_{sdqHF1} = \begin{bmatrix} \bar{V}^{r*}_{sdHF1} \\ 0 \end{bmatrix} = \begin{bmatrix} \left(R_{dHF} + j\omega_{dHF} L_{dHF}\right) \bar{I}^{r}_{sdHF1} \\ 0 \end{bmatrix} = \begin{bmatrix} \bar{V}^{r'}_{sdqHF1} \cos\left(\omega_{dHF} t + \varphi_{Z_d}\right) \\ 0 \end{bmatrix} \tag{40}$$

$$i^{r*}_{sdqHF1} = \frac{I^{*}_{HF}}{2} e^{j\omega_{dHF} t} + \frac{I^{*}_{HF}}{2} e^{-j\omega_{dHF} t} = i^{r*}_{sdqHF1pc} + i^{r*}_{sdqHF1nc} \tag{41}$$

$$v^{r'}_{sdqHF1} = \frac{\left|v^{r'}_{sdqHF1}\right|}{2} e^{j(\omega_{dHF} t - \varphi_{Zd})} + \frac{\left|v^{r'}_{sdqHF1}\right|}{2} e^{j(-\omega_{dHF} t + \varphi_{Zd})} \; v^{r'}_{sdqHF1pc} + v^{r'}_{sdqHF1nc} \tag{42}$$

$$Z_{dHF} = R_{dHF} + j\omega_{dHF} L_{dHF} = \frac{v^{r'}_{sdqHF1pc}}{i^{r*}_{sdqHF1pc}} = \frac{v^{r'}_{sdqHF1nc}}{i^{r*}_{sdqHF1nc}} \tag{43}$$

$$L_{dHF} = \frac{Im\left[Z_{dHF}\right]}{j\omega_{dHF}} \tag{44}$$

Figure 7 shows the inverter control block diagram and the signal processing needed for torque estimation when pulsating dq-axes HF current signals are used for dq-axes HF inductance estimation. Two HF resonant current controllers (45) are used to inject the HF currents, where K_p is the proportional gain, ω_{HF} is the resonant frequency, and c is the zero position of the controller.

$$PR(s) = K_p \frac{(s+c)^2 + \omega_{HF}^2}{s^2 + \omega_{HF}^2} \tag{45}$$

Inputs to the signal processing block are the commanded high frequency resonant currents i^{r*}_{sdqHF1} and i^{r*}_{sdqHF2}, the output voltage of the resonant controllers $v^{r'}_{sdqHF1}$ and $v^{r'}_{sdqHF2}$ and the fundamental commanded current I^{r*}_{sdq}. Two band stop filters, $BSF1$ and $BSF2$, are used to remove the negative sequence components of the HF currents and voltages. The d and q-axis HF impedances are estimated using (43), the d and q-axis HF inductances are estimated using (44), the PM flux is estimated using (17) and, the output torque, $\hat{T}_{out}$, is finally estimated using (18).

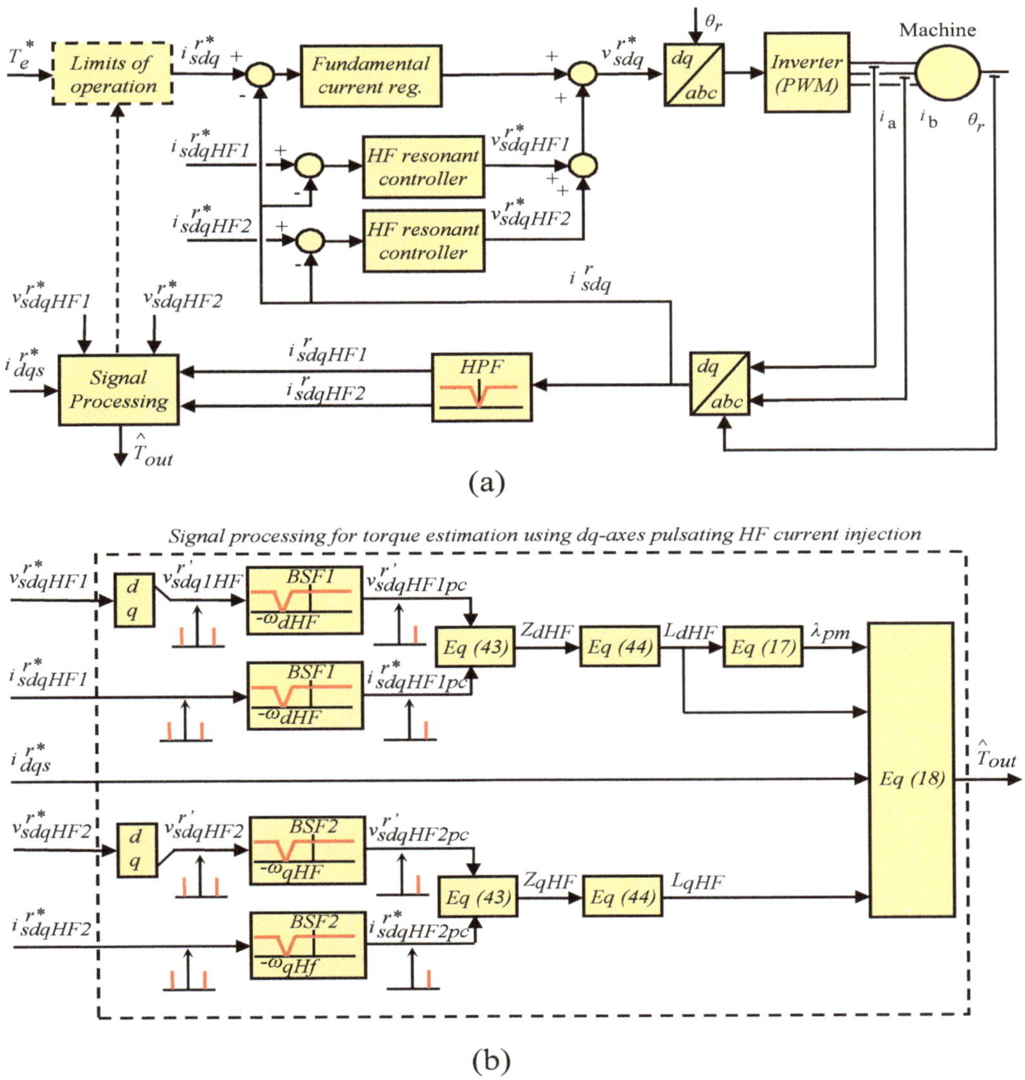

Figure 7. HF inductance estimation using HF pulsating current injection. (**a**) Power converter control. (**b**) Signal processing for torque estimation.

Note that in this case a pure inductive behaviour is not assumed, cross-coupling terms do not affect to the estimations either. However, the fact that there are two HF signals might have a larger impact on machine performance (noise, vibration, ...) compared to methods using a single HF signal, the computational burden also being larger. Finally, Table 2 summarises the main characteristics of the methods analysed in this section.

Table 2. Comparison between HF Signal injection methods for SM parameter estimation.

	Pulsating *dq*-axes HF Current Injection	**Pulsating HF Voltage Injection**	**Rotating HF Voltage Injection**
Pure inductive HF model can be assumed	✓	✗	✓
Sensitive machine speed	✓	✗	✗
Need of additional controllers	✗	✓	✓
Number of injected signals	2	1	1
Computational burden	✗	✓	✓

5. Experimental Results

The proposed methods have been tested in the machines designs shown in Figure 1; the corresponding parameters can be found in Table 3. A schematic representation of the test bench as well as pictures of the different elements are shown in Figure 8. The inverter feeding the machines under test (Inverter 2 in Figure 8a) uses 1200 V, 100 A IGBT power modules, with a switching frequency of 10kHz. The load machine is a 40kW axial PMSM machines (EMRAX 228 [36]) driven by a BAMOCAR-PG-D3 power converter [37] (Inverter 1 in Figure 8a). Torque is measured using a T5 Interface Torque transducer [13], with 12-Bit resolution, 10 kHz, ±100 Nm, 0.2% combined error. Currents in the machines being tested are measured using standard 1% Hall-effect based-current sensors and 12-bit analog-to-digital converters [38]. For the sake of completeness, a different type of HF excitation has been used with each machine design.

Table 3. Test machines parameters.

	P_{rated} [kW]	I_{rated}[A]	L_{d0} [mH]	L_{q0} [mH]	λ_{pm0}[Wb]	ω_{rated} [rpm]	Poles
IPMSM	4	14	10.5	23	0.64	1000	6
SPMSM	5 .3	15	5.54	6.81	0.59	1000	8
SynRM	1.5	3.9	410	100	-	1500	4

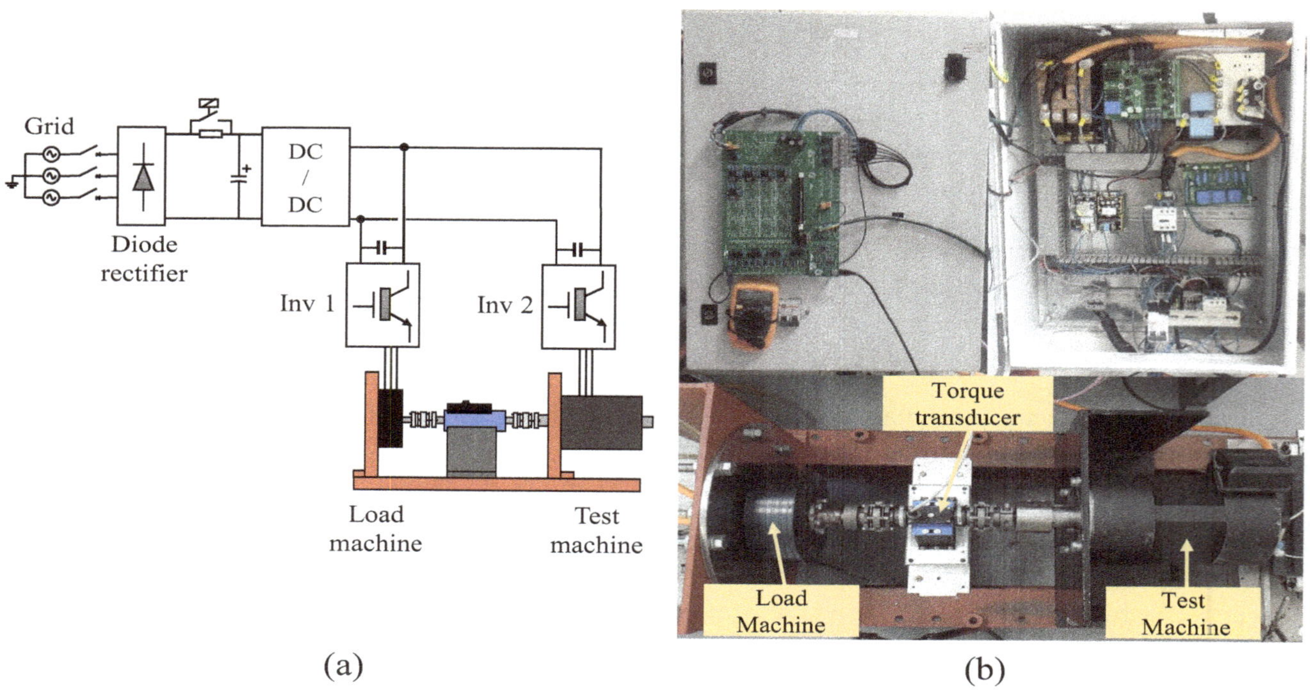

Figure 8. Experimental set-up: (**a**) Schematic representation. (**b**) Top left: Control board (Inverter 2). Top right: Power converter (Inverter 2). Bottom: Picture of the test bench.

5.1. IPMSM Torque Estimation

Two pulsating HF currents have been used in this case, of magnitude 0.05 pu and frequency of $\omega_{dHF} = 2 \cdot \pi \cdot 500$ rad/s and $\omega_{qHF} = 2 \cdot \pi \cdot 1000$ rad/s, respectively (38). The HPF needed to isolate the HF current components (see Figure 7) has a bandwidth of 5 Hz. Band stop filters $BSF1$ and $BSF2$ used to remove the negative sequence components of the HF currents and voltages have a bandwidth of $2 \cdot \pi \cdot 10$ rad/s.

Due to rotor magnets, the machine will work at high saturation levels even for no-load conditions. The coefficient linking the HF estimated inductances (incremental inductances) and the absolute inductances is assumed to be constant, $k_\mu = 14$ (see Figure 4).

Figure 9a shows the estimated torque from (2) ($T_{out_{conv}}$) assuming constant parameters; the estimated torque when the machine parameters are estimated from the injected HF current (18) ($\hat{T}_{out_{HF}}$); and the measured torque using the torque transducer shown in Figure 8 (T_{out}), when the magnitude of the fundamental current I^r_{sdq} changes from 0p.u. to 1p.u. following a Maximum Torque Per Ampere (MTPA) trajectory. Figure 9b shows the estimation error using both general torque equation and the proposed method. It can be observed that torque estimation error is reduced when the machine parameters are estimated using HF signal injection; the improvement being more relevant at higher current levels. This is an expected result since the dq-axes inductances values will differ more to their base values as the saturation level increases.

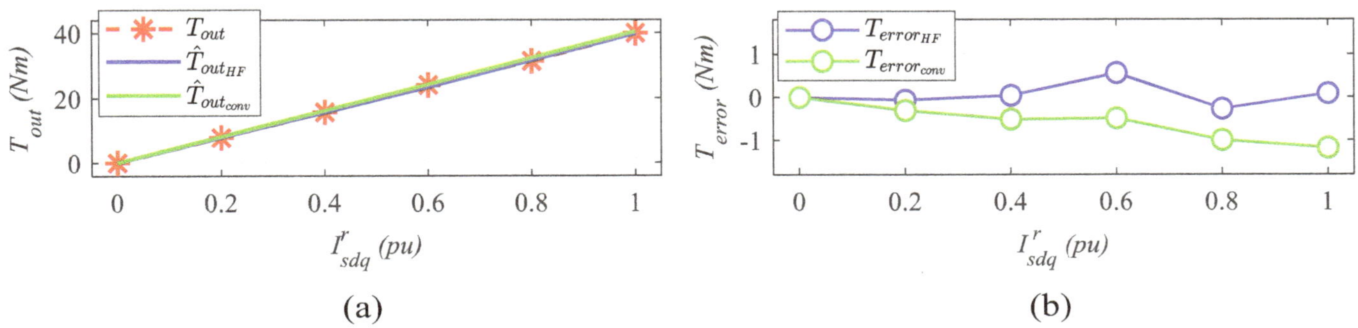

Figure 9. Experimental results: IPMSM & pulsating current injection: (**a**) Estimated and measured torque. (**b**) Estimated torque error. $0 < I^r_{sdq} < 1$ p.u., $\omega_r = 50Hz$, $I_{HF} = 0.05$ p.u., $\omega_{dHF} = 2 \cdot \pi \cdot 500$ rad/s, $\omega_{qHF} = 2 \cdot \pi \cdot 1000$ rad/s.

5.2. SPMSM Torque Estimation

HF inductances have been estimated in this case using pulsating voltage injection at 45° as described in Section 4.1. A HF voltage of 10 V and 250 Hz has been used (23). A band pass filter of 100 Hz was used to isolate the HF currents. Similar for the case of the IPMSM, the machine will work at high saturation levels even at no-load conditions due to the magnets. Therefore, also, in this case the coefficient linking the estimated HF inductances and the absolute inductances has been considered constant $k_\mu = 14$.

Similarly to Figure 9, Figure 10a shows the estimated torque assuming constant parameters and adapting machine parameters using HF voltage injection. The fundamental current,I^r_{sdq} was varied from 0 p.u. to 1 p.u. following a MTPA trajectory. Figure 10b shows the estimation error for both methods. As for the IPMSM case, torque estimation error is also reduced when the machine parameters are estimated using HF signal injection, the improvement being more noticeable than for the IPMSMs.

Finally, Figure 11 shows the actual and estimated torque when there is a step-like change in the q-axis current command from 0 to 1 p.u.. It can be observed from the error shown in Figure 11b that the torque estimator responds in the range of ms.

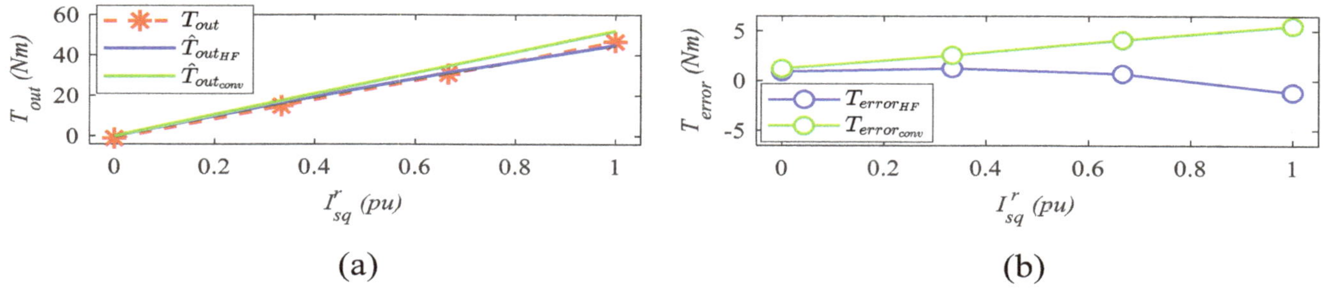

Figure 10. Experimental results: SPMSM & pulsating voltage injection: (**a**) Estimated and measured torque. (**b**) Estimated torque error. $0 < I^r_{sdq} < 1$ p.u., $\omega_r = 16$ Hz, $V_{HF} = 10$ V, $\omega_{HF} = 2 \cdot \pi \cdot 250$ rad/s.

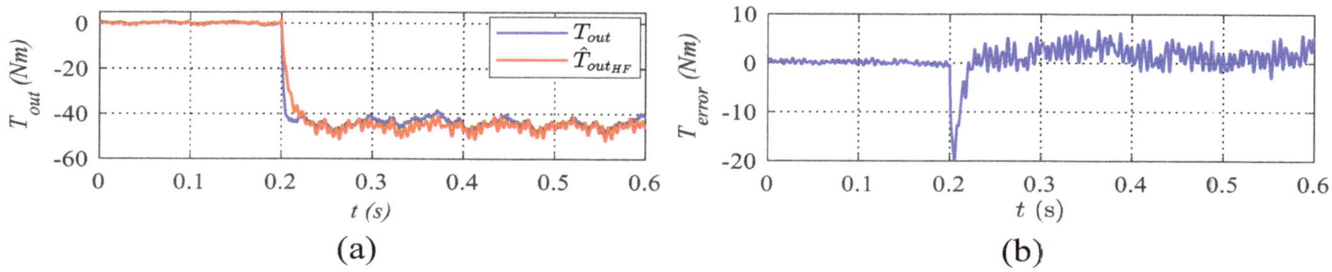

Figure 11. Experimental results: SPMSM & pulsating voltage injection: (**a**) Estimated and measured torque. (**b**) Estimated torque error. Transient response to a step-like change in q-axis current command from 0 to 1 p.u., $\omega_r = 16$ Hz, $V_{HF} = 10$ V, $\omega_{HF} = 2 \cdot \pi \cdot 250$ rad/s.

5.3. SynRM Torque Estimation

Torque estimation for the SynRM has been performed using rotating HF voltage injection. A HF voltage of 40 V and 500 Hz has been used (31). Figure 12 shows experimental results when the magnitude of the fundamental current, I^r_{sq}, changes from 0p.u. to 1p.u. following a MTPA trajectory. In a first approach, the machine has been considered to be working at low-middle saturation levels. Therefore, the coefficient linking the HF estimated inductances and the absolute inductances k_μ, has been considered to be constant and equal to 1 (i.e., the incremental inductance has been assumed to be approximately equal to the absolute inductance, see Figure 3). However, from the torque estimation error shown in Figure 12b, it is deduced that once the machine begins to saturate this approach is not longer valid. This suggests that the relation between the HF estimated inductances and the absolute inductances has to be adjusted using at least a second order polynomial. This is a subject of ongoing research.

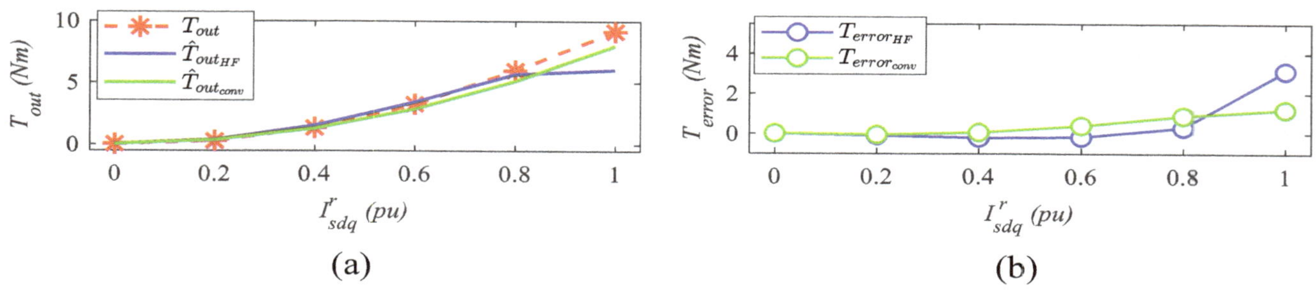

Figure 12. Experimental results: SynRM & rotating voltage injection: (**a**) Estimated and measured torque. (**b**) Estimated torque error. $0 < I^r_{sdq} < 1$ p.u., $\omega_r = 16$ Hz, $V_{HF} = 50$ V, $\omega_{HF} = 2 \cdot \pi \cdot 500$ rad/s.

6. Conclusions

Parameter estimation using HF signal injection aimed to improve the accuracy of torque estimation methods has been addressed in this paper, with the aim of making the estimation robust against variations in the operating conditions of the machine. This implies a reformulation of the torque equation, which will be function of the HF (incremental) inductances instead of the absolute inductances. Accurate modelling of the relationship between the incremental and absolute inductances will be therefore of paramount importance. Additionally, estimation of the PM flux is based on the linear relation with the d-axis inductance.

Three different types of HF signal injection have been considered: Pulsating Voltage Injection, Rotating Voltage Injection and Pulsating dq-axes Current Injection. In all the cases, the signal is superposed on top of the fundamental excitation applied by the inverter, not interfering therefore with the normal operation of the drive. It is concluded that pulsating current injection is advantageous as it is insensitive to the resistive components of the HF model and to cross-coupling effect. In change, its implementation is slightly more difficult due to the need of HF current controllers. In all the cases, no modification of the hardware is required.

Experimental verification using IPMSM, a SPMSM and a SynRM have been presented, which confirm the viability of the proposed methods.

Author Contributions: All the authors contributed in formulating the problem and designing the research proposal. M.M. wrote the paper, collected the experimental data and analysed the data. D.R. suggested the research topic, guided M.M. to complete the research and helped to analyse the data. D.F. developed the set up for the experimental tests. F.B. helped to write and finish the paper. All authors discussed the results and contributed to revised the final manuscript. All authors have read and agreed to the published version of the manuscript.

Abbreviations

The following abbreviations are used in this manuscript:

HF	High Frequency
IPMSMs	Interior Permanent Magnet Synchronous Machines
MTPA	Maximum Torque Per Ampere
SMs	Synchronous Machines
SPMSMs	Surface Permanent Magnet Synchronous Machines
SynRMs	Synchronous Reluctance Machines

References

1. Matsuo, T.; Lipo, T.A. Field oriented control of synchronous reluctance machine. In Proceedings of the IEEE Power Electronics Specialist Conference—PESC '93, Seattle, WA, USA, 20–24 June 1993; pp. 425–431. [CrossRef]
2. Maroufian, S.S.; Pillay, P. Torque Characterization of a Synchronous Reluctance Machine Using an Analytical Model. *IEEE Trans. Transp. Electrif.* **2018**, *4*, 506–516. [CrossRef]
3. Shi, R.; Toliyat, H.A.; El-Antably, A. A DSP-based direct torque control of five-phase synchronous reluctance motor drive. In Proceedings of the APEC 2001, Sixteenth Annual IEEE Applied Power Electronics Conference and Exposition (Cat. No.01CH37181), Anaheim, CA, USA, 4–8 March 2001; Volume 2, pp. 1077–1082, doi:10.1109/APEC.2001.912500. [CrossRef]

4. Shinke, A.; Hasegawa, M.; Matsui, K. Torque estimation for synchronous reluctance motors using robust flux observer to magnetic saturation. In Proceedings of the 2009 IEEE International Symposium on Industrial Electronics, Seoul, Korea, 5–8 July 2009; pp. 1569–1574. [CrossRef]
5. Yousefi-Talouki, A.; Pescetto, P.; Pellegrino, G.; Boldea, I. Combined Active Flux and High-Frequency Injection Methods for Sensorless Direct-Flux Vector Control of Synchronous Reluctance Machines. *IEEE Trans. Power Electron.* **2018**, *33*, 2447–2457. [CrossRef]
6. Senjyu, T.; Kinjo, K.; Urasaki, N.; Uezato, K. High efficiency control of synchronous reluctance motors using extended Kalman filter. *IEEE Trans. Ind. Electron.* **2003**, *50*, 726–732. [CrossRef]
7. Kim, S.; Im, J.; Go, S.C.; Bae, J.; Kim, W.; Kim, K.; Kim, C.; Lee, J. Robust Torque Control of DC Link Voltage Fluctuation for SynRM Considering Inductances With Magnetic Saturation. *IEEE Trans. Magn.* **2010**, *46*, 2005–2008. [CrossRef]
8. Gieras, J.F. *Permanent Magnet Motor Technology: Design and Applications*, 3rd ed.; CRC Press: Boca Raton, FL, USA, 2009.
9. Limsuwan, N.; Kato, T.; Akatsu, K.; Lorenz, R.D. Design and Evaluation of a Variable-Flux Flux-Intensifying Interior Permanent-Magnet Machine. *IEEE Trans. Ind. Appl.* **2014**, *50*, 1015–1024. [CrossRef]
10. Jung, H.; Park, D.; Kim, H.; Sul, S.; Berry, D.J. Non-Invasive Magnet Temperature Estimation of IPMSM Based on High-Frequency Inductance With a Pulsating High-Frequency Voltage Signal Injection. *IEEE Trans. Ind. Appl.* **2019**, *55*, 3076–3086. [CrossRef]
11. Reigosa, D.; Fernández, D.; Martínez, M.; Guerrero, J.M.; Diez, A.B.; Briz, F. Magnet Temperature Estimation in Permanent Magnet Synchronous Machines Using the High Frequency Inductance. *IEEE Trans. Ind. Appl.* **2019**, *55*, 2750–2757. [CrossRef]
12. Martinez, M.; Reigosa, D.; Fernandez, D.; Guerrero, J.M.; Briz, F. Enhancement of Permanent Magnet Synchronous Machines Torque Estimation Using Pulsating High Frequency Current Injection. *IEEE Trans. Ind. Appl.* **2019**, 1–9. [CrossRef]
13. Interface. Torque Sensors: T5. 2019. Available online: https://interfaceforce.co.uk/ (accessed on 1 September 2019).
14. HBM. HBM Torque Sensors. Available online: https://www.hbm.com/en/0264/torque-transducers-torque-sensorstorque-meters/ (accessed on 1 September 2019).
15. TE Connectivity. Measure Reaction and Rotating Torque. Available online: https://www.te.com/global-en/products/sensors/torque-sensors.html (accessed on 16 October 2019).
16. Futek. Futek Torque Sensors. Available online: https://www.futek.com/store/Torque%20Sensors (accessed on 16 October 2019).
17. Magtrol Motors Testing and Sensors. Magtrol Torque Transducers. Available online: https://www.magtrol.com/product-category/torque-transducers/ (accessed on 16 October 2019).
18. Heins, G.; Thiele, M.; Brown, T. Accurate Torque Ripple Measurement for PMSM. *IEEE Trans. Instrum. Meas.* **2011**, *60*, 3868–3874. [CrossRef]
19. Sue, P.; Wilson, D.; Farr, L.; Kretschmar, A. High precision torque measurement on a rotating load coupling for power generation operations. In Proceedings of the 2012 IEEE International Instrumentation and Measurement Technology Conference Proceedings, Graz, Austria, 13–16 May 2012; pp. 518–523. [CrossRef]
20. Zakrzewski, J. Combined magnetoelastic transducer for torque and force measurement. *IEEE Trans. Instrum. Meas.* **1997**, *46*, 807–810. [CrossRef]
21. Yeo, K.C.; Heins, G.; Boer, F.D. Comparison of torque estimators for PMSM. In Proceedings of the 2008 Australasian Universities Power Engineering Conference, Sydney, NSW, Australia, 14–17 December 2008; pp. 1–6.
22. Cheng, B.; Tesch, T.R. Torque Feedforward Control Technique for Permanent-Magnet Synchronous Motors. *IEEE Trans. Ind. Electron.* **2010**, *57*, 969–974. [CrossRef]
23. Jukic, F.; Sumina, D.; Erceg, I. Comparison of torque estimation methods for interior permanent magnet wind power generator. In Proceedings of the 2017 19th International Conference on Electrical Drives and Power Electronics (EDPE), Dubrovnik, Croatia, 4–6 October 2017; pp. 291–296. [CrossRef]

24. Xu, J.X.; Panda, S.K.; Pan, Y.J.; Lee, T.H.; Lam, B.H. Improved PMSM pulsating torque minimization with iterative learning and sliding mode observer. In Proceedings of the 2000 26th Annual Conference of the IEEE Industrial Electronics Society, IECON 2000, 2000 IEEE International Conference on Industrial Electronics, Control and Instrumentation, 21st Century Technologies, Nagoya, Japan, 22–28 October 2000; Volume 3, pp. 1931–1936. [CrossRef]
25. Lam, B.H.; Panda, S.K.; Xu, J.X. Torque ripple minimization in PM synchronous motors an iterative learning control approach. In Proceedings of the IEEE 1999 International Conference on Power Electronics and Drive Systems, PEDS'99 (Cat. No.99TH8475), Hong Kong, China, 27–29 July 1999; Volume 1, pp. 144–149. [CrossRef]
26. Chung, S.K.; Kim, H.S.; Kim, C.G.; Youn, M.J. A new instantaneous torque control of PM synchronous motor for high-performance direct-drive applications. *IEEE Trans. Power Electron.* **1998**, *13*, 388–400. [CrossRef]
27. Dong, X.; Tianmiao, W.; Hongxing, W. Comparison between model reference observer and reduced order observer of PMSM torque. In Proceedings of the 2011 6th IEEE Conference on Industrial Electronics and Applications, Beijing, China, 21–23 June 2011; pp. 663–667. [CrossRef]
28. Liu, Q.; Hameyer, K. High-Performance Adaptive Torque Control for an IPMSM With Real-Time MTPA Operation. *IEEE Trans. Energy Convers.* **2017**, *32*, 571–581. [CrossRef]
29. Mohamed, Y.A.I.; Lee, T.K. Adaptive self-tuning MTPA vector controller for IPMSM drive system. *IEEE Trans. Energy Convers.* **2006**, *21*, 636–644. [CrossRef]
30. Traoré, W.F.; McCann, R. Torque measurements in synchronous generators using giant magnetoresistive sensor arrays via the Maxwell stress tensor. In Proceedings of the 2013 IEEE Power Energy Society General Meeting, Vancouver, BC, Canada, 21–25 July 2013; pp. 1–5. [CrossRef]
31. Lin, Z.; Reay, D.S.; Williams, B.W.; He, X. Online Modeling for Switched Reluctance Motors Using B-Spline Neural Networks. *IEEE Trans. Ind. Electron.* **2007**, *54*, 3317–3322. [CrossRef]
32. Reigosa, D.; Kang, Y.G.; Martínez, M.; Fernández, D.; Guerrero, J.M.; Briz, F. SPMSMs Sensorless Torque Estimation Using High Frequency Signal Injection. In Proceedings of the 2019 IEEE Energy Conversion Congress and Exposition (ECCE), Baltimore, MD, USA, 29 September–3 October 2019; pp. 2388–2393. [CrossRef]
33. Martinez, M.; Laborda, D.F.; Reigosa, D.; Fernández, D.; Guerrero, J.M.; Briz, F. SynRM Sensorless Torque Estimation Using High Frequency Signal Injection. In Proceedings of the 2019 IEEE 10th International Symposium on Sensorless Control for Electrical Drives (SLED), Turin, Italy, 9–10 September 2019; pp. 1–5. [CrossRef]
34. Xu, W.; Lorenz, R.D. High-Frequency Injection-Based Stator Flux Linkage and Torque Estimation for DB-DTFC Implementation on IPMSMs Considering Cross-Saturation Effects. *IEEE Trans. Ind. Appl.* **2014**, *50*, 3805–3815. [CrossRef]
35. Hanselman, D.C. *Brushless Permanent-Magnet Motor Design*; McGraw-Hill Inc.: New York, NY, USA, 1994.
36. Emrax e-Motors. Standard Motors: Emrax 228. 2019. Available online: https://emrax.com/products/emrax-228 (accessed on 16 October 2019).
37. UniTek. Bamocar-D3. 2019. Available online: https://www.unitek-industrie-elektronik.de/bamocar-d3 (accessed on 16 October 2019).
38. Texas Instrument. Microcontrollers(MCU): TMS320F28335. 2019. Available online: http://www.ti.com/product/TMS320F28335 (accessed on 16 October 2019).

Permissions

All chapters in this book were first published in MDPI; hereby published with permission under the Creative Commons Attribution License or equivalent. Every chapter published in this book has been scrutinized by our experts. Their significance has been extensively debated. The topics covered herein carry significant findings which will fuel the growth of the discipline. They may even be implemented as practical applications or may be referred to as a beginning point for another development.

The contributors of this book come from diverse backgrounds, making this book a truly international effort. This book will bring forth new frontiers with its revolutionizing research information and detailed analysis of the nascent developments around the world.

We would like to thank all the contributing authors for lending their expertise to make the book truly unique. They have played a crucial role in the development of this book. Without their invaluable contributions this book wouldn't have been possible. They have made vital efforts to compile up to date information on the varied aspects of this subject to make this book a valuable addition to the collection of many professionals and students.

This book was conceptualized with the vision of imparting up-to-date information and advanced data in this field. To ensure the same, a matchless editorial board was set up. Every individual on the board went through rigorous rounds of assessment to prove their worth. After which they invested a large part of their time researching and compiling the most relevant data for our readers.

The editorial board has been involved in producing this book since its inception. They have spent rigorous hours researching and exploring the diverse topics which have resulted in the successful publishing of this book. They have passed on their knowledge of decades through this book. To expedite this challenging task, the publisher supported the team at every step. A small team of assistant editors was also appointed to further simplify the editing procedure and attain best results for the readers.

Apart from the editorial board, the designing team has also invested a significant amount of their time in understanding the subject and creating the most relevant covers. They scrutinized every image to scout for the most suitable representation of the subject and create an appropriate cover for the book.

The publishing team has been an ardent support to the editorial, designing and production team. Their endless efforts to recruit the best for this project, has resulted in the accomplishment of this book. They are a veteran in the field of academics and their pool of knowledge is as vast as their experience in printing. Their expertise and guidance has proved useful at every step. Their uncompromising quality standards have made this book an exceptional effort. Their encouragement from time to time has been an inspiration for everyone.

The publisher and the editorial board hope that this book will prove to be a valuable piece of knowledge for researchers, students, practitioners and scholars across the globe.

List of Contributors

Adrienn Dineva, Mate Gyimesi and Istvan Vajda
Institute of Automation, Kalman Kando Faculty of Electrical Engineering, Obuda University, 1034 Budapest, Hungary

Narjes Nabipour
Institute of Research and Development, Duy Tan University, Da Nang 550000, Vietnam

Timon Rabczuk
Department of Computer Engineering, College of Computer and Information Sciences, King Saud University, Riyadh 12372, Saudi Arabia

Amir Mosavi
Institute of Automation, Kando Kalman Faculty of Electrical Engineering, Obuda University, 1034 Budapest, Hungary
School of the Built Environment, Oxford Brookes University, Oxford OX3 0BP, UK
Queensland University of Technology (QUT), Centre for Accident Research Road Safety-Queensland (CARRS-Q), 130 Victoria Park Road, Queensland 4059, Australia
Institute of Structural Mechanics, Bauhaus University Weimar, D-99423 Weimar, Germany

Sina Faizollahzadeh Ardabili
Biosystem Engineering Department, University of Mohaghegh Ardabili, Ardabil 5619911367, Iran

Shahaboddin Shamshirband
Department for Management of Science and Technology Development, Ton Duc Thang University, Ho Chi Minh City, Vietnam
Faculty of Information Technology, Ton Duc Thang University, Ho Chi Minh City, Vietnam

Kwok-Wing Chau
Department of Civil and Environmental Engineering, Hong Kong Polytechnic University, Hung Hom, Hong Kong, China

Shuang Wang, Jianfei Zhao and Kang Yang
School of Mechatronic Engineering and Automation, Shanghai University, Baoshan District, Shanghai 200444, China

Xinmin Li, Guozheng Zhang and Qiang Geng
School of Electrical Engineering and Automation, Tianjin Polytechnic University, Tianjin 300387, China

Guokai Jiang
China Automotive Technology and Research Center Co. Ltd., Tianjin 300300, China

Wei Chen
School of Artificial Intelligence, Tianjin Polytechnic University, Tianjin 300387, China

Tingna Shi
College of Electrical Engineering, Zhejiang University, Hangzhou 310027, China

Dong Yu, Xiaoyan Huang, Lijian Wu and Youtong Fang
College of Electrical Engineering, Zhejiang University, Hangzhou 310027, China

Qingjiang Han, Ying Dai and Minqi Hua
School of Mechatronic Engineering and Automation, Shanghai University, Baoshan District, Shanghai 200444, China

GuangQing Bao and WuGang Qi
College of Electrical Engineering and Information Engineering, Lanzhou University of Technology, Lanzhou 730050, China
Key Laboratory of Gansu Advanced Control for Industrial Processes, Lanzhou University of Technology, Lanzhou 730050, China
National Demonstration Center for Experimental Electrical and Control Engineering Education, Lanzhou University of Technology, Lanzhou 730050, China

Ting He
Gansu Natural Energy Research Institute, Lanzhou 730046, China

Tingzhang Liu
School of Mechatronic Engineering and Automation, Shanghai University, Baoshan District, Shanghai 200444, China

Peter Bober and Želmíra Ferková
Faculty of Electrical Engineering and Informatics, Technical University of Kosice, 04200 Kosice, Slovakia

Junlei Chen, Shuo Chen, Xiang Wu, Guojun Tan and Jianqi Hao
School of Electrical and Power Engineering, China University of Mining and Technology, Xuzhou 221116, China

Grzegorz Utrata
Institute of Environmental Engineering, Czestochowa University of Technology, Brzeźnicka 60a, 42-200 Czestochowa, Poland

Jaroslaw Rolek and Andrzej Kaplon
Department of Industrial Electrical Engineering and Automatic Control, Kielce University of Technology, Tysiąclecia Państwa Polskiego 7, 25-314 Kielce, Poland

Si-Lu Chen, Yusheng Liang, Jinhua Chen, Chi Zhang and Guilin Yang
Zhejiang Key Laboratory of Robotics and Intelligent Manufacturing Equipment Technology, Ningbo Institute of Materials Technology and Engineering, Chinese Academy of Sciences, Ningbo 315201, China

Liming Yuan and Jie Zhang
Zhejiang Key Laboratory of Robotics and Intelligent Manufacturing Equipment Technology, Ningbo Institute of Materials Technology and Engineering, Chinese Academy of Sciences, Ningbo 315201, China
University of Chinese Academy of Sciences, Beijing 100049, China

Maria Martinez, David Reigosa, Daniel Fernandez and Fernando Briz
Electrical, Electronic, Computers and Systems Engineering, University of Oviedo, 33204 Oviedo, Spain

Index

www.ingramcontent.com/pod-product-compliance
Lightning Source LLC
Chambersburg PA
CBHW081216180926
51613CB00003B/254
9781639877478